Open Data and Energy Analytics

Open Data and Energy Analytics

Special Issue Editors

Benedetto Nastasi
Massimiliano Manfren
Michel Noussan

MDPI • Basel • Beijing • Wuhan • Barcelona • Belgrade • Manchester • Tokyo • Cluj • Tianjin

Special Issue Editors
Benedetto Nastasi
Sapienza University of Rome
Italy

Massimiliano Manfren
University of Southampton
UK

Michel Noussan
Fondazione Eni Enrico Mattei
Italy

Editorial Office
MDPI
St. Alban-Anlage 66
4052 Basel, Switzerland

This is a reprint of articles from the Special Issue published online in the open access journal *Energies* (ISSN 1996-1073) (available at: https://www.mdpi.com/journal/energies/special_issues/open_data_energy).

For citation purposes, cite each article independently as indicated on the article page online and as indicated below:

LastName, A.A.; LastName, B.B.; LastName, C.C. Article Title. *Journal Name* **Year**, *Article Number*, Page Range.

ISBN 978-3-03936-218-9 (Pbk)
ISBN 978-3-03936-219-6 (PDF)

Contents

About the Special Issue Editors

Benedetto Nastasi (PhD) is Senior Energy Planner and Lecturer at Sapienza University of Rome and Guest Researcher at TU Delft University of Technology. Previous affiliations include TU/e Eindhoven University of Technology, The Netherlands, and International Solar Energy Society and Guglielmo Marconi University, Italy. His work is related to Power-to-What solutions for energy systems design with a specific focus on the built environment. He has developed expertise on hydrogen technologies, energy efficiency, hybrid systems, energy efficiency in buildings, distributed generation, as well as micro and smart grids. He holds a PhD with Honors in Energy Systems Planning and Design at Sapienza University of Rome.

Massimiliano Manfren (PhD) is Lecturer in the Sustainable Energy Research Group (SERG), within the Faculty of Engineering and Physical Sciences of the University of Southampton (UK). His previous affiliations include Politecnico di Milano (IT) and University of Bologna (IT). His research focuses on analytics and predictive models for energy system design and operational optimization at multiple scales, from individual users to communities. His research aims to establish a convergence between scientific disciplinary knowledge in energy demand modelling at multiple levels; energy-efficient technologies; and advances in machine learning and operation research techniques, through an integrated use of simulation, optimization, statistics, and data mining on case studies. He holds a PhD in "Programming, Maintenance, and Rehabilitation of Buildings and Urban Systems"from Politecnico di Milano.

Michel Noussan (PhD) is Senior Research Fellow at Fondazione Eni Enrico Mattei (FEEM) Future Energy Research Program and Affiliate Professor of Sustainable Transport at Sciences Po's Paris School of International Affairs (PSIA). His current research activities are focused on the analysis and comparison of different mobility solutions in the framework of decarbonization and digitalization trends of the transport sector. He has developed expertise on energy systems analysis, combined heat and power, district heating, energy efficiency and local energy planning. He was a researcher and university lecturer at Politecnico di Torino in the domain of energy systems analysis, and he has a track record of several publications in international journals and conferences. He holds a PhD in Energy Engineering from Politecnico di Torino.

energies

MDPI

Editorial

Open Data and Energy Analytics

Benedetto Nastasi [1,2,*], Massimiliano Manfren [3] and Michel Noussan [4,5]

1 Department of Planning, Design and Technology of Architecture, Sapienza University of Rome, Via Flaminia 72, 00196 Rome, Italy
2 Department of Architectural Engineering & Technology, TU Delft University of Technology, Julianalaan 134, 2628BL Delft, The Netherlands
3 Faculty of Engineering and Physical Sciences, University of Southampton, Boldrewood Innovation Campus, Burgess Rd, Southampton SO16 7QF, UK; m.manfren@soton.ac.uk
4 Fondazione Eni Enrico Mattei, Corso Magenta 63, 20123 Milano, Italy; michel.noussan@feem.it
5 Department of Energy, Politecnico di Torino, Corso Duca degli Abruzzi 24, 10129 Torino, Italy
* Correspondence: benedetto.nastasi@outlook.com

Received: 10 March 2020; Accepted: 30 April 2020; Published: 7 May 2020

Abstract: This pioneering Special Issue aims at providing the state-of-the-art on open energy data analytics; its availability in the different contexts, i.e., country peculiarities; and at different scales, i.e., building, district, and regional for data-aware planning and policy-making. Ten high-quality papers were published after a demanding peer review process and are commented on in this Editorial.

Keywords: open data analytics; energy planning; smart cities; open energy governance; urban database; energy mapping; building dataset; energy modelling; data mining; machine learning

1. Overview

Open data and policy implications coming from data-aware planning require collection and the pre- and postprocessing as operations of primary interest. These procedures require that data are freely available to people and decision-makers. Openness is, therefore, the best way. Referring to the relationship between data and energy, public administrations, governments, and research bodies are promoting the construction of reliable and robust datasets (i) to pursue policies coherent with the sustainable development goals, as well as (ii) to allow citizens to make informed choices. Energy engineers and planners must provide the simplest and most robust tools to collect, process, and analyze data, to offer solid data-based evidence for future projections at building, district, and regional scales for an effective systems planning.

For all these reasons, researchers encouraged by the call for papers shared their original works in the field of "Open Data and Energy Analytics". Among the numerous submissions, the following 10 successfully passed the review process.

2. A Short Review of the Contributions to This Issue

Cutting-edge outcomes of ongoing and recently ended European research projects are published in this Special Issue. In detail, two H2020 projects, namely, PLANHEAT and HOTMAPS, are the sources of innovative results published in three original articles.

The paper authored by Fremouw et al. [1] deals with the role played by open data in supporting urban transition planning, thanks to the energy potential mapping within the H2020 project PLANHEAT. The aim of the paper is to identify the principal recurring issues in energy data acquisition and processing to overcome the existing barriers in data availability. An increase of the quality of energy mapping tools follows the relevance and availability of energy data. Thanks to the activities of the HOTMAPS project, Pezzutto et al. [2] present the design of an open-source toolbox to support urban planners, energy agencies, and public administrations for planning the heating and cooling supply at different scales. A bottom-up approach is used to collect and analyze market data related to space heating and domestic hot water systems and their performance in Europe. Within the same HOTMAPS project, Müller et al. [3] face the challenge of uncertainties coming from different databases and from large differences in available datasets among EU countries. A top-down approach is proposed, and a comparison between country-level and municipal-level building stock data is made for gross floor area and energy demand for space heating and domestic hot water. Transparency and regular update of datasets fostered by the increase of smart meters installation are crucial to support and effective energy planning.

Moreover, this Special Issue presents also different research works dealing with the potential of gathering useful information from available data in different fields, both for performance assessment and future scenarios design.

Korkovelos et al. [4] illustrate an overview of open-access geo-spatial data and GIS-based electrification models aiming to support SDG7, with a detailed discussion on their role in answering complex policy questions. Their research work presents an updated version of the Open-Source Spatial Electrification Toolkit (OnSSET-2018), which is described in detail and applied to a case study in Malawi, comparing the cost of different electrification options by 2030. The results highlight that the optimal mix includes off-grid PV systems for two-thirds of the population, and power grid extension for the rest. The sensitivity analysis provides additional insights on the crucial role of electricity demand projections in the optimal electrification solution.

Electricity data can also support a better evaluation of the distributors' performance, as described by Ganhadeiro et al. [5] in a case study in Brazil. The authors propose an improved methodology to better assess how environmental variables affect the energy efficiency of electricity distribution companies. The methodology presented by the authors can be extended to other countries where there is at least some influence of private sector in energy distribution, or any other regulated service.

Another interesting case for the potential of data in supporting energy analyses is presented by De Kok et al. [6], who focus on the use of user-generated contents in social media to understand and improve the energy consumption behavior of individuals. The authors highlight the interesting potential of social media content as a complementary support to other sources, thanks to the massive amount of data and the low cost of analysis. Thanks to an image and text processing pipeline, relevant information can be extracted to describe different energy-consuming activities. The strengths and weaknesses of this approach are presented, by applying the method to two case studies in Amsterdam and Istanbul.

Zipperle and Orthofer [7] present an innovative open-source interface for MESSAGEix model, named d2ix. MESSAGEix is an optimization model for strategic energy planning and integrated assessment of energy–engineering–economy–environment systems, including effects such as emissions, economic development, land and water use, and health implications. It can be linked also to the general-economy MACRO model to incorporate feedback between prices and demand levels for energy and commodities. The d2ix interface enables concise presentation and editing of model input data and increases the accessibility and transparency of the modelling processes, reducing barriers and simplifying collaborative working.

In the narrow field of energy efficiency in the built environment, Attanasio et al. [8] propose a methodology for the automatic estimation of building primary energy demand related to space heating and to the characterization of the relationship between the latter and the main building features. The methodology was tested using an energy performance certificate database with 90,000 flats in Piedmont region (Italy) and four machine learning algorithms. The methodology can be used for quick estimation of expected building energy demand as well as setting credible targets for improving building performance.

Another application of data analysis techniques in the built environment is presented by Manfren and Nastasi [9]. They describe an integrated workflow from parametric energy performance analysis to model calibration. A passive house building is a case study that seeks to show an effective and transparent way to link design and operation performance analysis together with reducing the efforts in modelling and monitoring by providing parametric performance boundaries. These performance boundaries are used to ease monitoring process and to identify insights in a simple, robust, and scalable way.

Finally, Vialetto and Noro [10] present an application of Internet of Things (IOT) and Industry 4.0 concepts to the industrial energy efficiency. A clustering modelling approach for the short-term forecasting of energy demand in industrial facilities is shown. The forecasting model is applied to an industrial facility (wood processing industry) with simultaneous heat and electricity demand, where it proves to be effective, with a very small error in the order of 3%.

Author Contributions: Conceptualization, B.N.; writing—original draft preparation, B.N., M.M., and M.N.; writing—review and editing, B.N., M.M., and M.N. All authors have read and agreed to the published version of the manuscript.

Funding: This research received no external funding.

Conflicts of Interest: The authors declare no conflict of interest.

References

1. Fremouw, M.; Bagaini, A.; De Pascali, P. Energy Potential Mapping: Open Data in Support of Urban Transition Planning. *Energies* **2020**, *13*, 1264. [CrossRef]
2. Pezzutto, S.; Croce, S.; Zambotti, S.; Kranzl, L.; Novelli, A.; Zambelli, P. Assessment of the Space Heating and Domestic Hot Water Market in Europe—Open Data and Results. *Energies* **2019**, *12*, 1760. [CrossRef]
3. Müller, A.; Hummel, M.; Kranzl, L.; Fallahnejad, M.; Büchele, R. Open Source Data for Gross Floor Area and Heat Demand Density on the Hectare Level for EU 28. *Energies* **2019**, *12*, 4789. [CrossRef]
4. Korkovelos, A.; Khavari, B.; Sahlberg, A.; Howells, M.; Arderne, C. The Role of Open Access Data in Geospatial Electrification Planning and the Achievement of SDG7. An OnSSET-Based Case Study for Malawi. *Energies* **2019**, *12*, 1395. [CrossRef]
5. Ganhadeiro, T.G.L.; Christo, E.D.S.; Meza, L.A.; Costa, K.A.; Souza, D.P.M. Evaluation of Energy Distribution Using Network Data Envelopment Analysis and Kohonen Self Organizing Maps. *Energies* **2018**, *11*, 2677. [CrossRef]
6. De Kok, R.; Mauri, A.; Bozzon, A. Automatic Processing of User-Generated Content for the Description of Energy-Consuming Activities at Individual and Group Level. *Energies* **2019**, *12*, 15. [CrossRef]
7. Zipperle, T.; Orthofer, C.L. d2ix: A Model Input-Data Management and Analysis Tool for MESSAGEix. *Energies* **2019**, *12*, 1483. [CrossRef]
8. Attanasio, A.; Savino Piscitelli, M.; Chiusano, S.; Capozzoli, A.; Cerquitelli, T. Towards an Automated, Fast and Interpretable Estimation Model of Heating Energy Demand: A Data-Driven Approach Exploiting Building Energy Certificates. *Energies* **2019**, *12*, 1273. [CrossRef]
9. Manfren, M.; Nastasi, B. Parametric Performance Analysis and Energy Model Calibration Workflow Integration—A Scalable Approach for Buildings. *Energies* **2020**, *13*, 621. [CrossRef]
10. Vialetto, G.; Noro, M. Enhancement of a Short-Term Forecasting Method Based on Clustering and kNN: Application to an Industrial Facility Powered by a Cogenerator. *Energies* **2019**, *12*, 4407. [CrossRef]

energies

MDPI

Article

Energy Potential Mapping: Open Data in Support of Urban Transition Planning

Michiel Fremouw [1,*], Annamaria Bagaini [2] and Paolo De Pascali [2]

1 Faculty of Architecture and the Built Environment, Department of Architectural Engineering + Technology, Delft University of Technology, 2628 BL Delft, The Netherlands
2 Department of Planning, Design and Technology of Architecture, Sapienza University, 00185 Rome, Italy; annamaria.bagaini@uniroma1.it (A.B.); paolo.depascali@uniroma1.it (P.D.P.)
* Correspondence: M.A.Fremouw@tudelft.nl

Received: 15 November 2019; Accepted: 25 February 2020; Published: 9 March 2020

Abstract: Cities play a key role in driving the transition to sustainable energy. Urban areas represent between 60% and 80% of global energy consumption and are a significant source of CO_2 emissions, making energy management at the urban scale an important area of research. Urban energy systems have a strong influence on the environment, economy, social dimensions and urban spatial planning. Energy consumption affects the urban microclimate, urban comfort, human health, and conversely, urban physical, economic and social characteristics affect the energy urban profile. In order to improve the quality of energy strategies, policies, and plans, local authorities need decision support tools, like energy potential mapping, which have risen significance in the last decades. Energy data are crucial for those tools. They can increase the quality and effectiveness of energy planning but also support the integration between energy and spatial planning. Energy data can also stimulate citizen engagement as well as encourage sustainable behaviours and CO_2 emission reduction. This paper aims to increase the practice of data-aware planning, through the study of problems in energy data acquisition and processing observed in European projects focused on developing energy mapping tools. The problems observed attend to two main areas: technical and socio-economic issues. Those were derived from a comparison of energy mapping tools, and the work conducted for the PLANHEAT development. The scope of the research is to understand the main recurring issues in energy data acquisition and processing, in order to overcome the barriers in data availability. Increasing awareness of the relevance of energy data can foster the use of energy mapping tools, increasing the quality of energy policies and planning.

Keywords: energy planning; energy potential mapping; urban energy atlas; urban energy transition; energy data; data-aware planning; spatial planning

1. Introduction

Human society is facing an unprecedented challenge. The extended use of fossil fuels as a source of CO_2 emissions has been the main driver of global climate change, and the built environment is playing a significant part in this. Changing urban planning practices is considered a primary component of the pathways towards climate mitigation [1,2]. In order to plan for the transition towards residual and renewable energy within this built environment, the nature of the urban fabric and local circumstances are highly important, yet often poorly mapped and quantified.

The availability of open data [3] contributes to the political, social and economic development of a country. Public administrations are often not aware of the value data can bring to societies, quality of life, environmental protection and energy turn. In order to accomplish this, data must be available, accessible, user-friendly, and reusable [4,5]. In this sense, energy data (e.g, data about energy demand

and local renewable energy sources) are crucial to innovate and increase the efficacy of urban energy policy and urban energy planning.

The urgency of countering climate change [1] spurs local governments to define their energy strategies, emission targets, as well as sustainability agendas, spurs local governments to define their energy strategies, emission targets, as well as sustainability agendas, but in many cases, these are lists of well-meant intentions, rather than operative actions, and not easy to translate into specific interventions. The reason behind this is that urban energy policies are not directly based on real data and information, but on top-down, aggregated estimations of the city energy profile (i.e., the urban energy performance in terms of energy demand, energy local sources, and the future trend of energy demand and supply). The city energy profile is however strongly related to urban shape, geometry and physical characteristics [6–11]. Energy demand and energy sources are rarely distributed homogenously within the city. Some areas show high energy consumption, and waste, whereas others exhibit energy poverty issues; and there are areas with high resource availability, while in others, locally available resources cannot satisfy the demand [12].

Many studies [7,8,13–15] show that, in the long term, the most suitable opportunities to influence energy consumption and the related CO_2 emissions are represented by the decisions taken in the field of urban planning. These decisions concern the land use, the design of public and private mobility, the urban waste system (i.e., collection and recycling process), the water system (i.e., supply and water treatment), the energy production from renewable sources (Spatial planners must find optimal locations for windmills, biomass and solar power plants as well as energy storage systems. At the same time conflicts with competing uses and the environment have to be minimized.) and its distribution; the green system and of course the design of buildings. The irregular spatial distribution of energy demand and supply sources within cities show a need for understanding the different relationships between energy and urban characteristics. These considerations find in the urban energy map a tool able to give effective support to the decision-making and planning process [11,16,17].

Energy mapping tools can give a spatial dimension to the energy issue and provide a means of understanding the relationships between urban and energy factors. The energy map aims to clarify the characteristics of a specific energy pattern (demand, supply, production and distribution), through the spatial visualization of energy data, which can support "informed" interventions, suggest strategies, and identify priorities and the most suitable locations for energy district developments. The energy map may be also a good basis to trigger an informed debate on the choices to take: an ideal catalyst for discussions and for defining shared objectives.

The term "energy mapping" is not officially defined, and because of the complexity of the subject, encompasses a wide range of implementations. Over the last decade, cities have developed different tools to increase their capacity to evaluate the availability of renewable energy sources (such as solar maps, geothermal maps) connected to the energy consumption and demand (heating and cooling maps). Several techniques, methods and objectives have been developed, some only in empirical form, others have been coded at the operational level, identifying steps of implementation [11,13,18].

All of these rely on sufficient availability of (geo) data. The literature on energy mapping instruments [11,13,16,19–23] offers a summary of the variables that an energy map should contain and integrate: the spatial distribution of the energy demand (electricity and heating/cooling energy demand); the spatial distribution of population; the land use; the characteristics of the building stock (use, year of construction, height of buildings, etc.); polluting emissions resulting from energy consumption; the spatial distribution of RES (Renewable Energy Sources, such as solar, solar thermal, wind, both in-land and off-shore, hydroelectric, geothermal, from biomass, from reconversion of excess heat, from waste-to-energy, etc.); heating degree hours (HDH); the presence and location of anchor loads; the design of energy grids; the location and size of expansion plans (residential, commercial, etc.), some specific barriers such as the presence of protected areas (frequently in the heritage and ecology categories); the mobility system; and possibly socio-economic indicators to identify the poorest or most degraded areas.

The innovative nature and the benefits of using an energy mapping tool have not necessarily resulted in swift implementation. One of the main difficulties is a lack of suitable data, that is also available to the urban planner. Access to energy data is essential to developing the appropriate tools and improve the ability to make decisions that merge physical-spatial issues with energy-environmental ones. "Making urban data accessible" is therefore becoming a fundamental prerequisite for urban innovation [24], and for increasing the quality of energy-related policies. Energy data should be accurate and based on real acquisition (in order to reduce the inaccuracies associated with estimation), geospatially referenced, and measured at short temporal intervals over at least a year (in order to account for both peaks and seasonal fluctuations).

Fortunately, the digital revolution is offering significant perspectives in terms of data acquisition, processing and use, providing new opportunities for urban investigation. New technologies improve the ability to analyse the urban energy profile (for instance smart meters–now also available for gas consumption), and can allow for a very detailed and real-time view of consumption (but also the use of mobile devices-mobile phones, tablets, etc.–for recording consumption and citizen habits, able to influence also people behaviour).

Aims and Objectives

This study builds upon the work conducted during the Horizon 2020 funded European PLANHEAT project, which started in 2016 [22]. The main objective of PLANHEAT is to develop an integrated tool, a potential energy map, which will empower public authorities (cities and regions) in the development of sustainable energy plans, with a special focus on distributed heat (cold) networks and energy district design. The PLANHEAT tool supports local authorities by providing:

- thermal energy (heat and cold) demand mapping;
- local potential mapping for distributed low carbon energy sources;
- forecasted demand mapping;
- a planning tool for defining scenarios which will be sustainable, feasible, and environmentally friendly based on the usage of renewable energy sources (as well as highly efficient cogeneration and district heating);
- a tool to understand the interactions of planned scenarios with already existing infrastructure such as district heating, gas and electricity networks and transport sector, etc.;
- a simulation module that calculates demand, supply and storage behaviour over a year and provides data on technical, environmental, economic and social impacts;
- a scenario evaluation instrument which allows for both a baseline and user defined scenario comparison [25]

Both the development of the PLANHEAT toolkit and its use require collecting energy related data. Thus, an assessment was made early on in the project, in order to discover the main issues related to the availability and quality of these types of data (both open and internal/proprietary data), as experienced by the toolkit's intended end users [25]. Through questionnaires and interviews, 26 cities in 8 countries (France, Belgium, Italy, Greece, Netherlands, Hungary, Croatia, and Spain) were asked to evaluate the rate of data openness in 7 data categories: heat demand, heat supply, transport sector, census information, energy audits, knowledge and motivation, and finally, the connection of local and national plans. The results showed that every country involved has more than 50% of the types of data considered available publicly [18–25], however local authorities highlighted structural difficulties in reaching and using these data [25]. Even if datasets exist, getting the required data (for example when the data owner is another stakeholder) sometimes proves difficult, and they may require processing or interpreting if the data was recorded for different purposes.

These results bring forward the need of investigating in greater detail the reasons behind those difficulties, with the aim of increasing the capacity of PLANHEAT to provide a useful toolkit and increase its usability by municipalities with varying levels of access to energy data. Thus, other projects with similar goals were analysed, with the aim to understand which type of data they used, how they collected them and which problems they had to deal with. The comparison between the difficulties emerged in these projects those that emerged in the PLANHEAT interviews made it possible to find similarities and common issues useful to determinate the main barriers in developing and using energy maps.

The scope of this paper is to understand the nature of commonly recurring problems in accessing and processing energy data. These problems can limit the usability of energy mapping tools by local authorities. Identifying the main issues assists in making these instruments more adaptable and therefore more effective, in terms of supporting energy turn and the decision-making process. Furthermore, this can also help improve the normative framework related to open data policies, and the elaboration of data standards and licenses for publication, which in turn may increase the future availability of suitable energy data.

2. Materials and Methods

This paper points out two main problems related to energy data availability and usability: technical issues (spatial and temporal resolution) and socio-economic issues (privacy, financial costs, ownership, concurrence, etc.). These two categories of problems raise from (1) the interviews conducted during the preliminary phase of the PLANHEAT project [25] and (2) from a comparative analysis of energy mapping projects and experiences, listed in Table 1.

In order to have a better understanding of the shared and common difficulties occurring in the energy data access and processing for developing energy maps, both a literature review and a study on European projects have been conducted. The intention of studying projects only in Europe comes from the need to remain under the same normative energy efficiency and data legislation. The literature review is built based on material collected through searching scholarly databases, mainly Scopus.com and Sciencedirect.com, using keywords including: "Urban energy maps"; "Urban energy mapping tools"; "Urban energy atlas"; "Energy web maps"; "Energy decision-support tools". For the selection of EU (European Union) projects, the research has been conducted on the European Commission web site, which collects all projects funded by topic. At the section Intelligent energy Europe, the projects have been selected in the categories: "Energy efficiency"; "Integrated initiatives"; "Heating and cooling" (https://ec.europa.eu/energy/intelligent/projects/). From this initial, wide range of energy mapping projects (EU projects and academic/institution ones) related to urban energy mapping tools, a selection was made of those deemed most suitable for comparison, from the perspective of evaluating the difficulties in collecting and processing data, as shown in Table 1. These projects provided sufficient information on the applied methodologies, the type of data used, and the results achieved for evaluation. Other projects found did not allow sufficient in-depth study for the analysis conducted, either because data collection was not a significant element, or because the underlying model is proprietary.

For each project, an analysis was made of the aim and type of the project; the steps of implementation; its status (ongoing, ended); the type of tool developed and its usability; the spatial scale(s) used; and the type of (geo) data used. The study focuses on building bound energy data availability and processing, intending to raise the most recurring problems, which can reduce the implementation of energy potential maps by local authorities. The aim of the paper is to increase data-aware planning and policy-making in the field of energy planning and urban energy policies, by identifying opportunities and solutions for problems with data acquisition and processing. This helps to improve the effectiveness of energy mapping tools and makes their usage more affordable for the large number of smaller local authorities.

In the majority of these projects either open data is used, but the level of detail is low (city/region/country), or the detail level is high (below city level), but the user is required to input significant amounts of private data to get to the planning stage.

Table 1. List of Projects Focused on Energy Potential Mapping Development.

Project Name	Reference	Start	End	Category	Result and Usability	Spatial Scale(s)	Types of (Geo) Data Used
PLANHEAT	[16,22,23,25]	2016	2019	EU project	open source plug-in Qgis	city/district	open databases for most maps
STRATEGO	[26,27]	2014	2016	EU project	open web-based GIS map	city/country	open databases
Scotland Heat Map	[28]	2014	ongoing	Institutional project	open web-based GIS map	region	databases updated by local authorities
Amsterdam Energy Atlas	[29]	2013	2015	EU project	open maps	city	datasets provided by local authorities and private sector
MUSIC iGUESS	[30]	2009	2014	EU project	open maps	city	datasets provided by local authorities and private sector
NL 3D heat maps	[31,32]	2009	2011	Institutional research	methodology–open map	city/country	new datasets production (data estimation)–Data provided by private sector and public bodies
London Heat Map	[33,34]	2009	2019 upgraded	Institutional project	open web-based GIS map	city	data provided by the 23 London Boroughs–Dataset production
POP Groningen	[19,35]	2006	/	Institutional project	methodology–open maps	province	public open base map and new datasets production (data estimation)
PlanVision	[11,36,37]	2009	2011	Academic project	methodology–Energy Zone Maps	city	datasets production (survey and public data collection)
PlanETer	[38]	2013	2015	Institutional project	open web-based GIS map	city	datasets production (private and public data collection)
ESTMAP3	[39]	2015	2016	EU project	open web-based GIS map	region/country	open databases
Elas calculator	[11]	2009	2011	Academic project	open web-based tool (calculator)	city	Data provided by local authorities in the calculator tool

3. Energy Planning Data: An Overview of Problems and Issues

Although all these projects are intended for energy planning, and supporting local and regional authorities, their specific implementation varies. In some cases the end result was an energy atlas, in others a spatial and/or quantitative decision support tool for the built environment. They do all share a requirement for and ability to use (geo) data in order to provide their potentials and assessments.

Furthermore, they benefit from more accurate data to be able to represent the real energy profile of the city [12,26,27,40]. Open energy data can create both economic as well as social value [5]. Those are largely driven by the level of openness and the cost of availability.

The first step for increasing the energy relevance into a planning process is the definition of which energy data are relevant at which steps and phases [26,41]. It allows more directed collection of data and avoids loss of time and resources. The second step should be a general overview of data owners and stakeholders. It is necessary to clarify which stakeholders are crucial for which elements [42], to increase the participation and understand which type of data is available and who can provide them (private bodies, public offices, European or international agencies). If the data needed are still not available, local authorities could consider building a new dataset. This process incurs a cost however, which may be a significant hurdle for small organisations.

In most cases, energy related datasets are available [43]. Municipalities, for example, have data about the floor space, year of construction, and building function (office, residential etc.). Energy and infrastructure companies have billing data that refers to the energy consumption of their clients. Standard renewable energy potentials, like for example solar (photovoltaic or thermal), are increasingly available at a high detail level due to their relatively low input requirements. In this case it is possible to use information on roof surface, slope and orientation, all covered by a high-resolution DEM that is an open-source dataset, with solar radiation data acquired from meteorological institutes (usually available as open data) to calculate roof potential [25].

However, in many cases, the use of datasets faces several challenges. From the comparative study of energy mapping and energy analysis tools (Table 1) we discovered two main categories of data acquisition problems:

- Technical issues: spatial resolution and temporal resolution problems;
- Socio-economic issues: privacy and data ownership; the financial cost of collecting and processing data, market competition, awareness.

3.1. *Technical Issues*

As the process of urban energy planning is relatively new, most of the relevant data has been collected for other purposes. Because of this, the exact definition of the values represented in the dataset determines if, and if so, how suitable these are for energy planning purposes.

Furthermore, available data are usually only (publicly) released in an aggregated form, if at all. This applies both to spatial and temporal aspects.

3.1.1. Spatial Resolution

An issue that was frequently encountered with geospatial dataset availability during the PLANHEAT project, is that either the spatial resolution is too low for suitable projection or further analysis, or that only a single figure is available for the area under consideration.

An example is open data on residential building gas consumption in the Netherlands, which is collected by energy supply companies (ESCOs) and made available publicly in aggregated form at the neighbourhood level, through Statistics Netherlands (CBS) [25]. As a result of privacy regulations, figures in some neighbourhoods were also occasionally reported as censored ('afgeschermd') when the number of houses or companies fell below a threshold. Experiences in other countries have been mixed, where sometimes the local ESCO did not provide even aggregated numbers at all, reportedly for competitive or political reasons. Section 3.2 goes into greater detail on this and other socio-economic issues.

Data may also not be available in the same type of spatial division. Statistics agencies for example tend to collect their data using administrative divisions, whereas energy supply potential maps may be based on environmental data which uses rasters. Although GIS (Geographic Information System) software allows these to be projected over one another, using different source formats may introduce additional inaccuracies during both the analysis and subsequent processing steps.

A low spatial resolution may also frustrate the analysis of demand and supply geodata and matching of sources and sinks, especially if there is a strong relation with the urban fabric for the categories under consideration. Although a planner would be able to use low resolution data to do an initial, fast assessment of the possibilities in their city, designing energy transition plans requires not only (geo)data on demand, retrofitting (i.e., demand reduction), and (residual and renewable) supply potentials, but also on the possibilities of existing, and space for new, infrastructure.

In some cases, commercial organisations provide high resolution energy potential maps themselves. For the same competitive reasons mentioned however they are rarely transparent about the exact calculation steps followed in order to produce the values displayed and may include internal assessments on suitability and cost that deviate from the considerations of other users. A lack of transparency decreases the level of confidence of these datasets in itself.

3.1.2. Temporal Resolution

A second issue is with the available temporal resolution. Datasets encountered during the PLANHEAT project that were suitable for energy planning, were usually either geospatial (annual figures) or temporal (for one or few buildings or areas) in nature, however rarely both (the one notable exception within the project being 1 to 5 km^2 resolution environmental satellite data, used for surface water potentials).

For mapping and planning purposes, annual data are usually sufficient, as at this stage of the planning process, quantities and concentrations are more important than temporal patterns. However, the fluctuating nature of both demand and some forms of residual and renewable energy supply might mean that the total potential figures require more than simply providing sufficient generation capacity, as there may be periods where supply vastly outpaces demand requirements (therefore effectively losing available energy), or conversely, demand outpaces supply.

An annual simulation can therefore be run to consider the impact of an energy transition plan. This however requires high resolution temporal, rather than spatial data for the components that are part of the planned energy system, in order to determine if these components are dimensioned to cope with both mismatches and peaks in demand and supply. As with high resolution spatial data, suitable high-resolution temporal data (at the hourly or smaller level) is frequently not available publicly in some cases, or at all in others.

More than with the spatial dimension, the technical and privacy considerations of data acquisition can be an issue here. Registering an hourly profile for household electricity demand over a year firstly requires the presence of a smart meter, and secondly permission of the household to read and store more than just the annual balance (which is required for billing purposes, the primary reason an electricity meter is installed). Accurately assessing the residual heat potential of an industrial process requires both temperature and volume monitoring, especially when used in a simulation. Even if monitoring is already active, commercial entities may be reluctant to (publicly or privately) release these figures, because they might subsequently be analysed by competitors.

3.2. Socio-Economic Issues

The analysis conducted on the PLANHEAT interviews' results and the literature review (European and academic researches related to energy mapping tools) highlights several socio-economic difficulties in dealing with energy data: data ownership and privacy, market competition, irregular updating of datasets, discrepancies between data formats, problems of data aggregation and disaggregation, public administration incapacity to treat and process data. Privacy (Article 8 of the Charter of Fundamental Rights of the European Union and Article 16 of the Treaty on the functioning of the European Union guarantee the protection of personal data. This means that acquisition, transfer and publication of personally identifiable data are subject to restrictions [44].) is always mentioned as one of the key challenges. The EU has several framework policies that protect the privacy of individuals (for instance the new GDPR (General Data Protection Regulation) directive in 2018 [45]), and therefore provide barriers both in the acquisition, transfer between actors (sometimes even between departments within the same civil authority) and publication [25]. When using statistical division-based datasets, sometimes even neighbourhoods and districts may cause confidentiality or privacy issues, if very few addresses are located there. For privacy issues in publication, a simple solution is available: aggregation, which can provide acceptable anonymity. But data aggregation may also reduce the quality of the final output. In many cases, local authorities possess data at a low spatial resolution, for example, city level energy consumption (or even regional scale statistical data). In this case a disaggregation process, using spatial indicators (like city cadastre or land use), would be needed to estimate the spatial distribution of energy consumption in the city.

Energy data are often in the hands of actors who may be unwilling to share it (energy companies, other private companies that for some reasons collect useful data). The competitive market is also a challenge when publishing open data is seen as potentially disadvantageous for companies. Sometimes (commercial) energy companies may have relevant data, but either do not provide this at all or only share at a low resolution, unsuitable for planning purposes. In the case is possible to run into problems of data combining and harmonization, because datasets come from different data owners, with different standards and aggregation methods, due to the liberalization of the energy production and sales market.

Sometimes new data acquisition is needed. The cost of creating and publishing datasets is high and take a long time. Understanding which types of data are available makes it easier to discover which data are missing, so data collection efforts can be specifically targeted for these, avoiding extra costs. Interviewing suitable stakeholders is also useful, especially when data collection projects are already planned, and the user simply must wait for it to become available. In this case, it will be more about cooperation than the associated cost.

Therefore, sometimes useful information might be derived from unexpected sources, and communicate any benefits coming from data sharing is beneficial to make it more interesting for companies to cooperate.

A lack of awareness is also a key challenge. The difficulty of accessing energy data is connected to the complexity of the energy production and supply system, increased with the liberalization of the energy market and the entry of different and new stakeholders. Commercial companies often have little interest in processing and sharing data. Public administrations have huge difficulties in managing data, as a result of differing standards and aggregation methods used by the operators. Municipalities and commercial companies are also not motivated and incentivised, and sometimes not aware of the potential benefits of sharing their energy data. When data owners are not aware of the potential value and the possibilities of open energy data to themselves, the step towards publishing their data is not likely to happen [5]. Now commercial companies (especially related to the energy market) see the opening of energy data as an added cost, without any benefit. The benefits of opening data would become much clearer if there would be direct incentives (financial incentives, government support) for private parties to publish energy data.

Lastly, data quality (including age) and the completeness of datasets is also crucial. Inaccurate or incomplete datasets can lead to misinterpretation and liability issues. For example, final consumption data that is a decade old may not be representative anymore, as not just the size and thermal efficiency of the building stock may have changed, but internal heat load (for example more electronics) and user habits will be different as well.

An interesting element from the Scotland heat map [28] is the methodology used to classify the quality of collected data: the confidence level. Data are categorized respect the level of detail provided. The confidence level 5—the most accurate one—represents data with high resolution, for example, data that come from the family bills. Meanwhile, at confidence level 1, there is the footprint of buildings from which to estimate or disaggregate the urban energy consumption. The goal of the Scotland heat map is to increase over time the quality of all the data collected, to obtain a real representation of the energy profile of the Scottish territory as accurately as possible [28].

Other problems linked to the quality of data are related to consumption habits and certain energy carriers. In some regions, part of the heat demand is fulfilled with unmetered sources, for example, wood (pellets) or fuel oil. The use of electric resistance heating means that some heat demand is effectively concealed as electricity consumption. This also applies to cooling, which is largely supplied by air conditioning. Cities where cooling demand plays a significant role or is expected to in the future because of climate change, are lacking methods to extract actual cooling demand from the data they have [25].

Finally, the lack of a governance framework, guidance, and regulation specifically on open energy data may also form a problem.

4. PLANHEAT: Using Public Open Data to Overcome Problems in Data Acquisition

Addressing the issues with data availability forms the basis of the PLANHEAT project [25]. In this Horizon 2020 funded project, a toolkit was developed that both integrates a wide range of open datasets and allows the user to replace these with higher resolution private data, and add own data for which there is no public substitute yet.

The PLANHEAT integrated toolkit is open source and QGIS3-based. The use of GIS software is becoming commonplace in urban planning practices but requires significant amounts of high-resolution data in order to produce useful results. A Heat/Cold (HC) potential map, as in this case, needs a large amount of data (geospatial, temporal and other), which municipalities may not have or cannot use. The main innovation of the PLANHEAT toolkit is that it can produce useful results, even if a local administration only has a limited amount of information available. Conversely, the same toolkit can use rich datasets as well, in order to be attractive for large metropolitan areas, where detailed data is frequently available to produce more accurate maps, plans and simulations. The objective of PLANHEAT is to offer a support tool, based as much as possible on public and open databases, for example, European ones, automatically loaded in the device settings. Initiatives like the EU Open Data Portal [13] aim to unlock data already collected by governments and institutions free of charge and without copyright. The PLANHEAT toolkit only asks for a relatively small amount of initial information from municipalities, like the municipal and district boundaries, the location of monuments or historic buildings with high cultural and aesthetic value, or particular consumption habits or resources used. Municipalities can replace the data provided in the toolkit when more detailed local data becomes available, in order to increase the accuracy of the results. The ability to use generic data at a lower detail level makes it possible to get started quickly without having to spend significant effort on data collecting. This will also reveal which layers to concentrate efforts for additional data collection, and which ones can be investigated later. For example, a map coming from the disaggregation process based on NUTS3 datasets (Nomenclature of territorial units for statistics) will allow visualizing the peaks of demand or the greatest territorial potentialities, which represents a crucial starting point for increasing the quality of urban energy strategies and planning.

The PLANHEAT tool consists of three modules: the mapping module (mapping local demand and supply sources), the planning module (plan new scenarios based on residual and renewable energy sources), and the simulation module (simulation of the new scenarios and Key Performance Indicator (KPI) evaluation). At the side of the energy consumption and supply mapping module, the tool provides two approaches.

The first one is the City Mapping Module (CMM), which applies a top-down or disaggregation method (Figure 1) [16,22,23]. For this method, the input required from the user is very limited. Only aggregated final consumption figures (GWh, Gigawatthour) per sector are required, and the boundary or boundaries to which they apply (city, districts). Subdivision within each boundary is made by applying the appropriate geospatial indicator, some of which are available directly through the PLANHEAT web database. Examples of geospatial indicators are the CORINE land use map (Coordination of Information on the Environment) and OpenStreetMap (specifically building footprints) [22].

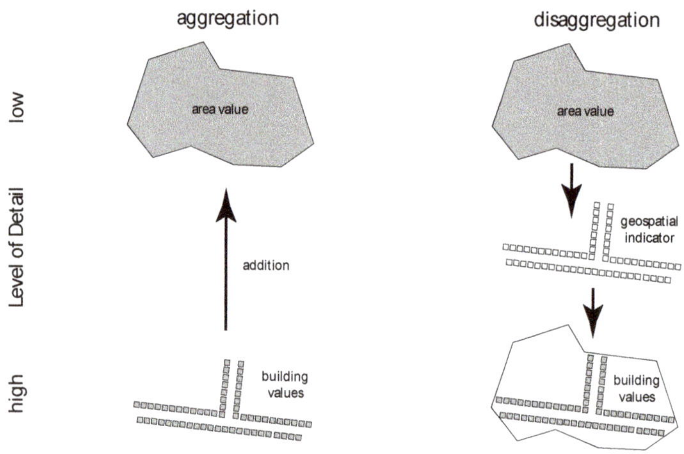

Figure 1. 'City' (top-down or disaggregation) and 'District' (bottom-up or aggregation) methods, as used in PLANHEAT [14].

The second one, the District Mapping Module (DMM), uses a bottom-up approach (aggregation, [16,22,23], Figure 1), by deriving heat demand from building characteristics (Table 2). Although more complicated, it provides two calculation routes: 'complete', which requires 11 variables and 'simplified', which will be less accurate, but requires only little input (a unique identifier for each building; gross floor area; age and use of each building) (Table 2) [22] shows a comparison between the input data for the 'complete' and 'simplified' methods. Both methods result in individual building values for use by the simulation module, that can also be aggregated for mapping purposes.

Table 2. Input data requirement comparison of the DMM calculation methods [22].

Method:	ID Building	Centroid	Age	Use	Total Height	Gross Floor Area	Number of Floors	Roof Area	Facade Area	Volume	Protection Degree
Complete	Y	Y	Y	Y	Y	Y	Y	Y	Y	Y	Y
Simplified	Y	Y	N	N	N	Y	N	N	N	N	N

The DMM was tested on a district close to Lecce, Italy, where little data was available. With this tool was possible determinate the district energy demand and identify opportunities for developing energy strategies based on RES and alternative energy resources. The validation test was incredibly useful for supporting the planning activity of the municipality.

On the supply side, both environmental and anthropogenic sources were included. Input data can be divided into technical parameters and geospatial indicators. An example is residual heat from wastewater treatment plants (WWTPs), for which the European Environment Agency had a base dataset that covered the EU28 [46,47]. Input data used are shown in Table 3.

Table 3. Geospatial indicators and technical variables used for residual heat potential from WWTPs.

Variable [Unit]	Data Specification	Default Maps
Sewage treatment plant locations	-	https://www.eea.europa.eu/data-and-maps/data/waterbase-uwwtd-urban-waste-water-treatment-directive-5
Sewage treatment plant, number of p.e. treated annually [p.e./yr]	(1) uwwLoadEnteringUWWTP: Number of person equivalents treated (note: actual use, not capacity). (2) If not available, use the national average uwwCapacity (based on known uwwCapacity and uwwLoadEnteringUWWTP in the UWWTP database)	https://www.eea.europa.eu/data-and-maps/data/waterbase-uwwtd-urban-waste-water-treatment-directive-5
Effluent of waste water treatment plants	Flow of the effluent of the waste water treatment plant	http://www.eea.europa.eu/themes/water/water-pollution/uwwtd/interactive-maps/urban-waste-water-treatment-maps-1
Sewage network temperature [°C]	Monthly average effluent temperature (12 values per year).	User required
Max cooling temperature [°C]	User specified, 2 °C will be the maximum realistic delta T.	2
Heat capacity [J/kg/K]	Heat capacity of sewage	4.2
Person equivalent [l/day]	EEA figure applicable to all of Europe, user may specify a more precise national or local figure	200

Although explicit levels of confidence were not implemented project wide, because in many cases a single data flow and corresponding continentally uniform dataset was chosen, the sewage heat recovery dataset contains two Confidence Levels (CLs). Measured flow is not available for some facilities, therefore residual heat potential was in some cases estimated based on capacity multiplied by the average national load (see Table 3), and therefore assigned a lower CL [14]. In future versions of the tool, which may contain a wider range of (higher or lower input data resolution) calculation methods catering to specific European regions, this feature will likely be expanded upon.

The full list of input data used in the PLANHEAT project is extensive, and can be found in the reports released by the project [16,18,22,24,46,47]. These represent a balance between accuracy, public availability, and full coverage of the EU28, and can be accessed through the PLANHEAT web database.

Experiences gained during the PLANHEAT project show that is possible to start an energy mapping process with relatively little data input. This opportunity is crucial for data-aware planning and to support policy-making both from an energetic and spatial perspective.

5. Conclusions and Recommendations

The energy transition will only be successful if it is integrated into the urban planning process [13,20]. As the need to increase the sustainability of the built environment is widely acknowledged, there is a clear need to rethink and implement new urban planning procedures in order to meet these expectations. A better understanding of and interactions between urban planning and energy issues are useful, not only for the planners themselves but also for the private sector, local communities and citizens who may take appropriate decisions and receive benefits from related economic and social added values [42].

The economic value could come from the increased economic activity and employment, while the social value could come from the improved social conditions within different communities of a city. Sharing responsibility makes the communities involved (Local energy communities).

Alhamwi et al. [21] and Manfren et al. [48] argue that the most innovative and advanced planning practices include communication strategies and actions for activating the participation of all urban actors. Supporting tools for visualizing urban phenomena and simulating future trends will play an increasing role in the coming years. The use of these tools supports inclusion and participation, because it allows the sharing of goals, shows the advantages of decisions, finds which actors will be involved in the processes, enables community participation (by the creation of local energy cooperatives), influences behaviour and increases overall awareness.

The energy sector produces economic benefits and attracts investments. In the transition from a centralized energy system to a distributed one, local economic opportunities increase. Energy potential maps are a useful tool to increase investment in clean energy and energy efficiency, providing the geospatially quantified information required to guide urban transformation processes. An energy potential map can also influence investor choices, attract external financing and demonstrate potential economic income, increasing the attractiveness and competitiveness of the city. The renewable energy sector is an expanding market, which rewards the most virtuous cities, communities and companies. In the long term, a city's ability to deal with climate change and offer high-quality, healthy environments will depend on its capacity to understand these complex phenomena. Supporting tools can represent a winning means to face new urban challenges and boost sustainable development.

The opportunity of applying innovative methodologies (and the tools that facilitate these) does not necessarily result in immediate concrete application of the opportunities identified. One of the problems that cause this is a lack of availability of required and suitable data. Using appropriate energy data in the planning process (both spatial and energy related) allows for a more effective analysis, as well as more effective interventions and strategies. Collecting data issues can be divided into two main categories: technical issues and socio-economic issues. Research on open data maturity in Europe (EU28+) [5] shows that countries completed over 55% of their open data journey with the development of basic open data policies and open data portal. Even if suitable datasets exist, the development of supporting tools is complicated. For this reason, the development of an incremental tool, like PLANHEAT, applicable in every context and based on public open data, permits a quick start and increases the usability.

From an open data perspective, the strategy should be the creation of open data policies, the increase of policy quality, and the setup of standards and licenses for the publication, all of which will in turn improve the availability of suitable energy data. Sometimes, a lack of standardization is related to a desire to facilitate data publication from data owners. Setting up standards would require a bigger effort from them, which might also translate to higher related costs. However, standardization is crucial for making the interoperability of datasets possible and increasing their use (or re-use) by local authorities.

Energy issues and spatial planning are tightly connected. For this reason, facilitating the vertical and horizontal coordination between urban and energy stakeholders is very important, in order to provide energy demand and supply data, to invest in integrated projects and to actively participate in the energy transition [42].

Author Contributions: Conceptualization, M.F. and A.B.; Methodology, A.B. and M.F.; Validation, M.F., P.D.P.; Formal Analysis A.B.; Resources, M.F. and A.B.; Data Curation, M.F. and A.B.; Writing-Original Draft Preparation, A.B., M.F.; Writing—Review & Editing, A.B., M.F.; Supervision, M.F. and P.D.P.; Project Administration M.F. and A.B. All authors have read and agreed to the published version of the manuscript.

Funding: Part of the research leading to this paper has received funding from the European Research Council (ERC) under the European Union's Horizon 2020 research and innovation programme (grant agreement n° 723757).

Acknowledgments: The authors would like to thank all PLANHEAT partners: RINA Consulting; Delft University of Technology; University of Zagreb; National Observatory of Athens; Tecnalia; Vito; Lecce local authority; Antwerp local authority; Velika Gorica local authority; the Regional environmental center for central and eastern Europe; Euroheat & Power; Geonardo; Artelys.

Conflicts of Interest: The authors declare no conflict of interest. The funding sponsors had no role in the design of the study; in the collection, analyses, or interpretation of data; in the writing of the manuscript, and in the decision to publish the results.

References

1. Masson-Delmotte, V.; Zhai, P.; Pörtner, H.-O.; Roberts, D.; Skea, J.; Shukla, P.R.; Pirani, A.; Moufouma-Okia, W.; Péan, C.; Pidcock, R.; et al. IPCC Special Report. In *Global Warming of 1.5 °C*; Summary for Policymakers; IPCC: Geneva, Switzerland, 2018; ISBN 978-92-9169-151-7.
2. Ann, J.; Mills, G. The role of urban form as an energy management parameter. *Energy Policy* **2013**, *53*, 218–228.
3. European Legislation on Open Data and the Re-Use of Public Sector Information. Available online: http://ec.europa.eu/digital-agenda/en/european-legislation-reuse-public-sector-information (accessed on 28 February 2020).
4. Directive 2003/98/EC of the European Parliament and of the Council of 17 November 2003 on the Re-Use of Public Sector Information. Available online: https://eur-lex.europa.eu/legal-content/en/ALL/?uri=CELEX%3A32003L0098 (accessed on 28 February 2020).
5. Carrara, W.; Nieuwenhuis, M.; Vollers, H. Open Data Maturity in Europe. Report 2016. Available online: https://euagenda.eu/publications/open-data-maturity-in-europe-2016 (accessed on 28 February 2020).
6. Knowles, R. *Energy and Form: An Ecological Approach to Urban Growth*; MIT Press: Cambridge, UK, 1974.
7. Owens, S. *Energy, Planning and Urban Form*; Pion Limited: London, UK, 1986.
8. Jaccard, M.; Failing, L.; Berry, T. From equipment to infrastructure: Community energy management and greenhouse gas emissions reduction. *Energy Policy* **1997**, *25*, 1065–1074. [CrossRef]
9. Steemers, K. Energy and the city: Density, buildings and transport. *Energy Build.* **2003**, *35*, 3–14. [CrossRef]
10. Madlener, R.; Sunak, Y. Impacts of urbanization on urban structures and energy demand: What can we learn for urban energy planning and urbanization management? *Sustain. Cities Soc.* **2011**, *1*, 45–53. [CrossRef]
11. Stoeglehner, G.; Neugebauer, G.; Erker, S.; Narodoslawsky, M. *Integrated Spatial and Energy Planning: Supporting Climate Protection and the Energy Turn with Means of Spatial Planning*; Springer: Berlin/Heidelberg, Germany, 2016.
12. Pereira, I.M.; Sad de Assis, E. Urban energy consumption mapping for energy management. *Energy Policy* **2013**, *59*, 257–269. [CrossRef]
13. Hemis, H. Integrating ENERGY in Urban Planning Processes—Insights from Amsterdam/Zaanstad, Berlin, Paris, Stockholm, Vienna, Warsaw and Zagreb. Report Urban Learning. Available online: http://www.urbanlearning.eu/fileadmin/user_upload/documents/D4-2_Synthesis-report_upgraded_processes_final_170807.pdf (accessed on 28 February 2020).
14. Owens, S. Land-use planning for energy efficiency. *Appl. Energy* **1992**, *43*, 81–114. [CrossRef]
15. Anderson, W.P.; Kanaroglou, P.S.; Miller, E.J. Urban form, energy and environment: A review of issues evidence and policy. *Urban. Stud.* **1996**, *33*, 7–55. [CrossRef]
16. PLANHEAT. D1.2 Report on End-Users' Current Status, Practices and Needs in H&C Plans. Report 2017. Available online: https://ec.europa.eu/research/participants/documents/downloadPublic?documentIds=080166e5aff3c1c4&appId=PPGMS (accessed on 28 February 2020).
17. Cornelis, E.; Meinke-Hubeny, F. Stratego Project. Local Action: Methodologies and Data Sources for Mapping Local Heating and Cooling Demand and Supply. Report 2015. Available online: https://ec.europa.eu/research/participants/documents/downloadPublic?documentIds=080166e5ba5e00b8&appId=PPGMS (accessed on 28 February 2020).
18. PLANHEAT. D1.4 Common IT Framework Specifications. Report 2017. Available online: http://planheat.eu/project-documents (accessed on 28 February 2020).
19. Stremke, S.; van den Dobbelsteen, A. *Sustainable Energy Landscapes. Designing, Planning, and Development*; Stremke, S., Ed.; Taylor & Francis: Copenagen, Denmark, 2013.
20. Hemis, H. Review of Current Governance Processes of Urban and Energy Planning in Amsterdam/Zaanstad, Berlin, Paris, Stockholm, Vienna, Warsaw and Zagreb. Available online: https://ec.europa.eu/research/participants/documents/downloadPublic?documentIds=080166e5adf1c1d3&appId=PPGMS (accessed on 28 February 2020).
21. Alhamwi, A.; Medjroubi, W.; Vogt, T.; Agert, C. GIS-based urban energy systems models and tools: Introducing a model for the optimisation of flexibilisation technologies in urban areas. *Appl. Energy* **2017**, *191*, 1–9. [CrossRef]
22. PLANHEAT. D5.1 Presentation on PLANHEAT Integrated Tool Functionalities. Report 2017. Available online: http://planheat.eu/project-documents (accessed on 28 February 2020).

23. PLANHEAT. Project Brief. Available online: http://planheat.eu/project-brief (accessed on 28 February 2020).
24. Delponte, I. Achieving Smart Energy Planning Objectives. The Approach of the Transform Project. *TeMA* **2014**. [CrossRef]
25. PLANHEAT. D1.7 Overcoming Barriers in Data Collection. Report 2018. Available online: http://planheat.eu/project-documents (accessed on 28 February 2020).
26. Cornelis, E.; Meinke-Hubeny, F. Stratego Project. Local Action: Methodologies and Data Sources for Mapping Local Heating and Cooling Demand and Supply. Available online: https://www.google.com/url?sa=t&rct=j&q=&esrc=s&source=web&cd=1&ved=2ahUKEwjN_OSy1ozoAhVU-WEKHQwZBbAQFjAAegQIBBAB&url=https%3A%2F%2Fwww.euroheat.org%2Fwp-content%2Fuploads%2F2016%2F04%2FD3.7_Methodologies-and-data-sources-for-mapping.pdf&usg=AOvVaw1nVLcUafGDsEk7urnjYJn4 (accessed on 1 November 2019).
27. Cornelis, E.; Holm, A.B.; Lauersen, B.; Lygnerud, K. Stratego Project. Insights from Drafting Local Heating and Cooling Action Plans. Available online: https://www.google.com.hk/url?sa=t&rct=j&q=&esrc=s&source=web&cd=1&ved=2ahUKEwiF4ZXn14zoAhWLHHAKHR9vCrEQFjAAegQIBxAB&url=https%3A%2F%2Fwww.euroheat.org%2Fwp-content%2Fuploads%2F2016%2F04%2FD3.d-WP3-Final-Report.pdf&usg=AOvVaw3kn6iVDl_t2yuoa-GjYcTD (accessed on 1 November 2019).
28. Scottish Government Scotland Heat Map. User Guide. 2.0 Methodology Report. Available online: https://www.gov.scot/binaries/content/documents/govscot/publications/advice-and-guidance/2018/11/scotland-heat-map-documents/documents/scotlands-heat-map-user-guidance/2.0-report-methodology/2.0-report-methodology/govscot%3Adocument/Scotland%2527s%2Bheat%2Bmap%2B2.0%2Breport%2Bmethodology%252C%2B17%2BNovember%2B2015.pdf (accessed on 27 February 2020).
29. TRANSFORM Open Energy Data: A Prerequisite for Cities to Become Low-Carbon. Available online: https://smartcities-infosystem.eu/sites/www.smartcities-infosystem.eu/files/transform_open_energy_data_-_a_prerequisite_for_cities_to_become_low-carbon.pdf (accessed on 1 November 2019).
30. De Sousa, L.; Eykamp, C.; Leopold, U.; Baume, O.; Braun, C. iGUESS–A Web-Based System Integrating Urban Energy Planning and Assessment Modelling for Multi-Scale Spatial Decision Making 2012. Available online: https://scholarsarchive.byu.edu/iemssconference/2012/Stream-B/293/ (accessed on 1 November 2019).
31. Broersma, S.; Fremouw, M.; van den Dobbelsteen, A. Heat mapping the Netherlands. Laying the foundations for energy-based planning. In Proceedings of the 6th World sustainable building Conference SB11, Helsinki, Finland, 18 –21 October 2011.
32. Broersma, S.; Fremouw, M.; van den Dobbelsteen, A. Energy potential mapping: Visualising energy characteristics for the exergetic optimisation of the built environment. *Entropy* **2013**, *15*, 490–506. [CrossRef]
33. Musco, F. *Rigenerazione Urbana e Sostenibilità*; FrancoAngeli: Milano, Italy, 2009.
34. ARUP Decentralised Energy Masterplanning. A manual for Local Authorities. Report 2011. Available online: https://www.google.com.hk/url?sa=t&rct=j&q=&esrc=s&source=web&cd=1&ved=2ahUKEwie_caxv4zoAhWwy4sBHTMVBJUQFjAAegQIBhAB&url=https%3A%2F%2Fwww.theade.co.uk%2Fassets%2Fdocs%2Fresources%2FDENet_manual_lo_v10.pdf&usg=AOvVaw0fReu8qQRPyLDGmlWSe9sR (accessed on 1 November 2019).
35. van den Dobbelsteen, A.; Jansen, S.; Vernay, A.L.; Gommans, L. Building within an energetic context: Low-exergy design based on local energy potentials and excess or shortage of energy. In Proceedings of the PLEA Conference, Singapore, 22–24 November 2007.
36. Stoeglehner, G.; Niemetz, N.; Kettl, K.H. Spatial dimensions of sustainable energy systems: New visions for integrated spatial and energy planning. *Energy Sustain. Soc.* **2011**, *1*, 1–9. [CrossRef]
37. Stoeglehner, G.; Narodoslawsky, M. Energy-Conscious Planning Practice in Austria: Strategic Planning for Energy-Optimized Urban Structures. In *Sustainable Energy Landscapes. Designing, Planning, and Development*; Stremke, S., Stremke, S., van den Dobbelsteen, A., Eds.; Taylor & Francis: Copenagen, Denmark, 2013; pp. 355–370.
38. Cherix, G.; Capezzali, M.; Rager, J. Territorial energy systems: A methodological approach and case study. In Proceedings of the 10th Conference on Sustainable Development of Energy, Water and Environment Systems, Dubrovnik, Croatia, 27 September–2 October 2015.
39. Van Gessel, S.; Bader, A.G.; Bialkowski, A.; Beccaletto, L.; Begemann, L. Energy Storage Data Collection. Report 2016. Available online: http://www.estmap.eu/downloads/ESTMAP-D3.04-v2016.12.14-Datacollection-report-public.pdf (accessed on 28 February 2020).

40. Meskel, E.; Weber, P. Review of Instruments and Tools Used for Energy and Urban Planning in Amsterdam/Zaanstad, Berlin, Paris, Stockholm, Vienna, Warsaw and Zagreb. Available online: http://www.urbanlearning.eu/fileadmin/user_upload/documents/D3.2_Synthesis_report_instruments_tools_170425_final.pdf (accessed on 28 February 2020).
41. Meshartility report D5.4 Recommendations for EU and National Policymakers on Improving the Collection and Access to Energy Data. Available online: http://www.meshartility.eu/images/documents/MESHARTILITY_deliverable_5.4.pdf (accessed on 28 February 2020).
42. Cajot, S.; Peter, M.; Bahu, J.M.; Guignet, F.; Koch, A.; Maréchal, F. Obstacles in energy planning at the urban scale. *Sustain. Cities Soc.* **2017**, *30*, 223–236. [CrossRef]
43. Stremke, S.; Kohn, J. Ecological concepts and strategies with relevance to energy-conscious spatial planning and design. *Environ. Plan. B Plan. Des.* **2010**, *37*, 518–532. [CrossRef]
44. Treaty on the Functioning of the European Union 2012/C 326/. Available online: http://data.europa.eu/eli/treaty/tfeu_2012/oj (accessed on 28 February 2020).
45. Regulation (EU) 2016/679 of the European Parliament and of the Council of 27 April 2016 on the Protection of Natural Persons with Regard to the Processing of Personal Data and on the free Movement of such Data, and Repealing Directive 95/46/EC (General Data Protection Regulation). Available online: https://eur-lex.europa.eu/eli/reg/2016/679/oj (accessed on 28 February 2020).
46. PLANHEAT. D2.5 Methods to Quantify and Map Unconventional Heating and Cooling Sources. Report 2018. Available online: http://planheat.eu/project-documents (accessed on 28 February 2020).
47. PLANHEAT. D2.6 Models for Quantifying and Mapping Energy Potential from Renewable Energy Sources. Report 2018. Available online: http://planheat.eu/project-documents (accessed on 28 February 2020).
48. Manfren, M.; Caputo, P.; Costa, G. Paradigm shift in urban energy systems through distributed generation: Methods and models. *Appl. Energy* **2011**, *88*, 1032–1048. [CrossRef]

energies
MDPI

Article

Assessment of the Space Heating and Domestic Hot Water Market in Europe—Open Data and Results

Simon Pezzutto [1,*], Silvia Croce [1,2], Stefano Zambotti [1], Lukas Kranzl [3], Antonio Novelli [1] and Pietro Zambelli [1]

1 Institute for Renewable Energy, European Academy of Bozen/Bolzano (EURAC Research), Viale Druso 1, 39100 Bolzano, Italy; silvia.croce@eurac.edu (S.C.); stefano.zambotti@eurac.edu (S.Z.); antonio.novelli@eurac.edu (A.N.); pietro.zambelli@eurac.edu (P.Z.)
2 Department of Civil, Environmental and Architectural Engineering, University of Padova, Via Marzolo 9, 35131 Padova, Italy
3 Institute of Energy Systems and Electric Drives, Energy Economics Group, TU Wien, Gusshausstrasse 25-29/370-3, 1040 Vienna, Austria; kranzl@eeg.tuwien.ac.at
* Correspondence: simon.pezzutto@eurac.edu; Tel.: +39-0471-055-622

Received: 13 April 2019; Accepted: 8 May 2019; Published: 9 May 2019

Abstract: The paper investigates the European space heating (SH) and domestic hot water (DHW) market in order to close knowledge gaps concerning its size. The stimulus for this research arises from incongruences found in SH and DHW market's data in spite of over two decades of scientific research. The given investigation has been carried out in the framework of the Hotmaps project (Horizon 2020—H2020), which aims at designing an open source toolbox to support urban planners, energy agencies, and public authorities in heating and cooling (H&C) planning on country, regional, and local levels. Our research collects and analyzes SH and DHW market data in the European Union (EU), specifically the amount of operative units, installed capacities, energy efficiency coefficients as well as equivalent full-load hours per equipment type and country, with a bottom-up approach. The analysis indicates that SH and DHW account for a significant portion of the total EU energy utilization (more than 20%), amounting to almost 3900 TWh/y. At the same time, the energy consumption provided by district heating (DH) systems exceeds the one of condensing boilers. While DH systems applications are growing throughout the EU, the replacement of elderly, conventional boilers progresses at a slower pace.

Keywords: space heating; domestic hot water; market assessment; EU28; district heating; open data

1. Introduction

While the member states (MS) of the European Union (EU) aim at reaching an integrated policy framework, especially directed at delivering sound market regulations to investors, national policies have been focused on a permanent improvement in efficiency by bringing the share of energy generated by renewable energy sources (RES) to 27% within 2030. Moreover, EU MS explicitly set a reduction in greenhouse gas (GHG) emissions goal of 40% compared to 1990 levels by 2030 [1], aiming to 80–95% by 2050 [2]. The achievement of the Paris Conference of the Parties 21 (COP21) accordance will require reaching at least the upper bound of this range [3].

In 2015, EU's primary energy consumption accounted for about 1600 Mtoe/y, of which a major contribution is provided by heating and cooling (H&C) applications (about 800 Mtoe/y, including also industrial heat), followed by transport and electricity (about 490 Mtoe/y and 310 Mtoe/y respectively) [4–8]. Buildings are responsible for approximately 640 Mtoe/y, which corresponds to 40% of the whole EU primary energy consumption [9,10]. The largest parts of energy utilization within the EU building stock (~75% in total), according to an order of magnitude, occur for space heating (SH), domestic hot water (DHW), and space cooling (SC) [11]. In the past three decades, EU MS invested massively in assessing the energy used by the different sectors [12–17]. In contrast to SC, the SH and DHW field is well researched in the scientific literature since more than two decades apart [7,18].

In particular, the EC supported a number of studies to provide quantitative data in this area, and inform related energy strategies and roadmaps (e.g., EC 2019 [19], Pardo et al. 2012 [20], and EC 2011 [21]). Further notable studies on the SH and DHW market in Europe were the result of various projects, such as the H2020 HRE4 [22], IEE STRATEGO [23], and Seventh Framework Programme (FP7) iNSPiRe [24]. Moreover, the following reports provide valuable insights into the investigated market: Patronen et al. 2012 [25], Boermans et al. 2012 [26], Von Manteuffel et al. 2016 [27], and Sanner et al. 2011 [28]. Finally, also scientific journal papers like Scoccia et al. 2018 [29], Balaras et al. 2007 [30], and Leurent et al. 2018 [31] contribute to the understanding of the SH and DHW market in Europe.

Carried out in the framework of the Horizon 2020 (H2020) Hotmaps project [32], our study generated a data repository for the SH and DHW market [33] (sources are available in the respective csv file [34]). The data are released as open data under the Creative Commons license CC-BY 4.0 [35]. This EU-funded project aims at designing an open source toolbox (released under the Apache 2.0 license [36]) to support urban planners, energy agencies, and public authorities in heating and cooling (H&C) planning at different scales (national, regional, and local), and in line with EU policies. As part of the project, the data analyzed in the present paper have been collected through a bottom-up approach. Based on the analysis carried out in [37], a more comprehensive investigation has been performed. In particular, we took a closer look at the uncertainty of generated results, provided an interpretation of the main outcomes, compared the main result with related findings of scientific literature as well as discussed its implications.

In order to create a high quality data set—characterized by completeness, accuracy, and reliability—in the framework of our analysis we place a special focus on the following aspects:

- Data inventory;
- Data reliability;
- Data definition and comparability [37–39].

1.1. Data Inventory

One of the main challenges of creating an inventory of SH and DHW market technologies consists of preparing an exhaustive list of all existing data. Generally, the use of data collected at EU-wide level offers unique advantages due to their extensive territorial scope (e.g., EurObserv'ER [40], EUROHEAT&POWER [41], and EHPA [42]). However, data completeness can never be fully ensured.

Attempts of closing data gaps require not only extrapolating and assembling data from large data sets available online (e.g., EU Buildings Database [43], EHPA's Online Stats Tool [44], and IGA [45]). To ensure a rigorous approach and address the lack of data, it also encompasses searching data source-by-source, especially by using individual scientific literature sources such as journal papers (e.g., Bertoldi et al. 2012 [46], Martinopoulos et al. 2018 [47], and Clay 2015 [48]).

One important aspect of the data inventory is to ensure that the information can be understood and interpreted correctly by any user. This requires a compilation of clear metadata description, annotation, contextual information, and documentation. The data documentation provides standardized structured information, indicating the creator, title, time references, access conditions, and terms of use of the data collection (please see Pezzutto and Zambelli 2019 [34] and Pezzutto and Zambelli 2019 [49]). The data repository is structured following the Frictionless data standards [50], to encode and describe the metadata and the main data set information using a data package.json file that is readable by both human and computers. A more detailed insight on the methodology that produced the data set is provided by the respective README.md file [51]. The license of the data repository is encoded using the Software Package Data Exchange (SPDX) format [52], in order to univocally identify the license. The Hotmaps' project selects a git repository to publish the data set, instead for example of a File Transfer Protocol (FTP) service, because a git service allows to: (i) preserve the history of the changes; (ii) perform an automatic versioning of the data that univocally identified the data set; and (iii) provide the functionalities to manage and discuss external contributions (e.g., open/assign/close issues, accept/comment/reject data modifications, etc.). Further details on the data inventory can be found in the Hotmaps Data Management Plan (DMP) [53].

1.2. Data Reliability

Much effort has been dedicated to analyze sources, assess the reliability of the gathered data, and fill existing gaps by in-depth investigations. We discern various types of information, by analyzing the different approaches applied for the collection of the identified data (e.g., amount of SH and DHW sold vs. operative units). In case of lacking or uncertain documentation, the data have not been considered for the development of the database.

All information collected on SH and DHW (i.e., amount of operative units, installed capacities, energy efficiency coefficients as well as equivalent full-load hours per equipment type and country) have been filtered and evaluated statistically; the methodology adopted is described in Section 2. Materials and Methods. Moreover, additional sources and types of information have been used to validate the outcomes obtained for the EU28 (see Section 4—Discussion) to assess their reliability.

1.3. Data Definition and Comparability

Although most data providers use standardized data formats and units, this does not necessarily mean that data are entirely comparable. In order to increase data comparability, the entire process of data elaboration requires adjusting differences and inconsistencies resulting from different methods, assumptions, measures, time references, and specifications [54].

Data have been collected for each EU MS using the most recent year available, while data over a decade old have been excluded (please see [34]). The developed data sets including the documentation are expected to improve data quality, add value to already existing data and provide data needed to monitor the progress of the SH and DHW field in Europe.

2. Materials and Methods

Our main data sources were derived from previous works. In particular, to those elaborated by AALBORG UNIVERSITY, HALMSTAD UNIVERSITY, and EUROPA-UNIVERSITÄT FLENSBURG in the context of several projects dedicated to the topic, including the data sets of the H2020 project Heat Roadmap Europe 4 (HRE4) [55], and the Intelligent Energy Europe (IEE) project STRATEGO (Multi level actions for enhanced Heating and Cooling plans) [56]. Another source, relevant for the data set compilation of the present investigation, is the data collection of the tender "Mapping and analyses of the current and future (2020–2030) heating/cooling fuel deployment (fossil/renewables)—ENER/C2/2014-641" led by the Fraunhofer Institute for Systems and Innovation Research—FH ISI [57].

Besides the deliverables achieved through the projects above (such as [58–60]), "Deliverable 2.1 Intermediate analysis of the heating and cooling industry" [61] was key in carrying out our work. The deliverable was produced within the tender "Support to key activities of the European technology platform on renewable heating and cooling"—PP-2041/2014.

Additional important information are provided by reports of Solar Heat Worldwide (e.g., [62,63]), EUROSTAT [64], and the TABULA WebTool [65]. Scientific publications have also been used as data sources, e.g., [66–68]. Given the large amount of references, in Section 3. Results and Table 1, only the major ones are indicated. Table 1 summarizes the most relevant data sources per type of information researched.

Table 1. Key data sources for amount of operative units, installed capacities, energy efficiency coefficients, and equivalent full-load hours per space heating (SH) and domestic hot water (DHW) equipment type and country (EU28), and information on public availability of data.

Source	Amount of Operative Units	Installed Capacities	Equivalent Full-Load Hours	Energy Efficiency Coefficients	Public Available
EurObserv'ER [40]	×				Yes
EUROHEAT&POWER [41]	×				Yes
HRE4 project [55]	×	×		×	No
STRATEGO project [56]	×	×		×	No
Tender ENER/C2/2014-641 [57]	×		×	×	Yes
Dengler et al. 2012 [58]	×	×		×	Yes
Persson et al. 2017 [59]		×	×	×	Yes
Connolly et al. 2016 [60]		×	×	×	Yes
Fedrizzi et al. 2016 [61]		×	×		No
Mauthner et al. 2017 [62]	×	×	×		Yes
Mauthner et al. 2016 [63]	×	×	×		Yes
TABULA project [65]	×			×	Yes
Nouvel et al. 2015 [68]				×	No
Pezzutto 2014 [69]		×	×	×	No

The analysis started by considering different SH and DHW technologies installed throughout Europe. The data were collected for each MS—as sources mainly provide information at country level—and were not subdivided by sector. The equipment typologies were categorized as found in [58,62,69]:

- Boilers:
  - Non-condensing;
  - Condensing;
- Stoves;
- Electric radiators;
- Heat pumps (HPs):
  - Aerothermal;
  - Geothermal;
- Solar thermal systems (STS):
  - Unglazed collectors;
  - Flat-plate collectors;
  - Evacuated tube collectors;
- Combined heat and power—Internal combustion (CHP-IC);
- District heating (DH).

In the list above, furnaces were classified in the category "Boilers, Non-condensing".

For each MS and type of equipment, data regarding number of units, installed capacity, yearly equivalent full-load hours, and energy efficiency coefficients were collected. With regard to energy efficiency coefficients, the absolute majority of the technologies identified were characterized by thermal efficiency. These include condensing and non-condensing boilers, stoves, electric radiators, CHP-IC units, and various solar thermal systems (unglazed, flat plate, and evacuated tube collectors). Aerothermal and geothermal HPs were instead described by the coefficient of performance (COP). In order to estimate the efficiency of DH systems, mean losses were included by considering DH network heat losses [70].

We also researched information on system types—in percentage at country level—as well as resources used to fuel each equipment considered. The latter are classified as proposed in [58,64,69]:

- Oil;
- Gas (natural gas);
- Coal;
- Renewables;
- Other fuels.

The category "Other fuels" included less dispersed combustibles (e.g., coke, peat, etc.) [64].

The data analysis was based on a bottom-up approach, which included an extensive literature analysis aimed at deriving reliable values. Data collected from scientific literature sources were filtered and statistically analyzed.

As a first step, for each MS and category of information (i.e., number of units, installed capacity, yearly equivalent full-load hours, and energy efficiency coefficients) at least three data were collected from different sources when possible. Then, their mean values were calculated. Depending on the amount of references, data that departed between a range of plus or minus one standard deviation around the mean of the respective data pool were excluded. The resulting numbers were utilized to calculate a more robust mean.

The sources of the data used as input cannot always be classified as open data, but the results of the statistical elaboration were released as open data. The published data set in [34] explicitly specifies when a value is the result of the statistical elaboration using more than one source (tagged as "Own calculation") or derived by a single source solely; for the latter, the source is specified.

Moreover, the coefficient of variation (CV) was utilized as a statistical indicator of uncertainty for generated values. The CV is the ratio of the standard deviation to the mean. The higher the CV the higher is the dispersion around the mean. Generally, it is indicated as a percentage [71,72] (displayed at the top of the columns in Figures 1–4). Unfortunately, due to missing data, it was not always possible to retrieve two or more data for each investigated value; in these cases, no statistical elaborations were carried out. In a minor amount of cases, data were extrapolated from one country, where data were available, to another, where data were missing—whenever in the presence of geographical, socio-economic, and historical similarities. Extrapolation of data was applied to the following countries:

- Czech Republic and Slovakia;
- Bulgaria and Romania;
- Estonia, Latvia, and Lithuania.

This approach was only applied with regard to mean installed capacities, efficiencies, and equivalent full-load hours. We did not put forward any specific assumption on mean installed capacities for DH systems. Zero values are present only when this was supported by one or more references, e.g., showing that no DH systems are available in Malta so far [73]. In a few specific cases, when information was available only at aggregated EU28 level, data were applied to all MS equally. As an example, this was the case of values regarding the mean installed capacity of stoves [61].

Based on the methodology proposed by Pezzutto et al. 2017 [7], once the data collection has been concluded, mean capacities installed per technology have been divided by their respective energy efficiency coefficients to obtain the work input (*W*) per equipment type. Then, in order to obtain energy consumption values per equipment type and sector, the number (*Nr.*) of units was multiplied by the equivalent full-load hours (*T*—time) in a year and by its work input (*W*) using the following Equation (1):

$$Energy\ Consumption_{SH\ \&\ DHW} = Nr._{units} \times T_{equivalent\ full\text{-}load\ hours} \times W \quad (1)$$

The utilized formula represents a simplified method to assess the energy consumption given by SH and DHW equipment at EU28 level, not differentiating between modulation and on-off equipment, as well as not considering partial load operation, efficiency of sources depending on its level of load, and accumulation of energy in buildings. Mainly due to not taking into consideration partial load operation, the used methodology thus might underestimate the assessed energy consumption.

To compare this investigation outcome with others in the scientific literature (please see Section 4. Discussion), we converted energy demand in energy consumption values by multiplying by 1.15. An important distinction is in place: What is meant with energy demand and energy consumption. The first is the net energy necessary to satisfy both SH and DHW needs. The second represents instead the input of energy at the level of devices necessary to cover the demand. On the basis of these definitions, the values of these quantities differ by a conversion factor. With these premises, since a boiler's efficiency is <1 (about 0.8–0.9 for those currently installed in Europe), energy consumption values are always higher compared to demand [37,66,69].

As the applied methodology relies on a number of assumptions, the main uncertainties associated with the final results were the following:

- The use of average data at the EU level, when data for each MS were not available, and of efficiency coefficients constants for all MS were necessary to fill the existing gaps in EU databases on SH and DHW and to perform the analysis for estimating the European SH and DHW market. However, these hypotheses result in inaccuracies related to the correctness of the final data relatively to each MS.
- Correspondence of full-load hours, efficiency, and mean installed capacity were assumed in case of missing data for countries with geographical, socio-economic, and historical similarities. However, SH and DHW systems are not always conforming and there might be differences between regions. This might also be influenced by diversities in climatic conditions.

At this point, it has to be stressed that the amount of data subject to assumptions accounted for approximately 4% of those needed to generate the results of the present investigation.

- The utilization of an EU-wide mean value to turn SH and DHW demand into energy consumption leads to imprecisions, given the energy efficiency level taken into consideration refers to boilers only. However, the considered equipment is the most diffused in Europe [57,58].

3. Results

The present paper displays the main results aggregated at the whole EU28 level and not for each MS individually. The entire data set, with detailed data for each MS, including sources, is available as open data in the Hotmaps git repository under [33] under [34]. The results at EU28 scale, per each type of equipment, regarded all the main data categories used for the estimation of the final energy consumption, i.e., number of installed units, equivalent full-load hours, mean installed capacity, and energy efficiency coefficients. Finally, the distribution of energy consumption per equipment type at EU28 level was presented and discussed.

In the column charts of Figure 1, the error bars indicated standard deviations, and above positioned percentages the coefficient of variation (CV).

With regard to installed units, Figure 1 shows the amount of SH and DHW units per equipment type at EU28 level (in millions—Mil.). Non-condensing boilers had the greatest diffusion, with about 80 Mil. installed devices, followed by stoves (60 Mil.). Other technologies, in order of distribution magnitude are electric radiators (approximately 30 Mil. units), condensing boilers, and aerothermal HPs, with about 10 Mil. units, respectively. They are followed by geothermal HPs (2 Mil. units) and STS flat-plate collectors (about 1 Mil. units). STS-evacuated tube collectors, CHP-IC, STS-unglazed collectors, and DH were less diffused equipment, with 0.14, 0.05, 0.03, and 0.02 Mil. units, respectively.

Looking at the average CV percentages per equipment type related to SH and DHW at EU28 level (indicated on the top of the columns over the bars in Figure 1) we inferred that the data building these bars were highly unequally distributed. The overall CV percentage was 34%. The highest variation was the one of condensing boilers (~66%), followed by aerothermal HPs (~52%). The lowest variation was the one of stoves (~8%). Other equipment types were characterized by variations around 30%.

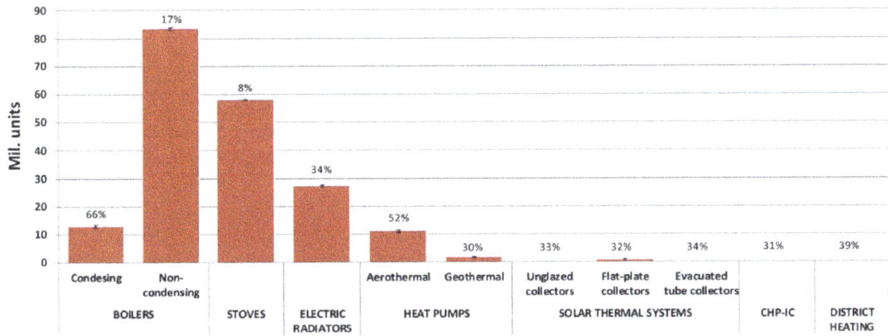

Figure 1. Number of operative units for SH and DHW per equipment type, EU28 [40,41,55–58,62,63,65] (As not visible in Figure 1: HP geothermal = 1.93, STS unglazed, flat-plate, and evacuated tube collectors = 0.03, 0.97, and 0.14, CHP-IC = 0.05, DH = 0.02 Mil. units.).

Figure 2 displays the annual distribution of equivalent full-load hours per equipment type. CHP-IC units had the highest mean value of full-load hours per year, nearly 1900 hours (h). Boilers (condensing and non) were second with over 1000 h. Equivalent full-load hours of electric radiators and DH amounted to 900 h each, closely followed by aerothermal HPs, with more than 700 h, and STS-flat-plate and STS-unglazed collectors with 400 h each. Geothermal HPs presented about 300 h, while stoves and STS-evacuated tube collectors were positioned last, with approximately 200 h each.

CV percentages included in Figure 2 indicated that the obtained data was rather dispersed. The mean value amounted to roughly 26%. The highest variation was given for STS-evacuated tube collectors (~53%), followed by the variation of STS-unglazed (~40%). The lowest variations related to condensing and non-condensing boilers (~6%). Other equipment types were characterized by variations between about 18% and 34%.

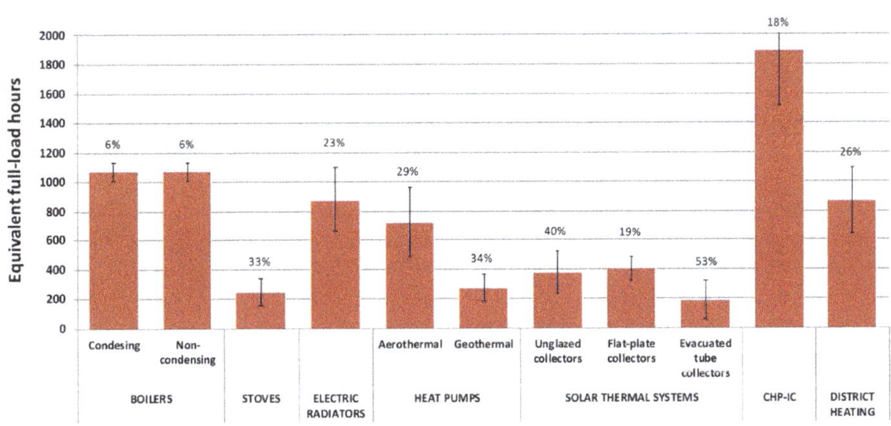

Figure 2. Distribution of mean SH and DHW units' equivalent full-load hours per equipment typology, EU28 [57,59–63,69].

The mean installed capacity per equipment type in kW is presented in Figure 3. DH's mean value exceeded Figure 3's axis indication, reaching a nominal number of almost 75,000 kW. CHP-ICs were characterized by means of about 200 kW. Next were STS-unglazed collectors with over 140 kW, followed by other STS types (i.e., flat-plate and evacuated tube collectors) with approximately 40 kW. Boilers (condensing and non) had a mean installed capacity of about 20 kW. Geothermal HPs and electric radiators came next with approximately 10 kW each. Conclusively, aerothermal HPs and stoves were positioned last with around 5 kW each.

In the case of average installed capacity per SH and DHW equipment type (EU28), the mean CV percentage was quite high, with an average of 38%. The highest variation related to stoves and aerothermal HPs (around 70%) followed by electric radiations (~60%). Geothermal HPs and CHP-IC followed with variations of about 40%. The lowest variations related to condensing and non-condensing boilers (~3%). Other equipment types were characterized by variations around 30%.

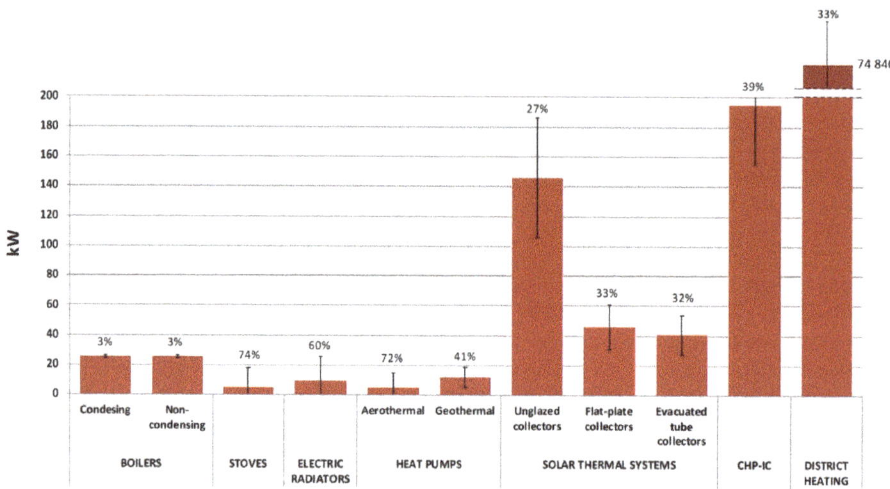

Figure 3. Mean installed capacity per equipment, EU28 [55,56,58–63,69].

Efficiency values at full-load (Figure 4) were evaluated by means of different indicators depending on the technology of the equipment type considered. Looking at technologies characterized by a thermal efficiency coefficient, we found that boilers and electric radiators had efficiency mean values near to 100%, with STS-unglazed collectors and non-condensing boilers being placed second and third with around 90% and 85%. Other STS systems, flat-plate and evacuated tube collectors, presented values around 60%; while, CHP-IC units and stoves of respectively 58% and 50%. For technologies characterized by a COP coefficient, geothermal HPs were significantly more efficient than aerothermal ones, the energy efficiency of the former amounting to 4.5 and of the latter to 3.5. Indicated values referred to nominal COPs and were not related to real operating conditions for building uses. To fully consider DH systems' efficiency, we included in the mean losses those deriving from DH network heat losses. The heat losses mean value for EU28 was found to be 13.70% [70].

For SH and DHW equipment, energy efficiency coefficients at full-load (EU28) had CV mean percentages with values around 10%. The highest variation was found for stoves (~36%). The lowest one for electric radiators (~2%). Other equipment types were characterized by variations between 5% and 12%.

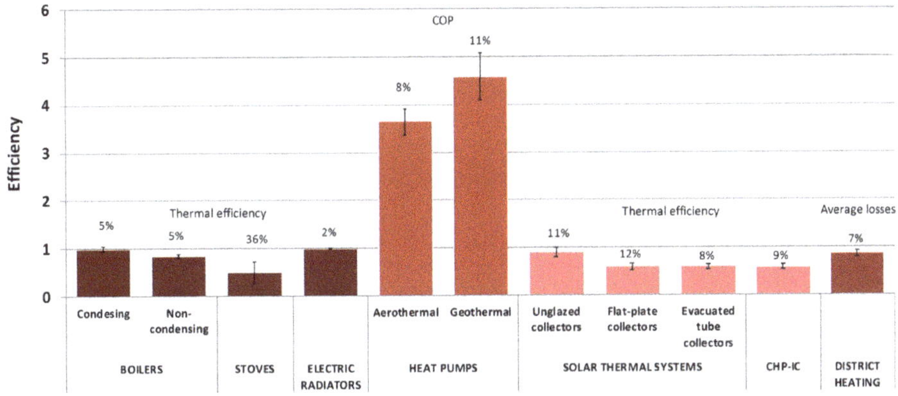

Figure 4. Energy efficiency coefficients at full-load per equipment type, EU28 [55–60,65,68–70].

Finally, including the data presented in previous figures in Equation (1), the results in terms of energy consumption per equipment type (Figure 5) were obtained. The entire EU28 energy use for SH and DHW technologies amounted to approximately 3880 TWh/y, and its largest share went to non-condensing boilers (over 2600 TWh/y, equaling to 67% of total). DH technologies came second with an energy use of about 500 TWh/y (13% of total). Condensing boilers' energy consumption corresponded to 350 TWh/y (i.e., 9% of total), while electric radiators consumed nearly 250 TWh/y (approximately 6% of total). These were followed by stoves, with about 130 TWh/y (approximately 3% of the above indicated 3880 TWh/y). CHP-IC, STS (flat-plate collectors), aerothermal HPs, STS (unglazed collectors), geothermal HPs, and STS (evacuated tube collectors) were last, accounting together for about 2% of total. A particularly striking feature was that the energy consumption deriving from DH systems exceeded the one of condensing boilers. The indicated difference was significant as the value for DH systems was approximately 25% higher.

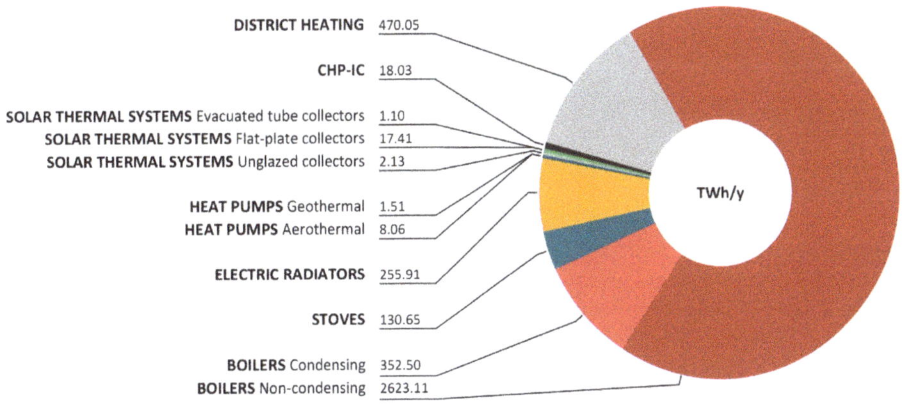

Figure 5. Energy consumption per type in TWh/y, EU28 [40,41,55–63,65,68–70].

Additionally, Table 2 displays the results in percentage with regard to various fuels utilization for SH and DHW equipment in the 28 EU MS.

Table 2. Fuels utilization at EU28 level for various SH and DHW equipment in percentage (NA—not available) [57,64,69,74,75] [1].

	Oil	Gas (Natural Gas)	Coal	Renewables	Other Fuels
Boilers-Condensing	33.77%	66.23%	NA	NA	NA
Boilers-Non-condensing	38.30%	54.30%	2.28%	5.12%	NA
Stoves	NA	NA	NA	100.00%	NA
CHP-IC	7.39%	43.73%	18.62%	22.84%	7.42%
DH	4.45%	38.35%	28.76%	26.02%	2.42%

[1] Electric radiators, aerothermal, and geothermal HPs, as well as STS—unglazed, flat-plate, and evacuated tube collectors were not considered in Table 2 due to not being fuel powered. A minor amount of coal and renewables driven condensing boilers, as well as not solely renewables (biomass) powered stoves are operative in Europe too [58,69,76].

As per Table 2, condensing boilers were mostly gas (natural gas) driven (~66%), followed by oil, with about 34%. The same ranking applies for non-condensing boilers. Gas (natural gas) was the first energy vector with nearly 54%. Oil was second (~38%), followed by RES and coal with about 5% and 2%, respectively.

Stoves were estimated to be for 100% RES (biomass) driven.

Regarding CHP-IC, gas (natural gas) again was the first (~43%). RES and coal followed with about 23% and 19%, respectively. In the last place we found "other fuels" and oil, with approximately 7% each.

DH systems were found to be mainly powered by gas (natural gas). Coal comes next, with about 29%. Close behind coal, what follows were renewables, with nearly 26%. Last positioned were oil and other fuels, with approximately 4% and 2%.

In conclusion, the outcome concerning centralized and individual boilers utilization showed individual technologies to be present in EU28 with slightly over half, nearly 54% [59,60,69].

4. Discussion

SH and DHW equipment's total energy consumption at EU28 level nearly reached 3900 TWh/y (approximately 3880 TWh/y) of which over 85% (about 3315 TWh/y) of this was provided by SH. Thus, only about 600 TWh/y (~580 TWh/y) accounted for DHW use. SH and DHW total consumption accounted for more than 20% of the EU's entire energy consumption [5]. If SC consumption was included, the latter amounted to only about 3% of the total energy consumption [37].

While notable studies in the field report SH and DHW values very close to ours (the H2020 project HRE4 [22], a publication of Patronen et al. 2012 [25], as well as the IEE STRATEGO project [23]), other studies differ greatly reporting values both falling short and exceeding ours. The three studies reporting values very close to ours differ by 3% to 6%.

The numbers falling short are provided by Boermans et al. 2012 [26], a report of the Seventh Framework Programme (FP7) iNSPiRe project [24], and by Von Manteuffel et al. 2016 [27]. In this case, differences with respect to our work range from 12% to 47%.

The detected values exceeding ours are given by Sanner et al. 2011 [28], Scoccia et al. 2018 [29], and Balaras et al. 2007 [30]. The indications provided by these authors vary compared to our own by 13% to 35%. Table 3 summarizes and analyzes in greater detail similarities and differences to our outcome found in scientific literature:

Table 3. Comparison of SH and DHW market quantifications at the EU level found in scientific literature and the present investigation's result.

Comparison to the Present Study's Result	Energy Consumption (TWh/y)	Deviation	Reference Year	Source
Close	~4025	3%	2015	[22]
Close	~3980	3%	2012	[25]
Close	~4150	6%	2010	[23]
Lower	~3400	12%	2012	[26]
Lower	~2760	29%	2014	[24]
Lower	~2070	47%	2015	[27]
Higher	~4450	13%	2006	[28]
Higher	~4715	18%	2006/2007	[29]
Higher	~5990	35%	2013	[30]

We wish to emphasize that even though only ~11% of the input data for our investigation derived from the HRE4 project, our result showed only a minor deviation, 3%.

Furthermore, concerning the IEE STRATEGO project result, we must recall that the output would be even lower if taking into consideration the decrease of energy use for SH and DHW in EU buildings in the past decade.

To complete our study, we calculated energy consumption for DHW per MS, using population and household data—by means of energy per person [77], number of inhabitants [78], and dwellings [79]. The values, expressed in TWh/y, were found to be approximately 510 TWh/y and 540 TWh/y, respectively. Thus, the differences with respect to the results shown above were 12% and 7%, respectively. Please find respective data set under [80] (details on all sources used are available in the respective csv file [81]).

As already mentioned in the Section 3. Results, a particularly striking result is given by the fact that energy consumption provided by DH systems exceeded the one of condensing boilers by 25%. Furthermore, comparing non-condensing boilers with condensing ones, we found that the replacement of conventional boilers with better performing SH and DHW technologies seemed to progress very slowly. As a plausibility check, we should first consider the share of biomass boilers (~2%), oil boilers (~19%), and natural gas boilers (~28%) [23]. In fact, while Regulation EU 813/2013 does not impose the use of condensing boilers over biomass boilers, it only came into place at the end of 2015, with significant exemptions. As a further confirmation of our results, if we suppose that all gas and oil boilers installed in the period 2015–2016 are substituted by condensing boilers, these would reach a total number of around 10.7 Mil. (assuming non-condensing boilers installed prior to 2015 and assuming a lifetime of 15 years) [82]. This is very much in line with the data presented in Figure 1. However, in the event of an enforcement of the Eco-design Regulation EU 813/2013, the share of condensing boilers should grow significantly in the coming years. This evidence is further reinforced by the indication that the thermal efficiency of currently installed boilers in Europe is approximately 85% [66], while condensing boilers are characterized by declared efficiency levels of approximately 99% [37]. On the other hand, a number of scientific resources confirm the steady growth of DH applications within the EU28 [83–85].

Concerning the fuels utilization at EU28 level for SH and DHW equipment (Table 2), it has to be stressed that gas (natural gas) dominated the ranking, while renewables were lower positioned, besides in the case of stoves. However, it is worth nothing that the indicated renewables value for stoves had been estimated due to a lack of sources. RES were mostly found at lower level positions (once again this was not valid for stoves and CHP-IC). Especially with regard to DH, the utilization for RES was characterized by a high potential [86–88].

5. Conclusions

This work presented a collection, statistical elaboration, calculation, and comparison of data that offered insights on the SH and DHW market for the EU28. The main aspects were the following:

- A significant portion of the total EU energy consumption went to SH and DHW (more than 20%), reaching almost the value of 3900 TWh/y.
- The energy consumption of DH systems covered around 13% of the total, and exceeded that of condensing boilers. While DH system applications are growing throughout the EU, the replacement of older, conventional boilers progresses at a slower pace.
- The investigation also included information on the use of fuels for SH and DHW technologies at the EU28 level. Gas (i.e., natural gas) was the most diffused, while renewables were ranking lower. This was different for stoves and CHP-IC. Especially DH offered a high potential for the use of RES.

The collected data per MS and the entire EU are also available as an open source data set, which allows for freely access and to retrieve information on SH and DHW consumption.

The data collection at the basis of the investigation and its insights presented certain limitations, which resulted from the assumptions indicated in Section 2. Materials and Methods.

All types of collected data (amount of operative units, installed capacities, energy efficiency coefficients as well as equivalent full-load hours per SH and DHW equipment type and country) were subject to not negligible variations, which resulted in pertinent CV values. This was especially true for collected data concerning the number of operative units for condensing boilers and aerothermal HPs (CV = 66% and 52%, respectively), the amount of equivalent full-load hours of STS—evacuated tube collectors (CV = 53%), and mean installed capacities of electric radiators (CV = 60%). Consequently, the performed analysis represented an assessment, and respective outcomes are to be interpreted with care.

The data collection and results of this work can form the basis for the collection and analysis of further data regarding Europe's building stock. Finally, our research indicates room for improvement in terms of data quality and completeness, as well as for extending the scope to areas such as industry and transportation.

Author Contributions: Conceptualization, S.P.; Data curation, S.C. and S.Z.; Supervision, L.K.; Validation, S.C., S.Z., L.K., A.N. and P.Z.

Funding: We are thankful to the Horizon 2020 Hotmaps project (Grant Agreement number 723677), for providing partial funding for the performance of the work.

Acknowledgments: We are thankful to Amy Segata (Eurac Research) for designing the figures of the study. In addition, our gratitude goes to Sonja Gantioler (Eurac Research) for the language editing of the text.

Conflicts of Interest: The authors declare no conflict of interest.

References

1. EC. 2030 Climate & Energy Framework. 2019. Available online: http://ec.europa.eu/clima/policies/strategies/2030/index_en.htm (accessed on 15 February 2019).
2. EC. 2050 Low-Carbon Economy. 2019. Available online: https://ec.europa.eu/clima/policies/strategies/2050_en (accessed on 15 February 2019).
3. EC. Paris Agreement. 2019. Available online: https://ec.europa.eu/clima/policies/international/negotiations/paris_en (accessed on 15 February 2019).
4. Benejam, G.M.; Mata, É.; Kalagasidis, A.S.; Johnsson, F. Bottom-up characterization of the Spanish building stock for energy assessment and model validation. In Proceedings of the Retrofit 2012 Conference, Manchester, UK, 24–26 January 2012.
5. EC. Consumption of Energy. 2019. Available online: http://ec.europa.eu/eurostat/statistics-explained/index.php/Consumption_of_energy (accessed on 16 February 2019).

6. Economidou, M. Energy performance requirements for buildings in Europe. *REHVA J.* **2012**, *92*, 16–21.
7. Pezzutto, S.; De Felice, M.; Fazeli, R.; Kranzl, L.; Zambotti, S. Status Quo of the Air-Conditioning Market in Europe: Assessment of the Building Stock. *Energies* **2017**, *10*, 1253. [CrossRef]
8. EC. Energy Efficiency Trends in Buildings in the EU. 2012. Available online: https://energiatalgud.ee/img_auth.php/6/68/Enerdata._Energy_Efficiency_Trends_in_Buildings_in_the_EU._2012.pdf (accessed on 17 February 2019).
9. EC. Buildings. 2019. Available online: https://ec.europa.eu/energy/en/topics/energy-efficiency/buildings (accessed on 18 February 2019).
10. Tronchin, L.; Manfren, M.; Nastasi, B. Energy efficiency, demand side management and energy storage technologies—A critical analysis of possible paths of integration in the built environment. *Renew. Sustain. Energy Rev.* **2018**, *95*, 341–353. [CrossRef]
11. IEA. Energy Efficiency Requirements in Building Codes, Energy Efficiency Policies for New Buildings. 2008. Available online: https://www.iea.org/publications/freepublications/publication/Building_Codes.pdf (accessed on 17 February 2019).
12. EEA. Final Energy Consumption by Sector and Fuel. 2019. Available online: https://www.eea.europa.eu/data-and-maps/indicators/final-energy-consumption-by-sector-9/assessment-4 (accessed on 18 February 2019).
13. IEA. Data Service. 2019. Available online: http://wds.iea.org/WDS/Common/Login/login.aspx (accessed on 18 February 2019).
14. Vinnova. H2020 Visualization. 2019. Available online: http://h2020viz.vinnova.se/#/ (accessed on 18 February 2019).
15. EC. Research and Innovation. 2019. Available online: https://ec.europa.eu/research/energy/eu/index_en.cfm?pg=projects&fp7page=10b (accessed on 18 February 2019).
16. Bointner, R.; Pezzutto, S.; Sparber, W. Scenarios of public energy research and development expenditures: Financing energy innovation in Europe. *Wiley Interdisciplin. Rev. Energy Environ.* **2016**, *5*, 470–488. [CrossRef]
17. Bointner, R.; Pezzutto, S.; Grilli, G.; Sparber, W. Financing Innovations for the Renewable Energy Transition in Europe. *Energies* **2016**, *9*, 990. [CrossRef]
18. Pezzutto, S.; Fazeli, R.; De Felice, M.; Sparber, W. Future development of the air-conditioning market in Europe: An outlook until 2020. *Wiley Interdisciplin. Rev. Energy Environ.* **2016**, *5*, 649–669. [CrossRef]
19. EC. An EU Strategy on Heating and Cooling. 2016. Available online: https://ec.europa.eu/energy/sites/ener/files/documents/1_EN_ACT_part1_v14.pdf (accessed on 29 April 2019).
20. Pardo, N.; Vatopoulos, K.; Krook-Riekkola, A.; Moya, J.A.; Perez, A. Heat and Cooling Demand and Market Perspective. 2012. Available online: http://publications.jrc.ec.europa.eu/repository/bitstream/111111111/26989/1/ldna25381enn.pdf (accessed on 29 April 2019).
21. EC. Energy Roadmap 2050. 2011. Available online: http://ec.europa.eu/smart-regulation/impact/ia_carried_out/docs/ia_2011/sec_2011_1565_en.pdf (accessed on 29 April 2019).
22. Fleiter, T.; Elsland, R.; Rehfeldt, M.; Steinbach, J.; Reiter, U.; Catenazzi, G.; Jakob, M.; Rutten, C.; Harmsen, R.; Dittmann, F.; et al. Profile of Heating and Cooling Demand in 2015. Heat Roadmap Europe, 2017. Available online: https://heatroadmap.eu/wp-content/uploads/2018/11/HRE4_D3.1.pdf (accessed on 24 February 2019).
23. Persson, U.; Werner, S. Quantifying the Heating and Cooling Demand in Europe. STRATEGO, 2015. Available online: http://www.heatroadmap.eu/resources/STRATEGO%20WP2%20-%20Background%20Report%204%20-%20Heat%20%26%20Cold%20Demands.pdf (accessed on 24 February 2019).
24. Birchall, S.; Wallis, I.; Churcher, D.; Pezzutto, S.; Fedrizzi, R.; Causse, E. D2.1a—Survey on the Energy Needs and Architectural Features of the EU Building Stock. iNSPiRe, 2014. Available online: http://inspirefp7.eu/wp-content/uploads/2016/08/WP2_D2.1a_20140523_P18_Survey-on-the-energy-needs-and-architectural-features.pdf (accessed on 25 February 2019).
25. Patronen, J.; Kaura, E.; Torvestad, C. Nordic Heating and Cooling. 2017. Available online: http://www.diva-portal.org/smash/get/diva2:1098961/FULLTEXT01.pdf (accessed on 24 February 2019).
26. Boermans, T. Building Renovation in Europe—What are the Choices? 2012. Available online: https://www.eurima.org/uploads/ModuleXtender/Publications/90/Renovation_tracks_for_Europe_08_06_2012_FINAL.pdf (accessed on 24 February 2019).
27. Von Manteuffel, B.; Petersdorff, C.; Bettgenhäuser, K.; Boermans, T. EU Pathways to a Decarbonised Building Sector. 2016. Available online: https://www.ecofys.com/files/files/ecofys-2016-eu-pathways-towards-a-decarbonised-building-sector.pdf (accessed on 25 February 2019).

28. Sanner, B. 2020–2030–2050 Common Vision for the Renewable Heating & Cooling Sector in Europe. 2011. Available online: http://www.rhc-platform.org/fileadmin/Publications/RHC_BROCHURE_140311_web.pdf (accessed on 25 February 2019).
29. Scoccia, R.; Toppi, T.; Aprile, M.; Motta, M. Absorption and compression heat pump systems for space heating and DHW in European buildings: Energy, environmental and economic analysis. *J. Build. Eng.* **2018**, *16*, 94–105. [CrossRef]
30. Balaras, C.A.; Gaglia, A.G.; Georgopoulou, E.; Sevastianos, M.; Sarafidis, Y.; Lalas, D.P. European residential buildings and empirical assessment of the Hellenic building stock, energy consumption, emissions and potential energy savings. *Build. Environ.* **2007**, *42*, 1298–1314. [CrossRef]
31. Leurent, M.; Da Costa, P.; Rämä, M.; Persson, U.; Jasserand, F. Cost-benefit analysis of district heating systems using heat from nuclear plants in seven European countries. *Energy* **2018**, *149*, 454–472. [CrossRef]
32. EU. Hotmaps. 2019. Available online: https://www.hotmaps-project.eu/ (accessed on 18 February 2019).
33. Pezzutto, S.; Zambelli, P. space_heating_cooling_dhw_bottom-up_SH+DHW.xlsx. 2019. Available online: https://gitlab.com/hotmaps/space_heating_cooling_dhw_demand/blob/master/data/space_heating_cooling_dhw_bottom-up_SH+DHW.xlsx (accessed on 18 February 2019).
34. Pezzutto, S.; Zambelli, P. space_heating_cooling_dhw_bottom-up_SH+DHW.csv. 2019. Available online: https://gitlab.com/hotmaps/space_heating_cooling_dhw_demand/blob/master/data/space_heating_cooling_dhw_bottom-up_SH+DHW.csv (accessed on 18 February 2019).
35. Creative commons. Attribution 4.0 International (CC BY 4.0). 2019. Available online: https://creativecommons.org/licenses/by/4.0/ (accessed on 18 February 2019).
36. The Apache Software Foundation. Apache License. 2019. Available online: https://apache.org/licenses/LICENSE-2.0https://apache.org/licenses/LICENSE-2.0 (accessed on 18 February 2019).
37. Pezzutto, S.; Zambotti, S.; Croce, S.; Zambelli, P.; Garegnani, G.; Scaramuzzino, C.; Pascuas, R.P.; Zubaryeva, A.; Haas, F.; Exner, D.; et al. D2.3 WP2 Report—Open Data Set for the EU28. Hotmaps, 2018. Available online: https://www.hotmaps-project.eu/wp-content/uploads/2018/03/D2.3-Hotmaps_for-upload_revised-final_.pdf (accessed on 18 February 2019).
38. Noussan, M.; Nastasi, B. Data Analysis of Heating Systems for Buildings—A Tool for Energy Planning, Policies and Systems Simulation. *Energies* **2018**, *11*, 233. [CrossRef]
39. Noussan, M.; Roberto, R.; Nastasi, B. Performance Indicators of Electricity Generation at Country Level—The Case of Italy. *Energies* **2018**, *11*, 650. [CrossRef]
40. EurObserv'ER. All Heat Pumps Barometers. 2018. Available online: https://www.eurobserv-er.org/category/all-heat-pumps-barometers/ (accessed on 18 February 2019).
41. EUROHEAT&POWER. District Heating and Cooling—Country by Country—2015 Survey. 2015. Available online: http://www.euroheat.org/wp-content/uploads/2016/03/2015-Country-by-country-Statistics-Overview.pdf (accessed on 18 February 2019).
42. Nowak, T.; Westring, P. European Heat Pump Market and Statistics Report. 2018. Available online: https://www.ehpa.org/market-data/market-report/ (accessed on 18 February 2019).
43. EC. EU Buildings Database. Available online: https://ec.europa.eu/energy/en/eu-buildings-database (accessed on 18 February 2019).
44. EHPA. The Online Stats Tool. 2019. Available online: http://www.stats.ehpa.org/hp_sales/country_cards/ (accessed on 18 February 2019).
45. IGA. OUR DATABASES. 2019. Available online: https://www.geothermal-energy.org/explore/our-databases/ (accessed on 18 February 2019).
46. Bertoldi, P.; Rezessy, S.; Lees, E.; Baudry, P.; Jeandel, A.; Labanca, N. Energy supplier obligations and white certificate schemes: Comparative analysis of experiences in the European Union. *Energy Pol.* **2010**, *8*, 1455–1469. [CrossRef]
47. Martinopoulos, G.; Papakostas, K.T.; Papadopoulos, A.M. A comparative review of heating systems in EU countries, based on efficiency and fuel cost. *Renew. Sustain. Energy Rev.* **2018**, *90*, 687–699. [CrossRef]
48. Clay, K. Power to the People: Energy in Europe over the Last Five Centuries. *J. Econ. Hist.* **2015**, *75*, 936–937. [CrossRef]
49. Pezzutto, S.; Zambelli, P. space_heating_cooling_dhw_demand. 2019. Available online: https://gitlab.com/hotmaps/space_heating_cooling_dhw_demand (accessed on 18 February 2019).

50. Open Knowledge International. Frictionless Data. Specifications and Software. 2019. Available online: https://frictionlessdata.io/ (accessed on 18 February 2019).
51. Pezzutto, S.; Zambelli, P. Space Heating, Cooling and DHW Demand—EU28. 2019. Available online: https://gitlab.com/hotmaps/space_heating_cooling_dhw_demand/blob/master/README.md (accessed on 18 February 2019).
52. SPDX Workgroup. Software Package Data Exchange®(SPDX®). 2019. Available online: https://spdx.org/ (accessed on 18 February 2019).
53. Zambelli, P.; Pezzutto, S.; Garegnani, G.; Pignatelli, A.; Lehtsalu, L.; Kranzl, L.; Fritz, S. Data Management Plan. Hotmaps, 2018. Available online: https://www.hotmaps-project.eu/wp-content/uploads/2019/03/Hotmaps_D2.1_Data-Management-Plan_2018-09-28_final.pdf (accessed on 18 February 2019).
54. Nouvel, R.; Zirak, M.; Coors, V.; Eicker, U. The influence of data quality on urban heating demand modeling using 3D city models. *Comput. Environ. Urban Syst.* **2017**, *64*, 68–80. [CrossRef]
55. EU. Heat Roadmap Europe. 2019. Available online: http://www.heatroadmap.eu/ (accessed on 18 February 2019).
56. EU. Stratego. 2014. Available online: http://stratego-project.eu (accessed on 19 February 2019).
57. EC. Mapping and Analyses of the Current and Future (2020–2030) Heating/Cooling Fuel Deployment (Fossil/Renewables). 2019. Available online: https://ec.europa.eu/energy/en/studies/mapping-and-analyses-current-and-future-2020-2030-heatingcooling-fuel-deployment (accessed on 19 February 2019).
58. Dengler, J.; Köhler, B.; Dinkel, A.; Bonato, P.; Azam, N.; Kalz, D. Work Package 2: Assessment of the Technologies for the Year 2012. Mapping and Analyses of the Current and Future (2020–2030) Heating/Cooling Fuel Deployment (Fossil/Renewables). 2016. Available online: https://ec.europa.eu/energy/sites/ener/files/documents/mapping-hc-final_report-wp2.pdf (accessed on 20 February 2019).
59. Persson, U.; Möller, B.; Wiechers, E. Deliverable 2.3: A Final Report Outlining the Methodology and Assumptions used in the Mapping. Heat Roadmap Europe, 2017. Available online: http://www.heatroadmap.eu/resources/HRE4_D2.3.pdf (accessed on 20 February 2019).
60. Connolly, D.; Hansen, K.; Drysdale, D.; Lund, H.; Vad Mathiesen, B.; Werner, S.; Persson, U.; Möller, B.; Wilke, O.C.; Bettgenhäuser, K.; et al. Enhanced Heating and Cooling Plans to Quantify the Impact of Increased Energy Efficiency in EU Member States. Stratego, 2016. Available online: http://stratego-project.eu/wp-content/uploads/2014/09/STRATEGO-WP2-Executive-Summary-Main-Report.pdf (accessed on 20 February 2019).
61. Fedrizzi, R.; Marchetti, R. iNSPiRe, Bolzano, Italy. Deliverable 2.1 Intermediate analysis of the heating and cooling industry. Support to key activities of the European technology platform on renewable heating and cooling. Unpublished work. 2016.
62. Mauthner, F.; Weiss, W.; Spörk-Dür, M. Solar Heat Worldwide. 2017. Available online: http://www.iea-shc.org/Data/Sites/1/publications/Solar-Heat-Worldwide-2017.pdf (accessed on 20 February 2019).
63. Mauthner, F.; Weiss, W.; Spörk-Dür, M. Solar Heat Worldwide. 2016. Available online: http://www.iea-shc.org/Data/Sites/1/publications/Solar-Heat-Worldwide-2016.pdf (accessed on 20 February 2019).
64. EC. ENERGY DATA. 2019. Available online: http://ec.europa.eu/eurostat/web/energy/data (accessed on 20 February 2019).
65. EU. TABULA WebTool. 2019. Available online: http://webtool.building-typology.eu/#bm (accessed on 21 February 2019).
66. Pezzutto, S.; Toleikyte, A.; De Felice, M. Assessment of the Space Heating and Cooling Market in the EU28: A Comparison between EU15 and EU13 Member States. *Int. J. Contemp. Energy* **2015**, *1*, 35–48.
67. Welch, T. Heat Pump Technology. 2009. Available online: https://www.cibsejournal.com/cpd/modules/2009-05/ (accessed on 21 February 2019).
68. Nouvel, R.; Cotrado, M.; Pietruschka, D. European Mapping of Seasonal Performances of Air-Source and Geothermal Heat Pumps for Residential Applications. In Proceedings of the CISBAT 2015, Lausanne, Switzerland, 9–11 September 2015.
69. Pezzutto, S. Analysis of the space heating and cooling market in Europe. Ph.D. Thesis, University of Natural Resources and Life Sciences, Vienna, Austria, 22 May 2014.
70. Werner, S. Possibilities with more District Heating in Europe. Ecoheatcool, 2006. Available online: https://www.euroheat.org/wp-content/uploads/2016/02/Ecoheatcool_WP4_Web.pdf (accessed on 21 February 2019).

71. Insee. Coefficient of Variation/CV. 2019. Available online: https://www.insee.fr/en/metadonnees/definition/c1366 (accessed on 23 February 2019).
72. Lee, C.F.; Lee, J.C.; Lee, A.C. *Statistics for Business and Financial Economics*, 2nd ed.; World Scientific Publishing: Singapore, 2000; pp. 103–104.
73. Malta's Office of The Prime Minister (Energy and Projects). Malta's National Energy Efficiency Action Plan. 2017. Available online: https://ec.europa.eu/energy/sites/ener/files/documents/mt_neeap_2017.pdf (accessed on 22 February 2019).
74. Andrews, D.; Riekkola, A.K.; Tzimas, E.; Serpa, J.; Carlsson, J.; Pardo-Garcia, N.; Papaioannou, I. Background Report on EU-27 District Heating and Cooling Potentials, Barriers, Best Practice and Measures of Promotion. 2012. Available online: https://setis.ec.europa.eu/system/files/1.DHCpotentials.pdf (accessed on 25 February 2019).
75. Werner, S. International review of district heating and cooling. *Energy* **2017**, *137*, 617–631. [CrossRef]
76. Gupta, A.; Bai, A.S. Europe Boiler Market Growth 2017–2024—Industry Size, Share Report. 2018. Available online: https://www.gminsights.com/industry-analysis/europe-boiler-market (accessed on 25 February 2019).
77. UNEP. Domestic Hot Water for Single Family Houses. 2015. Available online: http://www.estif.org/fileadmin/estif/content/publications/downloads/UNEP_2015/factsheet_single_family_houses_v05.pdf (accessed on 25 February 2019).
78. Index mundi. Demographics: Population. 2017. Available online: https://www.indexmundi.com/map/ (accessed on 25 February 2019).
79. Fuentes, E.; Arce, L.; Salom, J. A review of domestic hot water consumption profiles for application in systems and buildings energy performance analysis. *Renew. Sustain. Energy Rev.* **2018**, *81*, 1530–1547. [CrossRef]
80. Pezzutto, S.; Zambelli, P. space_heating_cooling_dhw_population_households.xlsx. 2019. Available online: https://gitlab.com/hotmaps/space_heating_cooling_dhw_demand/blob/master/data/space_heating_cooling_dhw_population_households.xlsx (accessed on 25 February 2019).
81. Pezzutto, S.; Zambelli, P. space_heating_cooling_dhw_population_households.csv. 2019. Available online: https://gitlab.com/hotmaps/space_heating_cooling_dhw_demand/blob/master/data/space_heating_cooling_dhw_population_households.csv (accessed on 25 February 2019).
82. EU. Commission Regulation (EU) No 813/2013 of 2 August 2013 Implementing Directive 2009/125 /EC of the European Parliament and of the Council with Regard to Ecodesign Requirements for Space Heaters and Combination Heaters. 2013. Available online: https://eur-lex.europa.eu/legal-content/EN/TXT/PDF/?uri=CELEX:32013R0813&from=EN (accessed on 25 February 2019).
83. Epp, B. Top District Heating Countries—EUROHEAT&POWER 2015 Survey Analysis. 2015. Available online: https://www.euroheat.org/news/district-energy-in-the-news/top-district-heating-countries-euroheat-power-2015-survey-analysis/ (accessed on 25 February 2019).
84. EUROHEAT&POWER. European Heating Sector Well Positioned for Renewables Integration. 2017. Available online: https://www.euroheat.org/news/european-heating-sector-well-positioned-renewables-integration/ (accessed on 25 February 2019).
85. Gustavsson, L. District heating systems and energy conservation—Part II. *Energy* **1994**, *19*, 93–102. [CrossRef]
86. IRENA. Renewable Energy Indistrict Heating and Cooling. A Sector Roadmap for Remap. 2017. Available online: https://www.irena.org/-/media/Files/IRENA/Agency/Publication/2017/Mar/IRENA_REmap_DHC_Report_2017.pdf (accessed on 28 February 2019).
87. Wangab, D.; Orehounigab, K.; Carmelieta, J. Investigating the potential for district heating networks with locally integrated solar thermal energy supply. *Energy Proc.* **2017**, *122*, 1057–1062. [CrossRef]
88. Olsthoorna, D.; Haghighata, F.; Mirzaeib, P.A. Integration of storage and renewable energy into district heating systems: A review of modelling and optimization. *Sol. Energy* **2016**, *136*, 49–64. [CrossRef]

energies

MDPI

Article

Open Source Data for Gross Floor Area and Heat Demand Density on the Hectare Level for EU 28

Andreas Müller [1,2,*], Marcus Hummel [1], Lukas Kranzl [2], Mostafa Fallahnejad [2] and Richard Büchele [2]

1 E-Think Energy Research, Zentrum für Energiewirtschaft und Umwelt, Argentinierstrasse 18, 1040 Vienna, Austria; hummel@e-think.ac.at

2 Institute of Energy Systems and Electrical Drives, Energy Economics Group, Technische Universität Wien, Gusshausstr. 25-27, 1040 Vienna, Austria; kranzl@eeg.tuwien.ac.at (L.K.); Fallahnejad@eeg.tuwien.ac.at (M.F.); buechele@eeg.tuwien.ac.at (R.B.)

* Correspondence: mueller@e-think.ac.at; Tel.: +43-158801-370362

Received: 4 November 2019; Accepted: 6 December 2019; Published: 16 December 2019

Abstract: The planning of heating and cooling supply and demand is key to reaching climate and sustainability targets. At the same time, data for planning are scarce for many places in Europe. In this study, we developed an open source dataset of gross floor area and energy demand for space heating and hot water in residential and tertiary buildings at the hectare level for EU28 + Norway, Iceland, and Switzerland. This methodology is based on a top-down approach, starting from a consistent dataset at the country level (NUTS 0), breaking this down to the NUTS 3 level and further to the hectare level by means of a series of regional indicators. We compare this dataset with data from other sources for 20 places in Europe. This process shows that the data for some places fit well, while for others, large differences up to 45% occur. The discussion of these results shows that the other data sources used for this comparison are also subject to considerable uncertainties. A comparison of the developed data with maps based on municipal building stock data for three cities shows that the developed dataset systematically overestimates the gross floor area and heat demand in low density areas and vice versa. We conclude that these data are useful for strategic purposes on aggregated level of larger regions and municipalities. It is especially valuable in locations where no detailed data is available. For detailed planning of heating and cooling infrastructure, local data should be used instead. We believe our work contributes towards a transparent, open source dataset for heating and cooling planning that can be regularly updated and is easily accessible and usable for further research and planning activities.

Keywords: open data; heating; building stock; heat map; spatial analysis; heat density map

1. Introduction

About 50% of the final energy consumption in Europe is spent on heating and cooling, including space heating and cooling, domestic hot water, and processing heat and cold [1]. The largest share of this demand is covered by fossil energy carriers [2]. Thus, the heating and cooling sector needs to be radically transformed in order to be in line with decarbonization targets. In contrast to other energy carriers (e.g., electricity and fuels), which are carried over hundreds (electricity) to thousands (oil, gas) of kilometers, the transmission of thermal energy (heat or cold) is still limited to local or regional systems [3]. Hence, the heating and cooling sector, per se, has a strong spatial dimension, which needs to be carefully considered in this transformation process. The European Energy Efficiency Directive [4,5] takes into account these considerations by requesting Member States of the European Union to provide a so called "comprehensive assessment of the potential for the application of high-efficiency cogeneration and efficient district heating and cooling". According to Annex VIII of the directive, this should include—among other things—"a map of the national territory, identifying (...) heating and cooling demand points". A preliminary version of these reports had to be delivered by the end of 2015, and an update will be due by the end of 2020. This comprehensive assessment is considered to support the strategic heating and cooling planning and mapping process on different spatial levels [6], which has a long tradition in countries like Denmark [7]. The existence of such datasets provides valuable support for assessments in this area of research. However, as pointed out by Noussan and Nastasi [8], Tronchin et al. [9], and others, the quality of the input data used for the analysis impacts the accuracy of the results. Therefore, an almost equally important step forward is ensuring the availability of such datasets to a broader group of users, such as policy makers and researchers, who foster increased data quality by identifying the weaknesses of datasets and subsequently improving their methodologies and underlying data foundations.

1.1. *Spacial Levels of Heating and Cooling Planning*

In general, three different spatial levels of analyses and regional detail may be distinguished for heating and cooling planning. These levels are very different based on the aims of the analyses, the required data, and the potential (research) questions that can be answered. At the first, or top level, analyses are performed at the strategic level of heat planning, either at a national, regional, or municipal scale, with the objective to identify possible areas of interest for district heating, excess heat integration, and the overall potentials for different decarbonisation options in the sector.

On a European level, the heat density map of Heat Roadmap Europe [10] (see also, e.g., Connolly et al., [11]; Persson et al., [12]; Möller et al. [13]) is one of the most relevant examples of this kind of analysis. Their methodology also uses a top-down approach but differs from our work insofar as they chose an econometric approach for the break-down of the demand data [14]. This project provides maps for downloading and use in the project platform. However, the datasets are not publicly available at the time of the submission of this paper. Moreover, the Heat Roadmap Europe project, a robust body of scientific literature (e.g., [15–25]) on similar yet more locally focused projects, is available. The heat density maps developed in the frame of the comprehensive assessments at the national level also belong to this category. While valuable because they visualize heat demands at a regional level, the synthesis report on the evaluation of the Comprehensive Assessments from the Joint Research Centre [26] finds that most of the maps developed and provided for the first round of Comprehensive Assessments are only choropleth maps at the level of predefined areas based on administrative borders (e.g., a county, region, or municipality) or other statistical sectors, which are shaded/coloured to indicate the heat demands within this area. This kind of map does not allow further planning and only advises upon which areas to look at closer. Only a few of the Member States provided interactive isopleth maps that allow one to zoom in or that show a disaggregation of the heat demands at a high resolution raster level with public access (e.g., Austria [27], Scotland [28], and the Netherlands [29]). Even fewer maps provide data or allow downloading of the heat demand density data for further assessments.

At the second level, analyses are done at the city or (larger) district level, aiming at developing regional development plans or evaluating the (pre-) feasibility of district heating. For example, a study performed by Brocklebank et al. [30] considers the initial stage of designing a district heating network, with energy mapping in the local area, using the case study of Darley Dale, England. The results of the mapping technique are compared to the heat mapping work carried out by the UK Government and are shown to be accurate enough for further analyses. Dorotić [31] presents an economic analysis to determine the actual demands and the potential energy supply by using GIS based heat demand mapping methods and applies the results to an analysis of district heating network expansion in the city of Velika Gorica. Wyrwa and Yi-kuang [32] developed and applied a methodology to generate the data needed for the development of district heating systems. Their article presents a combined bottom-up–top-down approach applied for the city of Krakow as a case study and calculates of the useful heat demand for space heating and hot water preparation. Further examples for this kind of analysis include the progRESsHEAT project [33], Čižman et al. [34], and Dochev et al. [35], which analyses the spatial heat demands for the City of Hamburg.

Detailed technical and local analyses at the district, building block, or even individual building levels constitute the third spatial level. At this level, different technological alternatives, technical designs, and planning options are assessed and compared in detail. For example, Törnros et al. [36] simulated the DH demand for a medium-sized DH network in a city in southern Germany and used a spatially explicit approach for the analysis by first geo-locating the buildings and their attributes obtained from various sources. Based on these results, the authors calculated the annual primary energy demands for heating and domestic hot water for all individual buildings and then aggregated these demands at the segment level of an existing DH network and simulated the water flow through the system to cover the demand.

1.2. Aims and Objectives of This Work

The objective of this paper, which builds on the work performed within the Hotmaps project, is to provide an open-data top-down derived dataset of the heated gross floor area and final energy demands for space heating and domestic hot water preparation for all EU28 countries (plus Iceland, Norway, and Switzerland) on a hectare (100 × 100 m) scale. The data are available as an open dataset and thus may be used by anybody to start analysing areas of their interest in the EU. By providing these data, we believe that we can substantially contribute to fulfilling the requirements defined by Annex VIII and support public authorities, energy agencies, and planners in strategic heating and cooling planning at the local, regional, and national levels.

Through the work and dataset presented in this paper, we contribute to the first two levels. The maps cover (heated) the gross floor area, as well as energy needs and final energy demands, for space heating and domestic hot water preparation in residential and non-residential buildings. The methodology can be classified as a top-down approach: Energy consumption data at the country level (NUTS (Nomenclature of Territorial Units for Statistics) 0) is broken down to the NUTS 3 level and subsequently to the hectare level based on different spatial indicators (for details, see the description of the methodology in Section 2). This generic top-down approach has very specific strengths and weaknesses. Its key strengths are the comprehensive availability of data for every region within the EU-28 (+ IS, NO, CH), the transparency of the applied method, and the consistency of the aggregated results with national statistics. This approach's weakness lies in the deviations between the developed data and the data generated based on a bottom-up approach using detailed data at the local level. This appears at a very high spatial resolution: The smaller the spatial selection of the heat density map, the higher the deviation between the concrete demand and building the stock data, which occurs naturally on the ground, and the statistical approach chosen for breaking down the data to the hectare level. Thus, it is important to be aware of this limitation and to apply the data for the purpose proposed earlier—i.e., for the strategic level of heat planning and regional energy planning but not for detailed technical design and planning (e.g., of district heating infrastructure).

After this introduction, we present in Section 2 the methods and approaches for developing EU-wide gross floor area and heat density maps. Section 3 presents the results, including a validation and comparison of the data for selected municipalities across Europe. We discuss the results in Section 4 and derive conclusions and provide an outlook for further work in this field in Section 5.

2. Materials and Methods

2.1. General Approach

The top-down heat density map developed in the Hotmaps project builds on a three-stage approach (see Figure 1). In the first stage (top-level), we derived the final energy demand (FED) and energy needs (EN) for space heating (SH) and domestic hot water preparation (DHW) based on extensive literature research for the individual countries considered in the study. These data sources include the following: energy consumption data from energy statistics (e.g., data derived from Eurostat and national energy balances [37]), statistical data on the number of buildings, households, and shares of those per construction period [38], statistical and project related data on the typical properties of different building types and construction periods per country, and average climate data for different regions. From these data we built a building dataset for each country [39], which consistently combines these different types of input data, starting with the u-values for different building components and associated typical heat transmitting areas on the one hand and, on the other, the energy consumption reported in the national energy balances for the energy services focused upon by this analysis. This step is done by applying the Invert/EE-Lab model ([40], see also [41]). The Invert/EE-Lab model is a dynamic building stock model that calculates the energy needs and final energy consumption for SH and DHW, as well as the space cooling for regions or countries based on an underlying building physics model described by national and international norms [42–46] alongside the energetic properties of archetype buildings and their building components, climate, and usage data [40].

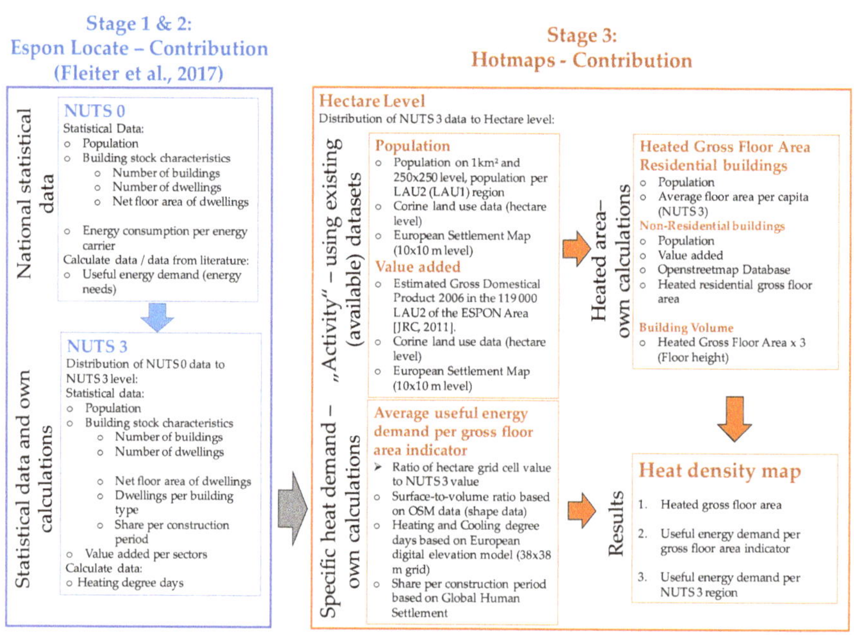

Figure 1. Schematic process of how we derived data maps at the hectare level for the EU-28 countries.

In the second stage, we distributed the EN for SH and DHW from the country level (NUTS 0) to the third territorial units at the statistics level (NUTS 3) based on an approach that we developed within the study, "Territories and low-carbon economy" (ESPON Locate) [47]. At this stage, we combine several indicators to estimate the share of EN for SH and DHW for the different NUTS 3 regions within each country.

For residential buildings, we consider the following indicators. Additional information on this approach is presented in the final report of the Espon Locate project [47].

- Data provided by the European Census Hub 2011 [48]:
    - Persons (population);
    - Number of dwellings;
    - Useful floor space per dwelling;

- Number of dwellings per period of construction;
- Number of dwellings per type of building;

- Heating degree days (HDD) at the NUTS 2-level are based on Eurostat [49]. Within the NUTS 2 level, the HDD at the NUTS 3 level are calculated based on the average HDD (18.5/18.5) calculated from the observed daily temperatures on a 25 × 25 km grid for the period 2002–2012 (see [50]).
- The final energy demand (FED) per m^2 gross floor area and building types are based on the Invert/EE-Lab model results derived by Fleiter et al. [37].

The following indicators are used for non-residential buildings (only services, excluding large industrial production facilities, etc.):

- Population, HDD, EN, and FED per m^2 gross floor area, building type, and construction period based on the Invert/EE-Lab building stock database [48];
- The share of dwellings per construction period of apartment buildings [48];
- The total value added of the service sector [51];
- The sectoral value added (VA) to the following sectors: (a) accommodation, restaurants, stores, and warehouses; (b) other private services; and (c) public buildings, research and education, art, culture, and the health sector [51].

The third level constitutes the distribution of the NUTS 3 results at the hectare level and thus derives a heat density map. We developed an approach that correlates information from the locally built environment with its EN for SH and DHW preparation within the Hotmaps project and applied it to the EU28 countries, plus Norway, Iceland, and Switzerland. This was done by using a spatial distribution function based on similar indicators used at to transfer country data to the NUTS 3 level. This approach builds on the central idea that the EN for space heating and domestic hot water preparation correlates with the population number within a plot area, as well as its economic activity, climatic conditions, and some building properties, such as the average construction period and volume-to-surface ratios of the buildings. The final energy consumption (delivered energy plus thermal energy provided by on-site renewable energy sources) is then calculated from the energy needs by applying the country specific national conversion factor between the energy needs and final energy consumption.

2.2. Population Distribution at the Hectare Level

Our main source for local population data is a work published by Gallego [52], where a dataset for the European population in 2006 at the level of 1 km^2 is given. In addition, we considered a dataset for the population in 2014 at the level of 250 × 250 m, developed by JRC [53], which is also publicly available. Although the latter source is newer, we decided to use the older population data as the primary input data source for the population distribution because a comparison of the population based on these data sources and data from the human settlement project (i.e., the share of the plot area sealed by buildings at a 10 × 10 m level) [54] revealed that a significant share of the population in rural regions is distributed in areas with no buildings (see Figure 2).

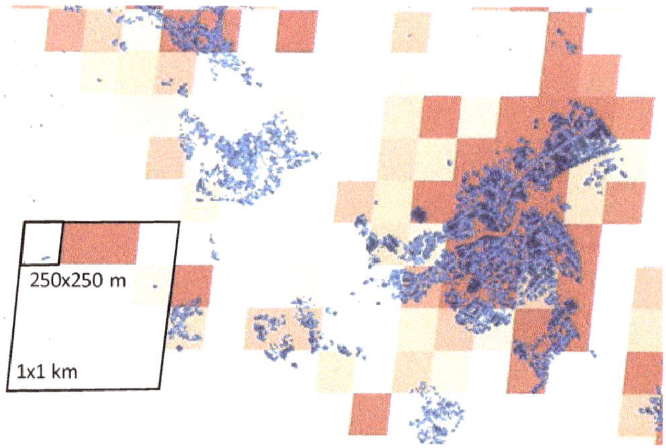

Figure 2. Comparison of the plot area covered by buildings on a 10 × 10 m level (blue) and the population in 2014 per 250 × 250 m for a small town in Carinthia, Austria (46°42′; 13°39′) with about 3000 inhabitants. Sources: [53,54].

An advantage of the new JRC population dataset [53] over the older dataset of Gallego [52] is that the new dataset partly covers areas that do not feature data in [52]. To combine these two datasets, we, therefore, calculated the population distribution on a 1 km^2 raster for Europe based on the following rules. If the primary population layer [52] does not contain data for a 1 km^2 grid cell, we use the data from a 1 km^2 grid cell layer derived from [53] as a fall-back option. An analysis of the resulting quality of the combined layer indicated that:

(a) The combination actually adds data in areas where the primary population layer does not cover all regions;
(b) However, it also introduces a bias towards higher populations in less densely populated areas since a (non-systematic) shift in the (1 × 1 km^2) grid cells between the two population layers can be observed in many regions. That is to say, for example, data source 1 locates the population in a neighbouring cell, similar to data source 2.

Consequently, we obtain an overestimation of the population in rural areas when we distribute the population of the NUTS 3/ local administrative units (LAU) regions at the hectare level. To reduce this undesired effect, we give the population of the JRC layer a weight factor of 30% (by multiplying the population data with a factor of 0.3). We have chosen this value by assessing the results that we derived for different weighting factors (in the range of 10–100%) for different regions (see Figure 3). This value, we believe, offers a good compromise that balances the two effects that accompany the data: first, the described effect of overestimating the population in rural areas and secondly, the underestimation of population in areas not covered by Gallego. We, however, have not performed any systematic analysis on the optimal level for the applied weighting factor.

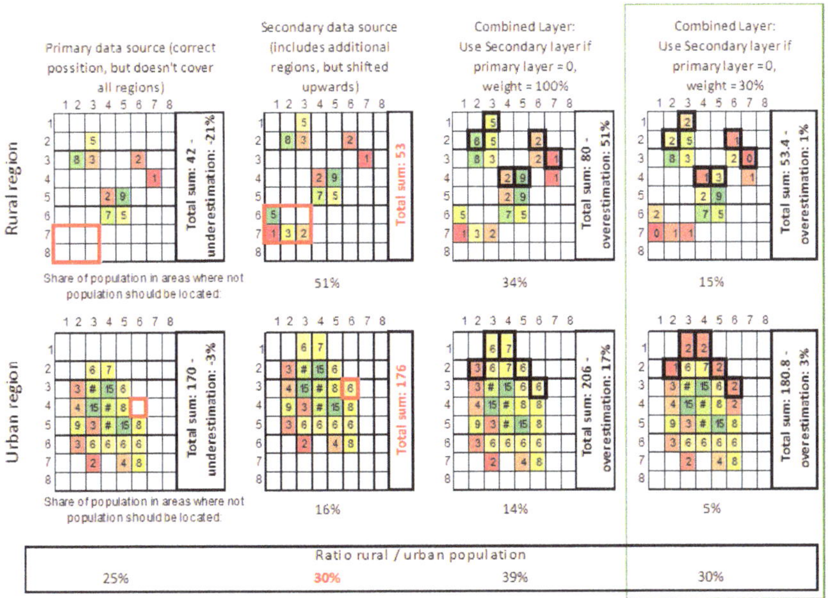

Figure 3. Schematic depiction of the method for combining the two data sources for population: Gallego [52] is the primary data source and places the population at the correct position but does not cover all areas (marked by red border lines), and JRC is secondary data source [53], which is shifted but covers additional areas.

Within the 1 km^2 grid cells, we distributed the population by considering the land usage type at the hectare level using the Corine land cover data [55] and the European Settlement Map layer [54], which depicts the share of the surface that is sealed by buildings on a 10 × 10 m grid. After distributing the population at the hectare level, we sum up the population for the local administrative units (LAU). This is done for the ~115 thousand regions (Eurostat [56], using the LAU 2, except for Greece and Denmark, where we used LAU 1, since the Census 2011 was performed at the LAU 1 level only). We then compared the population values with the population data in the local administrative units stated in the statistical data sources of Eurostat [48,57] and JRC [58]. To reduce the deviations, we then adjusted the population distribution to determine a compromise between the populations at the square kilometre level [52,53], the population per LAU region, and the upper limit for the population density per hectare. For this upper limit, we analysed the distribution of the indicator: the population per hectare divided by the population in the corresponding 1 × 1 km grid cell. For this analysis, we calculated the ratio between the population per hectare and the average population within the same square kilometre grid for each hectare cell. We than clustered the outcome by the population densities at the 1 km^2 grid level and removed all data points that did not exceed the average results for the density by a factor of 2. For the remaining data points, we calculated 95% and 99% for different population densities and defined an upper limit for the population at the hectare level (Figure 4), which is in the range of the 95–99% percentiles. The corresponding figure is read as follows: If 10 people live within a certain 1 × 1 km grid cell, then the population of each hectare cell within this 1 × 1 km^2 grid cell must not exceed ~5 inhabitants (50% of the total population of that square kilometre). If a 1 × 1 km^2 grid cell is populated by 10,000 people, the upper limit for inhabitants per hectare within that square kilometre must not exceed ~700 people (7%), and, for a population of 100,000 people per km^2, the upper limit for the population per hectare in that area is 2000 people (2%).

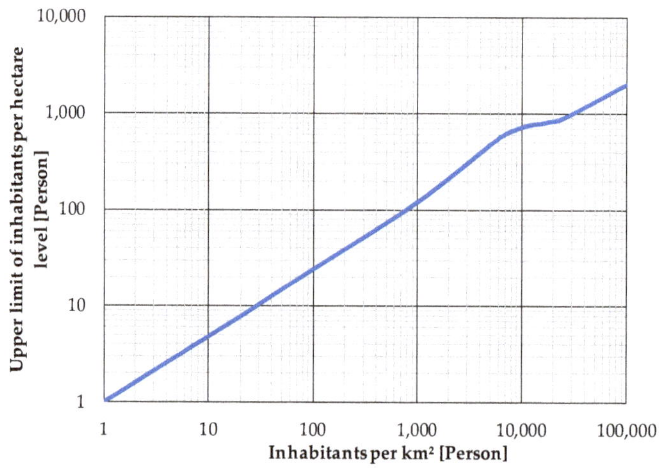

Figure 4. Implemented upper limit for population density at the hectare level.

2.3. Gross Floor Area of Buildings at the Hectare Level

Our methodology estimates the heated gross floor area of buildings at the hectare level based on two independent approaches. The first approach, which we call the "population-based" approach, builds on the population grid at the hectare level and estimates the gross floor area by multiplying the population on the hectare level with the average gross floor area per person in the corresponding NUTS 3 region. We derived this indicator from the data provided by the European Census Hub [48] for most European NUTS3 regions, namely the average floor area per dwelling and the average persons per household. This approach delivers reasonable results for the residential building stock. However, its predictive quality is poor in areas with a high share of non-residential buildings. To overcome this problem, we developed a second independent layer for the gross floor area of buildings.

The second approach derives the heated gross floor area by considering the building footprints and the estimated number of floors per buildings. We extracted the building footprints from the European Settlement Map [54] and data from the building layer of the OpenStreetMap (OSM) database [59]. The European Settlement layer contains the share of plot area that is sealed by buildings on a 10 × 10 m grid, while the OSM database contains the buildings as 2-dimensional vector data.

To calculate the gross floor area from the building's footprint, we required information on the building height or number of floors. In order to obtain estimates for these parameters, we developed a building height model that estimates the average building height from the footprint of the buildings using the OSM-building data. The developed approach applies a generic building height model (Figure 5), which is refined by the average regional (municipality-specific) building height-to-building footprint derived from buildings whose building height information is stored in the OSM database (~5 Mil. buildings spread over Europe).

As can be seen from Figure 5, the generic model for the number of floors matches well for buildings with a 60 to 1000 m^2 building footprint. It is assumed that buildings with footprints below 30 m^2 are mostly unheated (no floors) and are partially unheated if the footprint is between 30–45 m^2 (number of floors of 0.5). For buildings larger than 1000 m^2, the OSM-data show a further increase in the building height, while we, in contrast, keep the height constant until reaching buildings of 2500 m^2 and gradually reduce the number of floors to three for buildings with a footprint of 10,000 m^2 or more. The underlying reason for this method is as follows. The OSM-data mostly contain the building height, while the number of floors is calculated by us based on a constant floor height of 3 m (including structural elements). Based on personal experience, we believe that the ceiling height of very large buildings (industrial production halls, shopping centres, etc.) will be higher compared to smaller buildings and that the number of floors of such facilities rarely exceeds four. In the case of the OSM-data point for the largest building cluster (i.e., buildings with a footprint of more than 2500 m^2 (~20,000 buildings with a median footprint of about 6500 m^2)), the average floor height would correspond to about 5 m if we consider an average number of floors as 3.5. To us, this number appears plausible.

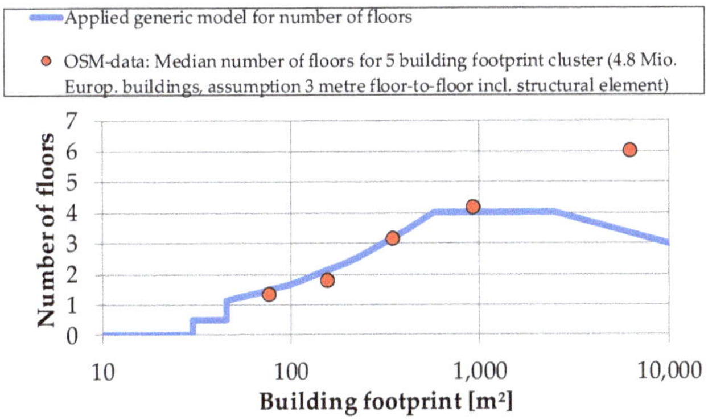

Figure 5. Generic building height model applied to the buildings covered in the OpenStreetMap database.

In order to estimate the average relationship between the building footprint and the building height, we calculated the number of floors for different building footprint sizes per municipality (~18,000 municipalities out of ~115,000 in the covered region, see Figure 6).

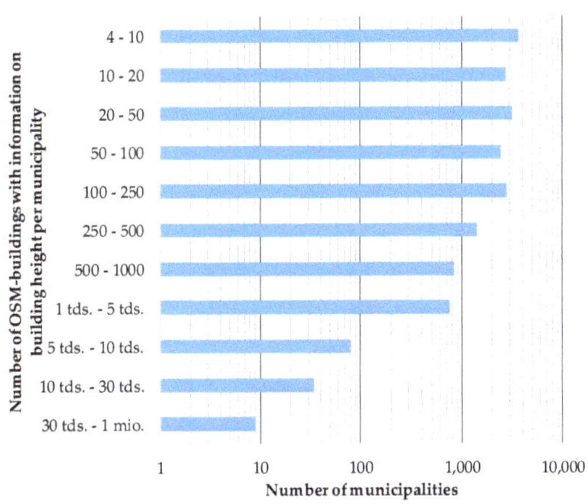

Figure 6. Number of buildings in the OpenStreetMap (OSM) database with information on the building height per municipality [59].

While each individual building from the OSM database with some information on their building height is given a weight of 1, the generic model for the number of floors is given a weight of 20. The resulting relationship between the footprint and number of floors for 150 randomly chosen municipalities (for which building height data are available) is shown in Figure 7.

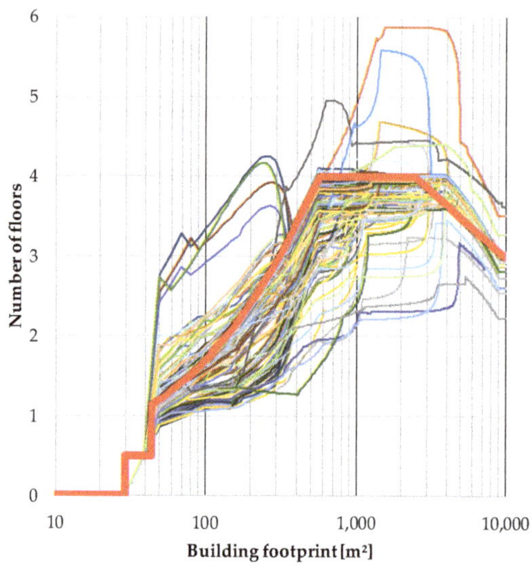

Figure 7. Calculated relationship (based on OSM data) between the average number of floors and the building footprint for 150 randomly chosen municipalities across Europe, as well as their generic functions (red line).

In the next step, we compare the gross floor areas derived from the OSM data with those from the population-based approach. If the outcome of the OSM-based approach is lower, then we scale-up the OSM data accordingly so that they match the outcome of the population-based approach. This is done up to a factor of four. If the gross floor area per inhabitant in a hectare cell is less than 15 m^2 according to the OSM-based data, then the OSM-quality indicator, which estimates the completeness of the OSM data (see Figure 8), is reduced. This process leads to a lower weight of the OSM-based heated floor area data in the final calculation of the heated gross floor area (see Table 1).

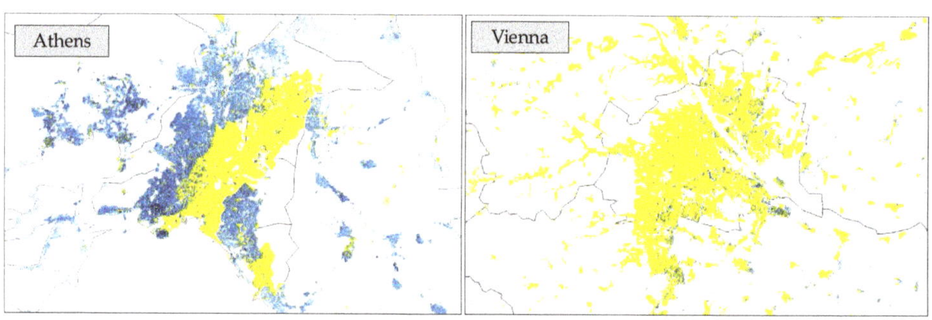

Figure 8. Completeness of the OpenStreetMap-building stock data: Comparison of the OpenStreetMap-data (yellow) with the European Settlement Map (blue) for the region of Athens (left map) and Vienna (right map). Sources: [54,59] (OSM: Planet dump May 2018).

Table 1. Weighting of the population and value added (VA) based approach versus the OSM based approach for areas in different Corine land cover classes used to calculate the heated gross floor area at the hectare level.

Corine Land Cover Class	Residential Gross Floor Area		Non-Residential Gross Floor Area	
	Weighting Factors * for Data from Approach Based on ...			
	Population Data w_{pop}	OSM Data w_{OSM}	Population & Value Added Data w_{pop}	OSM Data w_{OSM}
1: Continuous urban fabric	1	0.05	1	0.05
2: Discontinuous urban fabric	0.9	0.05	0.9	0.05
3: Industrial or commercial units	0.7	0.05	0.7	0.05
10: Green urban areas	0.1	0.05	0.1	0.05
11: Sport and leisure facilities	0.1	0.05	0.1	0.05
18: Pastures	0.5	0.05	0.5	0.05
20: Complex cultivation pattern	0.5	0.05	0.5	0.05
21: Land principally occupied by agriculture	0.5	0.05	0.5	0.05
Other classes	0.015	0.05	0.015	0.05

* The actual weighting factor is calculated, e.g., as $w_{pop}/(w_{pop} + w_{OSM})$.

Finally, the heated gross floor areas of the residential and non-residential buildings are calculated by combining the results of both methods. For residential buildings, we used a weight factor between 0.015 and 1 using the population-based approach (depending on the Corine land cover class (see Table 1), while the OSM-based approach is given a weight of 5%, if there is no indication that the OSM does not fully cover the given grid cell.

To estimate the heated floor area of non-residential buildings, we use (a) the value added per LAU region [45] instead of the residential floor area per inhabitant indicator and (b) the OSM-based approach. We calculate the non-residential heated gross floor area by subtracting the residential gross floor area from the calculated gross floor area using the OSM-building information. Compared to the residential gross floor area, we used a comparatively higher weight in the OSM-approach for non-residential buildings since the quality of the OSM data (degree of completeness) is estimated to be high. Based on own estimations, Table 1 depicts the detailed set of weighting factors that we used in our approach for different Corine land cover classes. These weighting factors are taken for areas where the OSM data quality is estimated to be high. If the data quality of the OSM data is considered to be low (see Figure 8), then the weight of the OSM approach is reduced accordingly.

2.4. Heating Degree Days at the Hectare Level

The starting point for the calculation of the local heating degree days is the observed average daily temperatures on the 25 × 25 km raster [50] for the period from 2002 to 2012. With a resolution of more than 600 km^2 per raster cell, this layer is too coarse to derive meaningful local heating degree days, as shown in Figure 9 (left figure) for the Alpine region, North Italy, and Croatia. To refine these data, we included in the calculation information on the local elevation using the digital elevation model over Europe (EU-DEM) layer at the 30 × 30 m grid level [60] and applied a temperature lapse rate of 6.5 °C per 1000 m elevation gain according to the specifications of the International Standard Atmosphere model [61] (see Figure 9, right).

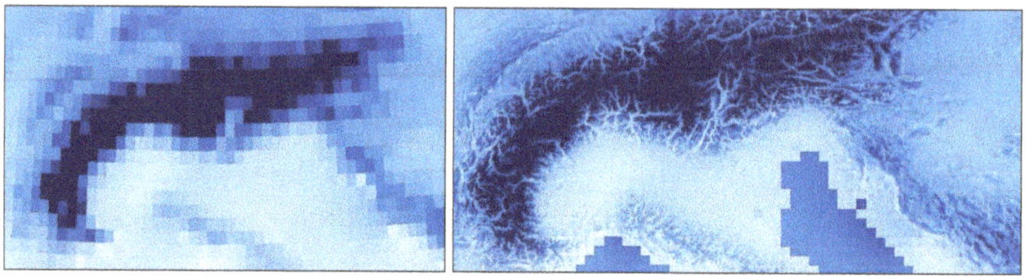

Figure 9. Heating degree days for the 25 × 25 km grid (left side) and the refined grid at the hectare level. Sources: [50,60] and our own calculations.

The energy needs for space heating (SH) on a local level are corrected by applying the ratio between the calculated site's specific HDD and the HDD at the NUTS 3 level using an elasticity of 0.5. We purposefully corrected the energy needs for the local climate conservatively, as we tried carefully to avoid "overshooting" our corrections and instead preferred results that are "too" uniform for different areas. With such low elasticity, we also covered the (plausible) assumption that buildings in colder areas (e.g., those at higher elevations), in general, might already have higher energy performance than similar buildings in warmer (lower) areas, even if they have to fulfil the same energy performance standards.

2.5. Surface-To-Volume Ratio of Buildings and Historical Construction Periods

In order to calculate the spatial distribution of EN and the final energy demand (FED) for SH from the heated floor area, we furthermore considered the surface-to-volume ratio of buildings and the share of buildings per construction periods. To obtain data on the surface-to-volume ratio, we built on the data from the OSM building layer: the building footprint (area and perimeter) and the estimated building height. For the share of buildings per construction period, we extracted information on the soil sealing data provided by the Global Human Settlement project [53] for 1975, 1990, 2000, and 2014 on a 38 × 38 m grid. Besides comparing the soil sealing ratio per grid cell for the time slots, we tried to correct the data for the soil sealed by other elements, such as roads, by considering the current share of soil sealed by buildings against the total share of sealed soil per grid cell (at the hectare level). We also generically considered building demolition. For the period after 2000, we considered an annual demolition rate of 0.2% for buildings constructed before 1975 and 0.1% for buildings constructed between 1975–1990. Thus, we assume that the stock of buildings constructed before 1975 is only 97% of that shown in the soil sealing map due to building demolitions between 2000 and 2014. Furthermore, we assume that at least 0.75% of the soil sealing share in each period (1975/1990/2000/2014) must stem from buildings constructed in the latest construction period. This means that if the soil sealing is 40% for a given grid cell in 1990, then the share of soil sealed by buildings constructed between 1975 to 1990 must be at least 40% × 0.75% = 0.3%. If the soil sealing map of 1975 already depicts soil sealing of 40%, we reduce that value by 0.3% and add that value to the construction period from 1975 to 1990. As an example, the outcome of this process is visualised in Figure 10 for the region of Vienna.

As with the local heating degree days, we tried to conservatively correct the energy needs for the surface-to-volume ratio and construction periods. Again, we manually performed checks for different regions to indicate that the outcomes are plausible on a general level. We needed to remain cautious of the significant uncertainties of this methodology. Therefore, we gave these two factors a rather low weight and considered an elasticity of only 33% for the surface-to-volume ratio. To buildings constructed after 2000, we assigned a specific EN of 80%, and to buildings constructed before 1990, we assigned an EN of 125%, which was compared to the specific energy needs per building (for buildings constructed from 1990 to 2000).

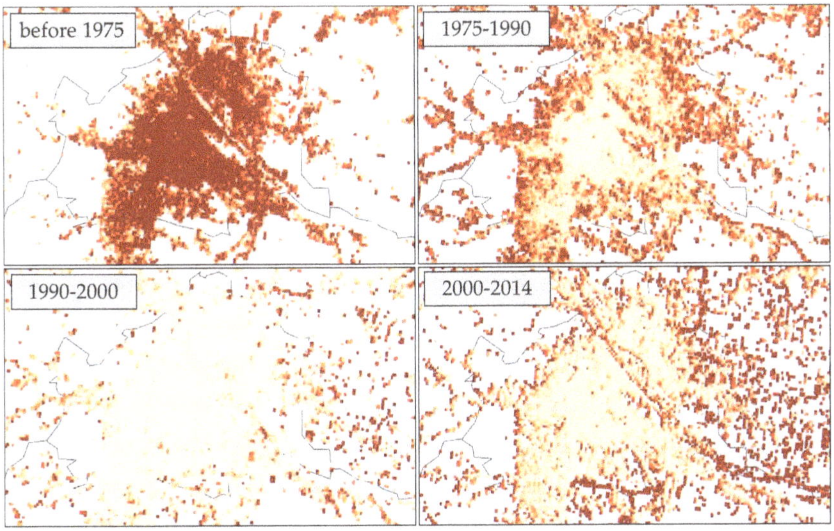

Figure 10. Estimated shares of buildings per construction period at the hectare level for Vienna and its surroundings. Red is used to color-code the high shares, whereas low shares are shown in beige.

2.6. Comparison of the Resulting Data with Data from Other Sources

In order to better understand the characteristics and quality of the developed gross floor area (GFA) density and heat demand (HD) density layers, we compared these with data from other sources. We did this at the following regional levels:

- For several NUTS 3 regions in Austria and Switzerland, a comparison of the overall gross floor area (GFA) of residential and service buildings and the respective final energy demand (FED) has been performed over the course of the project ESPON locate. The results can be found in the following report [47].
- For several LAU and NUTS 3 regions across Europe, we collected and compared the following data: number of inhabitants, GFA of residential buildings, GFA of service buildings, and FED for space heating and hot water preparation in residential and service buildings. We collected these data from local statistics on buildings and energy use and from reports of other projects (see Table 2).
- For three cities in Europe, we compared the developed GFA and the HD density maps with maps developed on the basis of building stock data from the city administrations. Then, we compared the values of both maps at the level of 100 × 100 m raster elements. These maps were developed according to the following methodology:
    - The basis for the bottom-up calculation of the GFA and the EN of the buildings in the three cities are shape files of the buildings containing the following information: shape and location of the building footprint, number of floors, building height and type, as well as age of the building. If data for certain buildings were missing, we filled these gaps using the average values of the other buildings with similar characteristics in the database.
    - In the second step, we joined these building stock data with data on specific EN values for space heating and for hot water generation from the Invert/EE-Lab model [40]. With this model, we calculated these values for typical buildings in the countries according to the type and construction period of the buildings. The resulting values applied to the overall building stock in the countries match the national energy balances. In joining the values

with the building stock databases of the cities, we performed climate correction from the average heating degree days (HDD) in the countries to the HDD in the cities. For this, we used the HDD data from the Hotmaps database (see Figure 9, available at [62]).

For the comparison, we used data from published and unpublished sources. At the NUTS 3 or LAU level for several locations, data can be found in the published literature. However, for many regions and municipalities, data on the energy demands of space heating and hot water generation, or on the number of buildings, are only available in unpublished databases or reports. Because public authorities usually do not publish their building inventories, for our comparison at the hectare level, we were only able to use unpublished data.

3. Results

3.1. Resulting Maps and Data

The developed raster files for gross floor area and heat demand for space heating and hot water preparation in residential and non-residential buildings are integrated into the Hotmaps database and toolbox and are accessible on their respective webpages [63]. On the website, these data can be visualised and analysed for selected locations and can be used in the different integrated calculation modules. Figure 11 shows a screenshot of the heat demand density total layer and the provided indicators for the NUTS 3 region of Vienna.

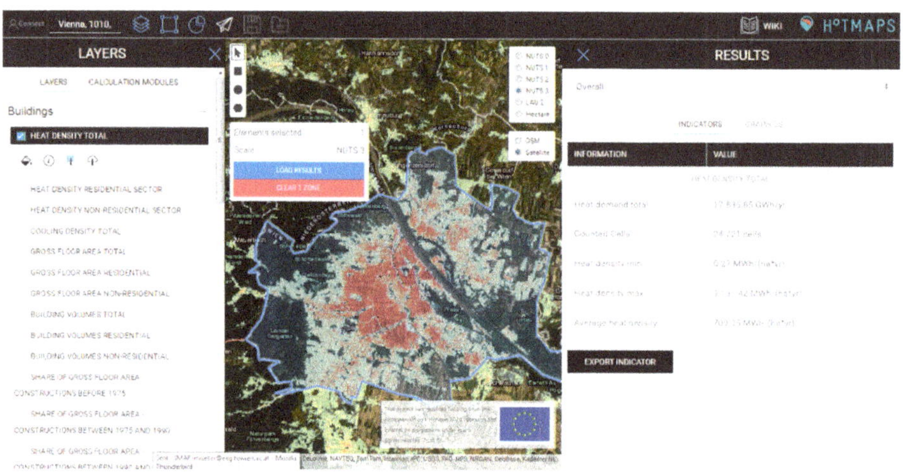

Figure 11. Screenshot of the developed heat demand density layer for the NUTS 3 region of Vienna, accessed via the Hotmaps database and toolbox. Source: [63].

The layers are also available for download from the Hotmaps database and toolbox or directly from a GitHub repository. The corresponding links are given in the supplementary materials section at the end of the manuscript.

3.2. Comparison of Results with Data from Other Sources at the NUTS 3, LAU 1, and LAU 2 Levels

To understand the quality of the developed gross floor area (GFA) and heat demand (HD) data, we first compare these at the levels of the NUTS 3, LAU 1 (local administrative units), and LAU 2 regions with data from other sources. We collected data on population, residential GFA, total GFA, and HD for space heating and hot water generation in residential and tertiary buildings for 20 different regions. Table 2 lists the data sources we used, and Figure 12 shows the results of this comparison.

Table 2. Data sources of the reference values for comparison at the LAU/NUTS level.

Location	Country	Data Sources
Ansfelden	Austria	[64]
Tralee	Ireland	[65]
Litomerice	Czech Republic	[66]
Skive	Denmark	[67]
Herten	Germany	[68]
Helsingor	Denmark	[69]
Bistrita	Romania	[70,71]
Hanau	Germany	[72]
Innsbruck	Austria	[73,74]
San Sebastian	Spain	[75,76]
Geneva City	Switzerland	[77]
Aalborg	Denmark	[67]
Milton Keynes	United Kingdom	[78]
Brasov	Romania	[79]
Aarhus	Denmark	[80,81]
Geneva Canton	Switzerland	[77]
Stuttgart	Germany	[82,83]
Frankfurt	Germany	[84,85]
München	Germany	[86]
Wien	Austria	[87,88]

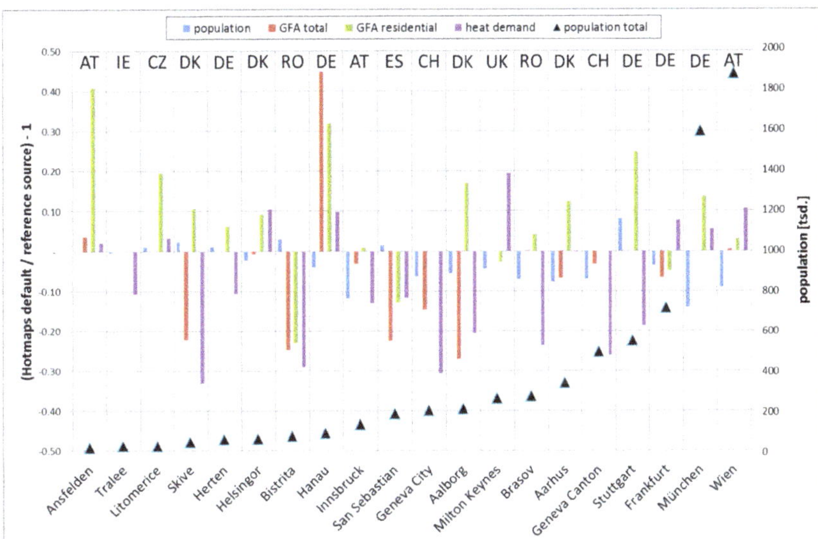

Figure 12. Comparison of the calculated values for the population, the residential gross floor area (GFA), the total GFA, and the heat demand with the values stated in other sources for selected locations.

This comparison of the developed data and other sources for the 20 selected regions shows an average difference of 12% in the mean values among the absolute values, with a standard deviation of 10% and a deviation of 8% for the median values. However, for some values, we found nearly no difference between the developed and other data. Some values show remarkable differences up to 45%. The values for the population and total GFA seem to better match the developed data and other sources, whereas residential GFA and HD show greater differences. Residential GFA, on average, is 9% higher in the developed data, and the HD, on average, is 8% lower. Both show a standard deviation of 16%. For regions with higher numbers of population, the differences seem to be lower. However, these statistics on the difference between the developed data and other data have a high uncertainty mainly due to the limited number of regions under comparison and the limited certainty of data from other sources. We discuss these uncertainties in Section 4.2.

3.3. *Comparison of the Results with Data from Other Sources at the Hectare Level*

For the cities of Bistrita, San Sebastian, and Frankfurt, we compared the developed gross floor area (GFA) and heat demand (HD) density maps (top-down) with maps developed from municipal building stock databases (bottom-up). Both types of maps were created with the same projection and raster size. For the comparison, we scaled the values in each hectare element of the top-down maps using the following ratio: the sum of the values of all hectare elements in the bottom-up map divided by the sum of values of all hectare elements in the top-down map. This allows us to focus on the difference of the distribution of the GFA and HD in the territories between the top-down and the bottom-up maps. For HD, we compare two maps: the top-down map and the bottom-up map. For GFA, we compare three maps: the top-down map, the bottom-up map of the GFA for all buildings in the region, and the bottom-up map showing only the heated GFA (HA) in residential and tertiary buildings.

In the first step, we compared the distribution of the GFA and HD over the GFA and HD density in the top-down and bottom-up maps. Figure 13 presents these distributions, showing the cumulated GFA and HD values for all hectare elements, from the elements with low density to the elements with high density. In order to compare the distributions for the three cities, we normalized both axes.

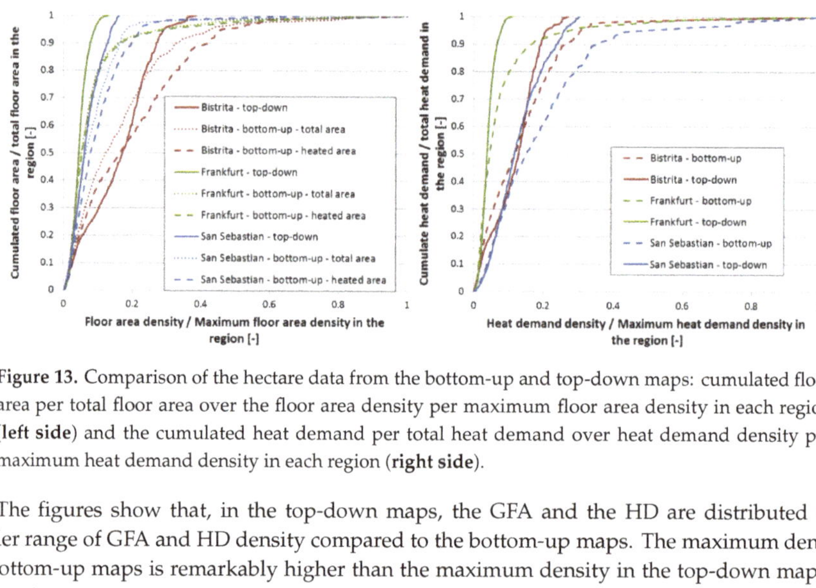

Figure 13. Comparison of the hectare data from the bottom-up and top-down maps: cumulated floor area per total floor area over the floor area density per maximum floor area density in each region (**left side**) and the cumulated heat demand per total heat demand over heat demand density per maximum heat demand density in each region (**right side**).

The figures show that, in the top-down maps, the GFA and the HD are distributed over a smaller range of GFA and HD density compared to the bottom-up maps. The maximum density in the bottom-up maps is remarkably higher than the maximum density in the top-down maps: e.g., in Frankfurt, in the top-down map, the highest GFA density is only 13% of the highest GFA density in the bottom-up map; for HD in Frankfurt, the GFA density is even lower at 11%. This is the same for all analysed cities but with a lower difference between the maximum values. Furthermore, the GFA and the HD are more evenly distributed in the top-down maps than in the bottom-up maps. The strong increase of density for the 5–10% GFA and HD in the areas of highest density visible in the bottom-up maps is remarkably underestimated in the top-down maps. Due to the fact that these characteristics can be found for all comparisons between the top-down and bottom-up groups, this seems like a systematic difference between the top-down and bottom-up results.

The figure also shows that the form of the distribution of the GFA and HD over the GFA and HD density is similar in the bottom-up and top-down maps. That is, the slope of the distribution curve for Frankfurt is steeper than that for San Sebastian, and the slope of the curve for San Sebastian is steeper than that for Bistrita. This is the same for both the bottom-up and top-down maps, as well as for the GFA and HD comparison. Finally, we also in the comparison of the GFA maps that the bottom-up maps showing only the heated area (HA) match better with the top-down maps for the three cities than with the bottom-up maps showing the entire GFA for all buildings in the region. This result seems logical, as the top-down maps reflect the heated area only.

For the second analysis, we compared the values from the top-down and bottom-up maps for each hectare element. Figure 14 shows the difference between the bottom-up and the top-down value in each hectare element for the three cities. In this way, the following maps are compared: (a) a top-down GFA map reflecting the heated area (HA) in the region, with the bottom-up GFA map containing all buildings in the region (including industrial and non-energy relevant buildings); (b) a top-down GFA map with the bottom-up HA map only containing the HA of the residential and tertiary buildings; and (c) a top-down HD map with a bottom-up HD map. For this comparison, the top-down values have been scaled so that the overall GFA, heated area, and HD in the area are the same. In the figure, each hectare cell of the selected municipality is represented by a single dot. Blue dots indicate that the Hotmaps' top-down data distribute a lower share of energy or gross floor area to a specific hectare cell then the bottom-data distribute. The hectare cells are shown in red dots if the bottom-up maps allocate a lower share.

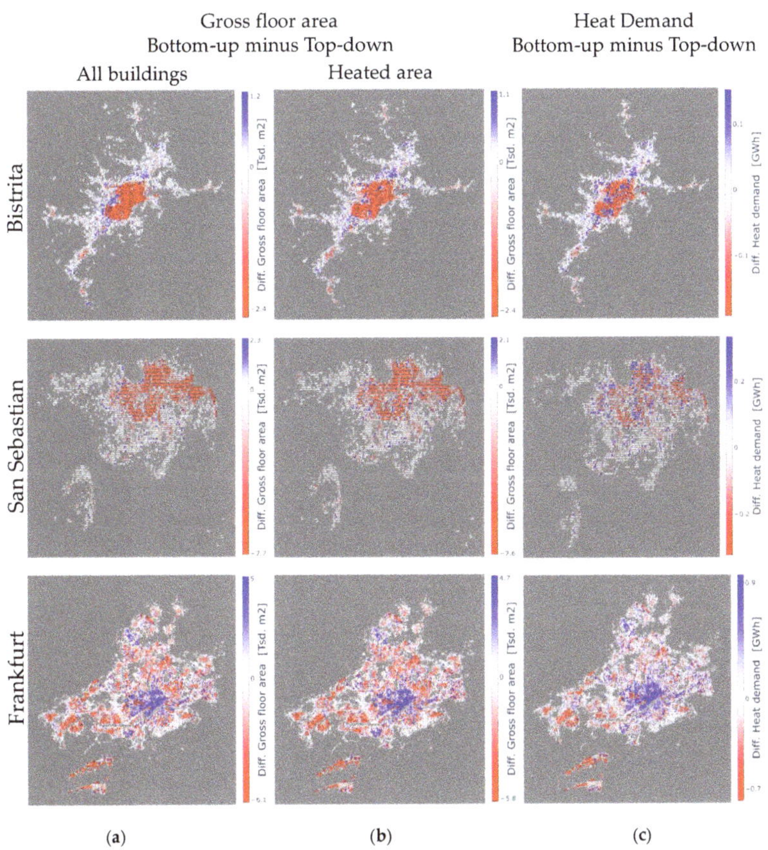

Figure 14. Difference between the top-down and bottom-up values for each hectare element in three cities: (**a**) gross floor area (GFA) of all buildings (including industrial and non-energy relevant buildings) in the bottom-up data vs. heated area (HA) in the top-down data (left column), (**b**) HA in the bottom-up data vs. HA in the top-down data (middle column), and (**c**) heat demand in the bottom-up and in the top-down data (right column).

4. Discussion

In this paper, we explained how we developed a gross floor area (GFA) density map and a heat demand (HD) density map at the level of 100 m × 100 m for the entire EU 28 (+ Norway, Iceland, and Switzerland) within the Hotmaps project. We also showed a comparison of the developed maps at the NUTS 3, LAU, and hectare levels with data and maps (developed) from other sources. In the following, we discuss the limitations of available data for the exercise and their effects on the results, as well as the uncertainty in the comparison of the results with data from other sources.

4.1. Limitations of the Data

The presented datasets and maps build on a statistical approach. This approach limits the accurateness of the data, as site specific or local conditions are not taken into account. Considering the input data, we believe that the population data are accurate up to a level between 250 × 250 m and 500 × 500 m. However, we must acknowledge that the input data are, on average, about 10 years old. In addition, the data are consistent with statistical data at the municipal level; given the limitation that statistical population data on LAU regions are not available for all census years and LAU regions (or contain inconsistencies), we used the average population data for the years 2008 to 2016. Checks, where we estimated the population of a given area using satellite images and estimations for the average number of persons per building, confirmed that the data are also plausible for higher resolutions.

Statistical data on the residential heated (net) floor area are available for most NUTS 3 regions. We again performed manual data quality checks, which indicated that our results are plausible at the hectare level for most regions. However, in the current dataset, we did not consider the observation that the heated area per inhabitant often decreases by increasing population density. For NUTS 3 regions with a strong urban versus rural area gradient, this might result in an overestimation of the heated residential gross floor area in urban areas. The heated gross floor area of non-residential buildings, however, remains very uncertain at the country level. Data quality checks indicate that the sum of the residential and non-residential heated gross floor area is in a plausible range. Also, the ratio between residential and non-residential gross floor area is plausible, although this indicator might not hold for grid cells, which contain few buildings. Furthermore, the comparison of regional building stock data indicates that we likely underestimated the floor area of non-residential buildings at the country level, which is an input in our model. The final energy demand for non-residential buildings, however, is in the correct order of magnitude, as well as the data for the heated gross floor area at the national level for building categories, such as offices, health and education, restaurants and hotels, retail and wholesale, and others. One possible reason for this result is that we overestimated the area-specific energy demand (e.g., due to the geometry of the buildings (small surface-to-volume ratio) and/or higher-than-estimated internal gains) or that a significant share of non-residential buildings is not fully heated (industrial production halls, warehouses, etc.). In order to systematically investigate this gap, further high-quality data on existing non-residential building stock is needed.

For the heat demand density map, we derived the local data from statistical data on energy consumption at the country level and national building stock characteristics, such as the average specific energy needs per construction period. To calculate the grid cell specific energy demand-per-floor area data, we assessed the surface-to-volume ratio of buildings based on the OpenStreetMap database, the share of floor area per construction periods, and the heating and cooling degree days. The impacts of the first two indicators are plausible but highly uncertain. We, therefore, give these indicators a low weight in our calculations. We belief that these last indicators, the heating and cooling degree days, are of higher accuracy, although we used a simple atmospheric temperature lapse rate model, which cannot account for local site-specific weather and climate conditions. Additional uncertainties exist, as we do not know if and how planers have already considered colder (or warmer) local climate conditions when the buildings were constructed in the past. Since we assume that this might be the case to some extent, we lowered the weight of the climate indicator compared to what is usually considered to be the actual thermodynamic degree of influence. Again, data quality checks indicate that our results are plausible. However, we recommend using individual data on the heated area-specific energy needs or final energy demands, whenever local data are available. Another uncertainty regarding the final energy consumption (which is not the case for the energy needs) arises from the lack of information on the applied heating systems and the corresponding efficiency. If, in a region or grid cell hiatus, biomass-based stoves are widely applied, then the final energy consumption will be higher than if electricity-based systems are commonly used, even if the actual energy needs of the buildings are identical.

4.2. Uncertainty in the Comparison of the Results with Data from Other Sources

To compare the developed maps with data and maps from other sources at different regional levels is important to understand the potential use and the credibility of the dataset. Although it was possible to find values for population, residential gross floor area (GFA), total GFA, and heat demand (HD) for 20 regions, the uncertainty in the statistics for the difference between the developed data and the data from other sources is high. There are two main reason for this uncertainty. First, the share of the regions for which we compared developed data with data from other sources on the overall territory covered in the developed data is very low. The 20 regions in the presented comparison cover 1.4% of the population of the entire analysed territory. Second, the reliability of the data from other sources is often unknown. Descriptions of the data stated in reports often lack detail to understand what exactly is being represented in the data, e.g., what type of heat demand is reflected, what types of buildings are taken into account, the year of reference, if the demand data are climate corrected, or what the regional borders of the analysis are. Therefore, no quantitative conclusions for the differences between the developed and other data at the NUTS and LAU level are possible.

We also compared the top-down developed maps with maps based on municipal building stock data (bottom-up) for three cities at the hectare level. The bottom-up maps were developed by estimating the HD via the average HD in typical buildings in the countries calibrated at the national level and climate corrected to the location of analysis. We found that the overall GFA and HD, as well as the split between residential and tertiary buildings, in the developed bottom-up maps match well with the data for buildings and the energy statistics from the cities. However, the data in the local statistics also imply uncertainty. Notably, the energy demand for space heating and hot water generation is not generally measured but developed based on the estimated shares of energy carriers used for different purposes. Bottom-up estimations, on the other hand, strongly depend on the input data—most importantly, energy demand per m^2, service factors, and building occupation. Furthermore, user behaviour is an important uncertainty when analysing HD at a very detailed regional level: Are people leaving houses for longer periods, are they used to having lower or higher indoor temperatures, or is a building or a flat really occupied or not?

5. Conclusions and Outlook

5.1. Conclusions

The developed GFA and HD density maps cover the entire territory of EU 28 + Norway, Iceland, and Switzerland. In addition, they are fully open source and therefore usable by everyone for every purpose. This is the first such dataset we know of.

A comparison of the developed data with data stated in other sources for selected cities and regions showed differences from very low up to 45%. The average difference of all compared values was 12% (median 8%), with a standard deviation of 10%. Differences in this range are also often experienced in comparing data from bottom-up estimations of GFA and HD based on local building stock data with values from buildings and energy statistics.

A comparison of the developed maps with maps based on municipal building stock datasets for three cities shows that, for these locations, the overall tendency of the distribution of GFA and HD over the GFA and HD density is similar in both approaches. This comparison also reveals the following systematic difference: The developed datasets seem to systematically overestimate the GFA and HD in low density areas and underestimate the GFA and HD in high density areas.

We conclude that the developed GFA and HD density maps allow a first analysis of GFA and HD distribution in all locations in Europe. Also, they can be used to identify areas that might be suitable for district heating. Especially for locations in Europe where detailed GFA and HD density maps are not available, the developed maps provide valuable data for initial and quick analyses. For the detailed planning of supply infrastructure, however, more detailed data from the local level should be used.

5.2. Outlook

Although the approach of developing heat density maps is not entirely new, and despite the achieved progress regarding the transparency, accessibility, and quality of the data presented in this paper, there is still a considerable need to enhance the work on heat density maps.

In the course of the Hotmaps project, the representatives of additional follower areas—beyond the cases presented above—will use the Hotmaps toolbox (www.hotmaps.eu). The creation of detailed, bottom-up heat density maps will provide the grounds for more data to be compared with the EU-28 default data map presented in this paper. The authors intend to use this process to continuously develop further calibration, a better understanding of possible deviations and biases, and regularly update the database via the toolbox and on the mentioned Git-Repository. This model, which creates the described heat density maps, will be published as open source at the end of the Hotmaps project; until then, the model is available on request.

High quality data on heating and cooling energy demands and consumption for entire regions or areas are rare and usually subject to substantial uncertainty. Thus, reference values for the plausibility checks of heat density maps are also rare and partly uncertain. In addition, even the assessment of the reliability of a source is often difficult since geographic system boundaries are not always clear, and the definitions, and indicators are often not fully documented. Thus, the improvement of data availability, reliability, documentation, and know-how on the municipal, regional, national, and European levels regarding energy demands and, in particular, heating and cooling, should be given higher priority. On the municipal level, this prioritization should also be integrated in the process of establishing strategic heating and cooling planning and mapping processes. We are convinced that a better data foundation and correspondingly trained persons are essential preconditions for more effective planning and mapping processes and thus a decarbonisation of the heating and cooling sector.

Cooling density maps have also been developed by the authors in the frame of the project Hotmaps. Due to the restricted space in this paper, we decided to limit the scope to heating only. The elaboration of cooling density maps has led to other types of uncertainties and issues that need be further analysed in future research work.

The validation and subsequent reliability of the heat density maps could be further improved by integrating other data sources. This refers, for example, to the data from the EPC databases. The project Enerfund (http://enerfund.eu/) has provided a rich map of EPC data, at least for some countries. Another important source and step to improve the quality of heat density maps is using real and measured energy consumption data. By increasing the roll-out of smart meters and devices that enhance the smart-readiness of buildings, a substantial amount of data will be available in the future. These data will have good potential to improve the reliability and real-time updates, adding higher time resolution and quality to the heat density maps. However, the availability of these data does not mean they are also accessible for the improvement of heat density maps. Clarifying data protection rules and the ownership of consumers' energy consumption data will be one of the key prerequisites.

Overall, we are convinced that with an increasingly stronger focus on the heating and cooling sector in achieving energy and climate policy targets, local work on decarbonising this sector will gain relevance. In this way, a better understanding of the spatial dimension of heating and cooling supply and demand will become progressively more important.

Supplementary Materials: The following developed datasets (maps) are available online at https://gitlab.com/hotmaps: Heated gross floor area density maps of buildings in EU28 + Switzerland, Norway and Iceland for the year 2015: Residential building: https://gitlab.com/hotmaps/gfa_res_curr_density, Non-residential buildings: https://gitlab.com/hotmaps/gfa_nonres_curr_density, all buildings: https://gitlab.com/hotmaps/gfa_tot_curr_density, Heat density (final energy demand for space heating and DHW) on a hectare level 100 × 100 m for EU28, Norway, Iceland and Switzerland in 2015: Residential building: https://gitlab.com/hotmaps/heat/heat_res_curr_density, Non-residential buildings: https://gitlab.com/hotmaps/heat/heat_nonres_curr_density, All buildings: https://gitlab.com/hotmaps/heat/heat_tot_curr_density.

Author Contributions: Conceptualization, A.M. and L.K.; methodology, A.M.; software, A.M., M.H., and M.F.; validation, A.M., M.H., and M.F.; formal analysis, A.M. and M.H.; data curation, A.M., M.H., and R.B.; writing—original draft preparation, A.M., M.H., and L.K.; writing—review and editing, A.M., M.H., L.K., M.F., and R.B.; visualization, A.M., M.H., and M.F.

Funding: This research was funded by the Horizon 2020 programme of the European Union, grant number 723677—Hotmaps.

Acknowledgments: This work has benefited greatly from the discussions with the Hotmaps project.

Conflicts of Interest: The authors declare no conflict of interest.

Abbreviations and Variables

CDD	Cooling degree days (K·days)
DHW	Domestic hot water
EN	Energy needs, defined by EN 13790, (kWh)
FED	Final energy demand (kWh)
HDD	Heating degree days (K·days)
LAU	Local administrative units
NUTS	Nomenclature of Territorial Units for Statistics. NUTS 0: Country Level. Below Nuts 0 three NUTS levels are defined and two levels of local administrative units (LAUs) below.
OSM	Open street map
SH	Space heating
VA	Value added (€)

References

1. Fleiter, T.; Elsland, R.; Rehfeldt, M.; Steinbach, J.; Reiter, U.; Catenazzi, G.; Jakob, M.; Rutten, C.; Harmsen, R.; Dittmann, F.; et al. *Profile of Heating and Cooling Demand in 2015*; Fraunhofer Institute for Systems and Innovation Research: Karlsruhe, Germany, 2017.
2. Eurostat Complete Energy Balance [nrg_bal_c]. Available online: https://ec.europa.eu/eurostat/web/energy/data/database (accessed on 1 December 2019).
3. Kavvadias, K.C.; Quoilin, S. Exploiting waste heat potential by long distance heat transmission: Design considerations and techno-economic assessment. *Appl. Energy* **2018**, *216*, 452–465. [CrossRef]
4. Directive 2012/27/EU of the European Parliament and of the Council of 25 October 2012 on Energy Efficiency, Amending Directives 2009/125/EC and 2010/30/EU and repealing Directives 2004/8/EC and 2006/32/EC Text with EEA Relevance. *Off. J. Eur. Union* **2012**, *L315*, 1–56.
5. Directive (EU) 2018/2002 of the European Parliament and of the Council of 11 December 2018 amending Directive 2012/27/EU on energy efficiency (Text with EEA relevance.). *Off. J. Eur. Union* **2018**, *L328*, 210–230.
6. Büchele, R.; Kranzl, L.; Hummel, M. Integrated strategic heating and cooling planning on regional level for the case of Brasov. *Energy* **2019**, *171*, 475–484. [CrossRef]
7. Djørup, P.S.; Bertelsen, N.; Mathiesen, B.V.; Schneider, C.A.; Sørensen, R.P.A.; Guddat, M.G.A. *Handbook I Definition & Experiences of Strategic Heat Planning*; Aalborg Universitet: Aalborg, Denmark, 2019; Volume 36.
8. Noussan, M.; Nastasi, B. Data Analysis of Heating Systems for Buildings—A Tool for Energy Planning, Policies and Systems Simulation. *Energies* **2018**, *11*, 233. [CrossRef]
9. Tronchin, L.; Manfren, M.; Nastasi, B. Energy efficiency, demand side management and energy storage technologies—A critical analysis of possible paths of integration in the built environment. *Renew. Sustain. Energy Rev.* **2018**, *95*, 341–353. [CrossRef]

10. Peta4—Heat Roadmap Europe. Available online: https://heatroadmap.eu/peta4/ (accessed on 23 September 2019).
11. Connolly, D.; Lund, H.; Mathiesen, B.V.; Werner, S.; Möller, B.; Persson, U.; Boermans, T.; Trier, D.; Østergaard, P.A.; Nielsen, S. Heat Roadmap Europe: Combining district heating with heat savings to decarbonise the EU energy system. *Energy Policy* **2014**, *65*, 475–489. [CrossRef]
12. Persson, U.; Wiechers, E.; Möller, B.; Werner, S. Heat Roadmap Europe: Heat distribution costs. *Energy* **2019**, *176*, 604–622. [CrossRef]
13. Möller, B.; Wiechers, E.; Persson, U.; Grundahl, L.; Lund, R.S.; Mathiesen, B.V. Heat Roadmap Europe: Towards EU-Wide, local heat supply strategies. *Energy* **2019**, *177*, 554–564. [CrossRef]
14. Möller, B.; Wiechers, E.; Persson, U.; Grundahl, L.; Connolly, D. Heat Roadmap Europe: Identifying local heat demand and supply areas with a European thermal atlas. *Energy* **2018**, *158*, 281–292. [CrossRef]
15. Andrews, D.D.; Krook-Riekkola, A.; Tzimas, E.; Serpa, J.; Carlsson, J.; Pardo-Garcia, N.; Papaioannou, I. *Luleå Tekniska Universitet; Institutionen för Ekonomi, Teknik och Samhälle Background Report on EU-27 District Heating and Cooling Potentials, Barriers, Best Practice and Measures of Promotion*; Publications Office of the European Union: Luxembourg, 2012; ISBN 978-92-79-23882-6.
16. Nielsen, S.; Möller, B. GIS based analysis of future district heating potential in Denmark. *Energy* **2013**, *57*, 458–468. [CrossRef]
17. Persson, U.; Werner, S. Heat distribution and the future competitiveness of district heating. *Appl. Energy* **2011**, *88*, 568–576. [CrossRef]
18. Müller, A.; Büchele, R.; Kranzl, L.; Totschnig, G.; Mauthner, F.; Heimrath, R.; Halmdienst, C. *Solarenergie und Wärmenetze: Optionen und Barrieren in Einer Langfristigen, Integrativen Sichtweise (SolarGrids)*; Energy Economics Group (TU Wien): Wien, Austria, 2014.
19. Fallahnejad, M.; Hartner, M.; Kranzl, L.; Fritz, S. Impact of distribution and transmission investment costs of district heating systems on district heating potential. *Energy Procedia* **2018**, *149*, 141–150. [CrossRef]
20. Dorfner, J.; Hamacher, T. Large-Scale District Heating Network Optimization. *IEEE Trans. Smart Grid* **2014**, *5*, 1884–1891. [CrossRef]
21. Eggimann, S.; Hall, J.W.; Eyre, N. A high-resolution spatio-temporal energy demand simulation to explore the potential of heating demand side management with large-scale heat pump diffusion. *Appl. Energy* **2019**, *236*, 997–1010. [CrossRef]
22. Chambers, J.; Narula, K.; Sulzer, M.; Patel, M.K. Mapping district heating potential under evolving thermal demand scenarios and technologies: A case study for Switzerland. *Energy* **2019**, *176*, 682–692. [CrossRef]
23. Leurent, M. Analysis of the district heating potential in French regions using a geographic information system. *Appl. Energy* **2019**, *252*, 113460. [CrossRef]
24. Pampuri, L.; Belliardi, M.; Bettini, A.; Cereghetti, N.; Curto, I.; Caputo, P. A method for mapping areas potentially suitable for district heating systems. An application to Canton Ticino (Switzerland). *Energy* **2019**, 116297, in Press, Corrected Proof. [CrossRef]
25. Lund, R.; Persson, U. Mapping of potential heat sources for heat pumps for district heating in Denmark. *Energy* **2016**, *110*, 129–138. [CrossRef]
26. Carlsson, J.; Jakubcionis, M.; Kavvadias, K.; Moles, C.; Santamaria, M. *Joint Research Centre Synthesis Report on the Evaluation of National Notifications Related to Article 14 of the Energy Efficiency Directive*; European Commission: Brussels, Belgium, 2018; ISBN 978-92-79-88815-1.
27. Austrian Heat Map. Available online: http://www.austrian-heatmap.gv.at/das-projekt/ (accessed on 1 December 2019).
28. Heat Map Scotland. Available online: http://heatmap.scotland.gov.uk/ (accessed on 1 December 2019).
29. Netherlands Enterprise Agency Nationaal Expertise Centrum Warmte—WarmteAtlas. Available online: www.warmteatlas.nl (accessed on 1 December 2019).
30. Brocklebank, I.; Styring, P.; Beck, S. Heat mapping for district heating. *Energy Procedia* **2018**, *151*, 47–51. [CrossRef]
31. Dorotić, H.; Novosel, T.; Duić, N.; Pukšec, T. Heat demand mapping and district heating grid expansion analysis: Case study of Velika Gorica. *E3S Web Conf.* **2017**, *19*, 01021. [CrossRef]
32. Artur, W.; Yi-kuang, C. Mapping Urban Heat Demand with the Use of GIS-Based Tools. *Energies* **2017**, *10*, 720. [CrossRef]

33. Hummel, M. *Supporting the Progress of Renewable Energies for Heating and Cooling in the EU on a Local Level (progRESsHEAT)*; Funded under H2020-LCE-2014-3, Grant agreement ID: 646573; Technische Universität Wien, Energy Economics Group: Vienna, Austria, 2017. Available online: www.progressheat.eu (accessed on 1 December 2019).
34. Čižman, J.; Staničić, D.; Česen, M. Use of Thermal Atlas and Heating Model for Strategic Municipal Energy Planning. In Proceedings of the 12th SDEWES Conference, Dubrovnik, Croatia, 4–8 October 2017; University of Zagreb: Zagreb, Croatia, 2017; p. 10.
35. Dochev, I.; Peters, I.; Seller, H.; Schuchardt, G.K. Analysing district heating potential with linear heat density. A case study from Hamburg. *Energy Procedia* **2018**, *149*, 410–419. [CrossRef]
36. Törnros, T.; Resch, B.; Rupp, M.; Gündra, H. Geospatial Analysis of the Building Heat Demand and Distribution Losses in a District Heating Network. *ISPRS Int. J. Geo-Inf.* **2016**, *5*, 219. [CrossRef]
37. Fleiter, T.; Marlene, A.; Ali, A.; Rainer, E.; Tobias, F.; Clemens, F.; Andrea, H.; Simon, H.; Michael, K.; Mario, R.; et al. *Mapping and Analyses of the Current and Future (2020—2030) Heating/Cooling Fuel Deployment (Fossil/Renewables)—Work package 1: Final energy consumption for the year 2012*; Fraunhofer Institute for Systems and Innovation Research (ISI): Karlsruhe, Germany, 2016.
38. ESS Census Hub. Available online: http://ec.europa.eu/eurostat/web/population-and-housing-census/census-data/2011-census (accessed on 1 December 2019).
39. EEG. Invert/EE-Lab European building stock database. In *Database on the Building Stock of the EU-28 Member States + Norway, Switzerland and Iceland*; Technische Universität Wien, Energy Economics Group: Vienna, Austria, 2019.
40. Müller, A. Energy Demand Assessment for Space Conditioning and Domestic Hot Water: A Case Study for the Austrian Building Stock. Ph.D. Thesis, Technische Universität Wien, Vienna, Austria, 2015.
41. The Invert/EE-Lab Model. Available online: www.invert.at (accessed on 1 December 2019).
42. ISO EN 13790:2008. *Energy Performance of Buildings—Calculation of Energy Use for Space Heating and Cooling*; European Committee for Standardization: Brussels, Belgium, 2008.
43. Austrian Standards. *ÖNORM B 8110-5: 2007 Wärmeschutz im Hochbau—Teil 5: Klimamodell und Nutzungsprofile*; Austrian Standards: Wien, Austria, 2007.
44. Austrian Standards. *ÖNORM B 8110-6, 2007. Wärmeschutz im Hochbau—Teil 6: Grundlagen und Nachweisverfahren—Heizwärmebedarf und Kühlbedarf*; Austrian Standards: Wien, Austria, 2007.
45. Austrian Standards. *ÖNORM H 5056, 2007 (Vornorm). Gesamtenergieeffizienz von Gebäuden—Heiztechnik-Energiebedarf*; Austrian Standards: Wien, Austria, 2007.
46. *Energy Performance of Buildings—Overall Energy Use and Definition of Energy Ratings*; ECS EN 15603:2008; European Committee for Standardization: Brussels, Belgium, 2008.
47. Schremmer, C.; Derszniak-Noirjean, M.; Keringer, F.; Raffaelm, K.; Michaelm, L.; Ursula, M.; Edith, S.; Tordy, J.; Lukas, K.; Mostafa, F.; et al. *Territories and low-Carbon Economy (ESPON Locate), Annex to the Final Report (Scientific Report)*; ÖIR GmbH: Vienna, Austria, 2017.
48. Eurostat CensusHub2. Eurostat, Luxembourg. Available online: https://ec.europa.eu/CensusHub2/query.do?step=selectHyperCube&qhc=false (accessed on 15 February 2018).
49. Eurostat Heating degree-days by NUTS 2 regions—Annual data [nrg_esdgr_a]. Eurostat, Luxembourg. 2013. Available online: https://appsso.eurostat.ec.europa.eu/nui/show.do?dataset=nrg_chddr2_a&lang=en (accessed on 9 December 2019).
50. Haylock, M.R.; van den Besselaar, E.J.M.; van der Schrier, G.; Klein Tank, A.M.G. A European daily high—Resolution observational gridded data set of sea level pressure. *J. Geophys. Res.* **2011**, *116*, D11110.
51. Eurostat Gross value added at basic prices by NUTS 3 regions [nama_10r_3gva]. Eurostat, Luxembourg. 2016. Available online: https://data.europa.eu/euodp/en/data/dataset/VhCfyrAU2sc2FmN0pneyuw (accessed on 1 December 2019).
52. Gallego, F.J. A population density grid of the European Union. *Popul. Environ.* **2010**, *31*, 460–473. [CrossRef]
53. European Commission, Joint Research Centre (JRC); Columbia University, Center for International Earth Science Information Network—CIESIN GHS population grid, derived from GPW4, multitemporal (1975, 1990, 2000, 2015), European Commission, Joint Research Centre (JRC). 2015. Available online: http://data.europa.eu/89h/jrc-ghsl-ghs_pop_gpw4_globe_r2015a (accessed on 8 December 2019).

54. Joint Research Center European Settlement Map, European Commission, Joint Research Centre, Institute for Protection and Security of the Citizen. 2017. Available online: http://land.copernicus.eu/pan-european/GHSL/european-settlement-map/esm-2012-release-2017-urban-green/view (accessed on 8 December 2019).
55. European Environment Agency (EEA) Corine Land Cover (CLC) 2012, Version 18.5.1 2012. European Environment Agency. Available online: http://land.copernicus.eu/pan-european/corine-land-cover/clc-2012/view (accessed on 8 December 2019).
56. European Commission, Eurostat (ESTAT), GISCO Communes. 2013—Administrative Unit. Eurostat, Luxembourg. 2013. Available online: http://ec.europa.eu/eurostat/web/gisco/geodata/reference-data/administrative-units-statistical-units/communes (accessed on 15 February 2018).
57. Eurostat. Correspondence Table LAU 2—NUTS 2010, EU-27. Eurostat, Luxembourg. 2010. Available online: https://ec.europa.eu/eurostat/documents/345175/501971/EU-27_2010.xlsx (accessed on 10 December 2019).
58. Joint Research Center. Estimation of the Gross Domestical Product 2006 in the 119 000 LAU2 of the ESPON Area. JRC; 2011. [Dataset] Provider: GISCO; ESPON Database 2013 Project, Date 01/02/2011 (access restricted to ESPON partners). Technical Report: Groza, O.; Rusu, A. Local & Regional Data. Producing Innovative Indicators at Local Scale. UAIC, CUGUAT-TIGRIS, Iasi, Romania. 2011. Available online: https://www.espon.eu/sites/default/files/attachments/3.4_TR_Local_data_innovative_indicators.pdf (accessed on 8 December 2019).
59. OSM OpenStreetMap Contributors. Planet Dump. March 2019. Available online: https://planet.osm.org/planet/2019/planet-190304.osm.bz2 (accessed on 8 December 2019).
60. EEA Copernicus Land Monitoring Service. EU-DEM v1.1. European Environmental Agency. 2016. Available online: http://land.copernicus.eu/pan-european/satellite-derived-products/eu-dem/eu-dem-v1.1/view (accessed on 8 December 2019).
61. De, M. Manual of the ICAO Standard Atmosphere, Third Edition. Doc 7488/3, International civil Aviation organization. 1993. Available online: https://tinyurl.com/rs7fozv (accessed on 10 December 2019).
62. Müller, A.; Fallahnejad, M. European Heating Degree Days (HDD) for the reference period 2002–2012. Hotmaps Open Data Set for the EU28. 2018. Available online: https://gitlab.com/hotmaps/climate/HDD_ha_curr (accessed on 8 December 2019).
63. Hotmaps Hotmaps Database and Toolbox. Available online: www.hotmaps.eu (accessed on 1 December 2019).
64. Büchele, R.; Hummel, M. *Factsheet of the Status Quo in Ansfelden*; TU Wien-Energy Economics Group: Vienna, Austria, 2016. Available online: http://www.progressheat.eu/IMG/pdf/d2-1-ansfelden_upload_2016-11.pdf (accessed on 10 December 2019).
65. XD Consulting. *Heat Mapping of Tralee Town in Course of the SmartReflex Project*; XD Sustainable Energy Consulting Ltd.: Clonakilty, Ireland, 2016.
66. Klusak, J.; Münster, M. *Factsheet of the Status Quo in Litomerice*; City of Litoměřice: Litoměřice, Czech Republic, 2016. Available online: http://www.progressheat.eu/IMG/pdf/d2-1_litomerice_upload_2016-11.pdf (accessed on 10 December 2019).
67. AAU. *Heat Atlas Denmark*; Aalborg University: Aalborg, Denmark, 2016.
68. Aydemir, A.; Münster, M. *Factsheet of the Status Quo in Herten*; Fraunhofer ISI: Karlsruhe, Deutschland, 2016. Available online: http://www.progressheat.eu/IMG/pdf/d2-1-herten_upload_2016-11.pdf (accessed on 10 December 2019).
69. Ben Amer-Allam, S.; Münster, M. *Factsheet of the Status Quo in Helsingor*; Technical University of Denmark: Copenhagen, Denmark, 2016. Available online: http://www.progressheat.eu/IMG/pdf/d2-1_litomerice_upload_2016-11.pdf (accessed on 10 December 2019).
70. Municipality of Bistrita. *Bistrita Municipal Building Inventory Bistrita*; Municipality of Bistrita: Bistrita, Romania, 2019.
71. INS. *Statistics on Natural Gas Demand*; Institutul National de Statistica: Bucharest, Romania, 2018.
72. Stadt Hanau Kommunales Klimaschutzkonzept Hanau—Im Rahmen der kommunalen Klimaschutzinitiative der Bundesregierung. Stabsstelle Nachhaltige Energien, Stadt Hanau. 2013. Available online: https://www.hanau.de/mam/Stadtentwicklung/energie_klima/klimaschutzkonzept/kommunales-klimaschutzkonzept-hanau_abschlussbericht.pdf (accessed on 12 December 2019).
73. Dobler, C.; Streicher, W. *Energieplan Innsbruck—Energieszenarien 2015–2050*; Universität Innsbruck: Innsbruck, Austria, 2017.

74. Pfeifer, D. *Entwicklung, Untersuchung und Bewertung von Berechnungsmodellen zur Erstellung von kommunalen Energiebilanzen im Gebäudebereich*; Universität Innsbruck: Innsbruck, Austria, 2017.
75. DSS. *Informe Anual de Sostenibilidad*; DSS: Donostia San Sebastian, Spain, 2018.
76. Fomento San Sebastian (FSS). *San Sebastian Municipal building inventory San Sebastian*; Fomento San Sebastian (FSS): Donostia San Sebastian, Spain, 2019; unpublished.
77. OCEN. *Data from OCEN*; Office Cantonal de l'énergie (OCEN): Geneve, Switzerland, 2018; unpublished.
78. Milton Keynes Energy Mapping Report, Milton Keynes Council, AECOM, Project Number: 60549497. 2018. Available online: http://www.milton-keynes.gov.uk/environmental-health-and-trading-standards/mk-low-carbon-living/energy-mapping-report (accessed on 12 December 2019).
79. Büchele, R.; Hummel, M.; Rata, C. Factsheet of the Status Quo in Brasov; D2.1 in course of the project progRESsHEAT, TU Wien, Vienna, Austria. 2016. Available online: http://www.progressheat.eu/IMG/pdf/d2-1-brasov_upload_2016-11.pdf (accessed on 12 December 2019).
80. PlanEnergi. *Personal Information from PlanEnergi*; PlanEnergi: Skørping, Denmark, 2019.
81. *Aarhus Municipal Building Inventory Aarhus unpublished*; Aarhus Kommune: Aarhus, Denmark, 2019.
82. LH Stuttgart Energieatlas Stuttgart. Available online: https://www.stadtklima-stuttgart.de/index.php?klima_kliks_energieatlas (accessed on 1 December 2019).
83. SLA Baden-Württemberg Wohnfläche je Einwohner in Stuttgat seit 1990. Available online: https://servicex.stuttgart.de/lhs-services/komunis/documents/10274_1_Wohnflaeche_je_Einwohner_1990_bis_2016.PDF (accessed on 1 December 2019).
84. *Frankfurt Municipal Building Inventory Frankfurt unpublished*; Energiereferat Frankfurt: Stadt Frankfurt, Germany, 2019.
85. Energiereferat Frankfurt. *Energiebilanzen der Stadt Frankfurt*; Stadt Frankfurt am Main, Der Magistrat, Energiereferat: Frankfurt, Germany, 2019; unpublished.
86. Kenkmann, T.; Hesse, T.; Hülsmann, F.; Timpe, C.; Hoppe, K.; Blanck, R.; Bürger, V.; Friedrich, A.; Sachs, A.; Winger, C. *Klimaschutzziel und Strategie München 2050*; Öko-Institut e.V.: Freiburg, Germany, 2017.
87. Statistik Austria. *Nutzenergieanalyse für Wien*; Statistik Austria: Vienna, Austria, 2018. Available online: http://www.statistik.at/wcm/idc/idcplg?IdcService=GET_NATIVE_FILE&dDocName=066287 (accessed on 8 December 2019).
88. Fritz, S. *Economic Assessment of the Long-Term Development of Buildings' Heat Demand and Grid-Bound Supply*; TU Wien: Vienna, Austria, 2016.

energies

MDPI

Article

The Role of Open Access Data in Geospatial Electrification Planning and the Achievement of SDG7. An OnSSET-Based Case Study for Malawi

Alexandros Korkovelos [1,*], Babak Khavari [1], Andreas Sahlberg [1], Mark Howells [1] and Christopher Arderne [2]

1 Division of Energy System Analysis, KTH Royal Institute of Technology, Brinellvägen 68, 10044 Stockholm, Sweden; khavari@kth.se (B.K.); asahl@kth.se (A.S.); mark.howells@energy.kth.se (M.H.)
2 The World Bank Group, Washington, DC 20433, USA; carderne@worldbank.org
* Correspondence: alekor@kth.se; Tel.: +46-735-843-613

Received: 10 March 2019; Accepted: 8 April 2019; Published: 11 April 2019

Abstract: Achieving universal access to electricity is a development challenge many countries are currently battling with. The advancement of information technology has, among others, vastly improved the availability of geographic data and information. That, in turn, has had a considerable impact on tracking progress as well as better informing decision making in the field of electrification. This paper provides an overview of open access geospatial data and GIS based electrification models aiming to support SDG7, while discussing their role in answering difficult policy questions. Upon those, an updated version of the Open Source Spatial Electrification Toolkit (OnSSET-2018) is introduced and tested against the case study of Malawi. At a cost of $1.83 billion the baseline scenario indicates that off-grid PV is the least cost electrification option for 67.4% Malawians, while grid extension can connect about 32.6% of population in 2030. Sensitivity analysis however, indicates that the electricity demand projection determines significantly both the least cost technology mix and the investment required, with the latter ranging between $1.65–7.78 billion.

Keywords: open data; electrification modelling; Malawi; OnSSET

1. Introduction

The 2030 Agenda for Sustainable Development has set the goal of universal access to electricity by 2030 (SDG7) [1]. The challenge is significant. It involves reaching populations with limited income, often living in sparsely populated areas, mostly in developing and least developed countries [2]. Selecting the optimal electrification approach is also difficult; grid vs. off-grid, fossil fuel vs. renewable, public vs. private investment are just a few examples. Coping with dilemmas of this nature—involving the deployment of big technological systems—requires thorough analysis of the social, technical, economic and political characteristics of the studied area or country [3]. This in turn, requires access to reliable data and information [4,5]; e.g., distribution and density of population settlements, electricity demand levels, resource availability, poverty rate and economic activity, distance from functional infrastructure (e.g., transmission and distribution network, roads, power stations) to name a few.

Despite progress, in most countries where universal electrification is still to be achieved, such official information is yet difficult to access [6]; these data are typically not covered by standard national energy statistics. The paucity of such information is one reason hampering electrification progress [7,8]. However, this situation is gradually being overcome with the increasing availability of new data and analytical tools, especially in the field of geospatial analysis. Geographic Information Systems (GIS) and remote sensing techniques are becoming openly available and can now provide a range of location-specific information that has not been previously accessible.

In the energy sector, the use of GIS data and associated analytical tools to conduct strategic planning remains at an early stage, yet such efforts have multiplied in recent years to further support both public and private stakeholders in prioritizing and rationalizing energy infrastructure investments [9]. From a public-sector perspective, GIS analytics are increasingly being used by governments and utilities to prioritize and sequence their grid extension efforts, as well as integrate off-grid solutions within national strategies aiming to achieve universal electricity access in a given timeframe (e.g., Tanzania [10], Afghanistan [11], Zambia [12], Madagascar [13]. From a private sector perspective, similar analytics are used to demonstrate the opportunity for supplying off-grid customers with decentralized energy services (market opportunity identification) and support subsequent operational roll outs (business models).

With this paper we aim to: (a) provide an overview of the main GIS data and modelling efforts aiming to support electrification planning and the achievement of SDG7; (b) discuss their role (especially if open) as providers of useful insights to difficult policy questions; (c) illustrate narrative through a case study of Malawi using an open-data-based and updated version of the Open Source Spatial Electrification Toolkit (OnSSET 2018), and (d) identify critical data/methodological gaps and suggest actions of future development.

2. GIS Based Electrification Planning

2.1. Open Access Data

The availability and quality of open access, publicly available GIS datasets has improved significantly over the past years; new datasets emerge conveying useful information regarding resource availability, status of infrastructure, social and economic characteristics of global populations. The following paragraphs present GIS datasets that have been (or can be) used in geospatial electrification analysis. A summarized list of useful GIS data for geospatial electrification modelling, providing status and gaps, is available in Appendix A.

2.1.1. Energy Infrastructure

The development of effective—GIS based—electrification plans depends greatly on the availability of credible and up-to-date records of existing infrastructure in the area of interest. The distribution of grid network for example, is an important input parameter. To illustrate, unelectrified settlements might find it more economical to connect to the national grid if in close proximity to service transformers or medium voltage (MV) lines. In contrast, areas that are located far from grid network might find off-grid technologies (mini-grids or solar home systems) are a better alternative. Therefore, low quality (erroneous or inadequate) datasets of the grid network may have a considerable impact on the results of electrification models. Other infrastructure (and thus for planning their datasets) such as the road network, are equally important; take for example remote villages without access to proper roads. They might experience high logistic costs for certain technologies e.g., high diesel prices. Several efforts have been recorded over the past few years aiming at reducing infrastructure data gap; few of them are briefly described below.

A noteworthy initiative recording power plants worldwide is the Global Power Plant Database by World Resource Institute [14]; the dataset contains geo-located entries of 28,500 power plants from 164 countries, including information on capacity, generation, ownership, and fuel type. It is open and frequently updated. The Global Roads Open Access Data Set (gROADS), v1 [15] provides a range of road data from the 1980s to 2010. Unfortunately, most country data is not 'date stamped' and spatial accuracy varies. OpenStreetMap (OSM) comes to fill data gaps in several instances [16]; OSM is a big—and growing—repository of open geospatial data including various elements of infrastructure, including roads [17]. The World Bank, has developed an online data explorer that records existing and planned transmission and distribution lines over Sub-Saharan Africa and Middle East [18]. The explorer draws from a comprehensive dataset [19] including power lines ranging from sub-kV to

700 kV. It should be noted however that there is large variation in the completeness of data by country. The ECOWAS Centre for Renewable Energy and Energy Efficiency (ECREEE) has provided a similar dataset for West Africa [20].

These efforts collect, organize and redistribute existing data. However, for many countries datasets are incomplete and in some cases of uncertain quality; for example, metadata describing the content, its source and how it was derived is often incomplete or missing. In order to overcome selected barriers, new methodologies have been developed. The energy access team at Facebook has released a remote sensing base predictive model for more accurate MV network mapping; the model is open source with output—as of the time of writing—being available for six countries in Sub-Saharan Africa [21]. Note that [22] provides an adaptation of this work available as an executable, open source code. Other initiatives consider the use of machine learning techniques and artificial intelligence. For example, Development Seed has developed an open source pipeline to efficiently map the high-voltage (HV) grid at a country-wide scale. The method uses high resolution satellite maps (0.5 m/pixel), from DigitalGlobe Platform to identify HV-towers. Then machine learning algorithms are applied to predict the distribution of transmission lines between the towers. Results are available for Nigeria and Zambia [23]. Finally, other initiatives [24–26] have also been developed in this area; some are and some are not focused in Sub-Saharan Africa. As they are open however, they can be applied globally and provide the potential to overcome important data shortages.

2.1.2. Resource Mapping

Natural resource availability—such as sunlight for PV panels—is a significant decision parameter when choosing electrification options. Electrification solutions should take into account local conditions in order to be achieve long term sustainability [27]. Remote areas with abundant solar irradiance, far from oil supply, for example might be better served by photovoltaic systems rather than diesel generators. Similar logic is applied to other resources. So called 'big data' from Earth-orbiting satellites have enabled scientists to better assess resource availability on a global scale. This body of data, if processed properly can provide useful information for electrification projects as well. For example, a Global Horizontal Irradiation (GHI) dataset is available by [28]. They provide information regarding solar availability in a location (usually in $kWh/m^2/year$). Other datasets such as wind speed [29,30], Digital Elevation Models (DEM) [31], land cover [32–35], river network [36], drainage basins [37], water discharge flows [38,39] are also highly useful. Combination of those, can yield very useful outputs such as wind power density [29] or capacity factors [40], hydro potential maps [41], which in turn can provide insights for the development of successful electrification projects.

2.1.3. Socio-Economic

A critical challenge in current electrification efforts is to construct sustainable business models. That is, electrification projects (both private and public) need to be able to recover investment and operational costs and be profitable—or at least break even [42]. Information regarding the socio-economic context under which such projects are developed, is thus important during the design phase. The following paragraphs describe how GIS can help identify some of these characteristics and incorporate them into electrification modelling.

2.1.4. Population Density & Distribution

Population density and distribution maps are used to indicate where population resides, thus where there is potential residential demand [43]. The map type as well as spatial resolution determines the detail (and sometimes accuracy) of information. Gridded population datasets (similarly to any other raster layer) represent information in the form of grid cells. In this case, grid cell values indicate population headcounts or density in a specific time.

Worldpop [44,45] has developed gridded population layers for many Sub-Saharan African countries at 100 m spatial resolution; 1 km resolution layers are available at continental level. These layers use interpolation techniques, which may give rise to inaccurate population estimates in certain cells; for example, some cells indicate population headcounts that have no physical meaning (e.g., less than 1). The Global Human Settlement layer (GHS) [46] suggests an alternative approach by indicating population values only in urban, peri-urban or rural areas; locations without population are eliminated. In similar manner, the Global Urban Footprint (GUF) [47–49] layer specifies in high spatial resolution (12 or 75 m) where settlements are located. The High Resolution Settlements Layer (HRSL) [50] provides population density maps in very high spatial resolution (30 m) but only for selected number of countries in Sub-Saharan Africa. Finally, [51] has developed a methodology that further processes the above datasets in order to provide more accurate vector type settlement layers for the case study of Tanzania.

2.1.5. Night-Time Lights

Night-time light (NTL) maps capture light sources on the surface of the Earth using satellite imagery. These can be a good proxy for assessing where electrified human settlements are, as they indicate light pollution. The Visible Infrared Imaging Radiometer Suite (VIIRS) dataset is available in raster format at 250 m spatial resolution; it provides the luminosity value in every cell; low value indicates that there is little visible light while higher values indicate high luminosity [52]. As of 2018, VIIRS provides annual composites for 2015 and 2016 as well as monthly composites for all years between 2012–2018. Its availability in monthly composites allows for detailed analysis of light sources and reduces the occurrence of false positives—areas that seem to be lit but in reality are not. Note that DMSP-OLS V4 [53] is VIIRS predecessor; it is available in raster format at 1 km spatial resolution and available for composites until 2012. It should be noted that DMSP-OLS V4 composites have been processed in order to provide stable light values over time series. Finally, Earth Observatory [54] also provides night light data at various spatial resolutions but without providing stable light composites.

2.1.6. GDP—Poverty Maps

Reference [55] developed a GIS layer presenting the Gross Domestic Product (GDP) in gridded format and on global scale for three intervals between 1990 and 2015 under 1 sq. km spatial resolution. The study uses primarily national GDP, PPP (purchasing-power-parity) values in constant 2011 international U.S dollars ($). In this instance, GDP illustrates the sum of gross value added by all resident producers in the economy plus any product taxes and minus any subsidies not included in the value of the products. It is calculated without making deductions for depreciation of fabricated assets or for depletion and degradation of natural resources. In addition, the study uses sub-national GDP, PPP values (for 82 countries) where available. The values were also converted in constant 2011 international U.S. dollars ($). These values were adjusted so that—when weighted by population—they total the GDP, PPP at the country level. By combining national and sub-national data, the global gridded GDP, PPP per capita maps were created. Poverty maps indicate the headcount ratio of population that lives below the poverty line (threshold usually being $1.25 or $2 per day) in an administrative area that can range from high level districts to lower level wards and municipalities. High resolution poverty maps (1 sq. km) have used geo-statistics in combination with GPS-located household survey data; such maps are however limited to a few countries. A combination of the aforementioned maps can provide very useful insights in electrification planning activities; they can be used as a proxy for economic activity or well-being in an area; or to create some sort of "heat map" indicating a better suited electricity access target per location.

2.1.7. Other

The list of open, energy related geospatial data is big and growing together with online GIS data platforms, map catalogues and repositories that make such datasets publicly available such as Energydata.info [56], OpenStreetMap [17], Google Earth Engine [57], IRENA global atlas [58], World Resource Institute [59], UN biodiversity lab [60], NREL GIS data & OpenEI [61,62], Earth Data [63].

Country specific GIS platforms have also been developed to support open data dissemination such as in Bolivia [64], Brazil [65], Kenya [66], Malawi [67], Uganda [68] and Namibia [69].

2.2. GIS Based Electrification Modelling Frameworks

The advent of geospatial information stimulated the development of modelling tools, methodologies and user interfaces that leverage on them so as to better support electrification planning decisions. The first tools that used GIS information in order to assess local resources and support techno-economic optimization included HOMER, RETScreen, SWERA UNEP, PVGIS, HOGA, DER-CAM [9]. Such tools are out of the scope of this paper as they mostly focus on the assessment of individual projects. The focus here is on what we shall term, the "second generation" of GIS modelling frameworks. The latter, utilize geospatial information and GIS software in order to support higher level electrification planning efforts. A short description of those most commonly used in electrification planning efforts is presented below. We shall also turn much of our attention to open access efforts, as they allow for reproducibility and thus can form the basis for scientific expansion.

IMPROVES-RE program (2007–2009) is one of the first efforts to support rural electrification activities with the use of a geospatial information. It is an open web-based platform aiming to support rural electrification projects and increase their impact on sustainable development and poverty alleviation in Burkina Faso [70]. It should be noted that IMPROVES-RE can be considered the predecessor of GEOSIM, a commercial tool that has been used for rural electrification planning in some countries (e.g., Tanzania [10]). GEOSIM is only marginally included in this review since it is not open source and relies on proprietary GIS software (Manifold).

Network Planner is a GIS based, open source [71] modelling framework for planning electricity infrastructure projects. Its underlying model identifies the optimal electrification technology mix for currently unserved demand centers; those include demand for households and other productive uses of electricity. Network planner uses a modified version of Kruskal's algorithm (minimum spanning tree) in order to find the maximum length of medium voltage lines for which grid extension is cheaper than the available off-grid options (solar home systems, diesel mini-grids) [72]; it does not however include biomass, wind and hydro as potential energy sources. Network Planner has been applied to Liberia [73], Ghana, [74] Kenya [75] and Senegal [76].

RE^2nAF is an open access web mapping application that enables geographically based exploratory analysis for off-grid electricity systems in the African continent. It overlays population settlements, infrastructure features (grid network, power plants and roads) and solar resources indicators (kWh/m^2) aiming to provide a comparison between diesel and PV based electricity costs for electrification [77]. All underlying GIS datasets have been made publicly available; results have been analysed and discussed in [27,78,79]. It shall be noted however that the underlying model is not available in the form of a customizable tool that could allow replication or modification by a broader user base, thus less capable of capturing specificities associated with individual projects.

The Reference Electrification Model (REM) [80,81] is an optimization tool designed to provide detailed engineering designs for electrification projects. It combines geospatial information with electricity demand and technology costs in order to estimate and compare different combinations of electrification modes (grid, mini-grids and stand-alone systems). Using satellite imagery, deep learning-based computer vision and clustering algorithms, REM can provide high level of granularity ranging from country to village level analysis. The model also offers the possibility to assess the impact of various factors such as demand levels, grid reliability, fuel and technology costs and cost of non-served-energy. REM has been used for electrification planning in India [82], regions of Rwanda and Uganda [83], Kenya and Colombia. REM (in liaison with its sibling tools GridForm and uLink [84]) offer a comprehensive modelling approach to rural electrification challenges. However, as in the time of writing the model is not yet open source.

IntiGIS is a plug-in application for ArcGIS that uses geospatial information in order to assess and compare the techno-economic performance of several electrification technologies; these include (a) stand-alone systems (PV, wind, diesel), (b) mini-grid systems (diesel, hybrid—wind/PV/diesel) or (c) connection to grid MV lines. Results include numerical and cartographic values of each of the selected technologies, including the optimal levelized cost of electricity at each point of demand and sensitivity of various technical parameters. IntiGIS is distributed freely however its operation is dependent on proprietary software (ArcGIS). Results of its application are available for Ghana [85].

The Open Source Spatial Electrification Toolkit (OnSSET) [85] is a GIS based tool developed to identify the least-cost electrification option(s) between seven alternative configurations; grid connection/extension, mini grid systems (solar PV, wind turbines, diesel gensets, small scale hydropower) or stand-alone systems (solar PV, diesel gensets). OnSSET combines geospatial information related to infrastructure, resources, topology and socio-economic characteristics over a modelled area, in order to inform a tree search algorithm. The algorithm traverses iteratively through a sub-set of the tree nodes (un-electrified population settlements) using Locality-Sensitive Hashing (LHS) to identify the nearest neighbor and optimal electrification technology. Results indicate the optimal technology mix, capacity and investment requirements for achieving electricity access goals under pre-defined time series (This may include multiple time steps; minimum duration of a time step is one year). The model also considers a prioritization algorithm, which defines how electrification progresses over time. Findings can be presented in various GIS compatible formats such as interactive maps, graphs and tables. OnSSET has informed IEA's energy access outlook publications [2,86], UN estimates for all Latin American and African countries [87], as well as country studies for Ethiopia [88], Nigeria [89], Kenya [90], Afghanistan [11], Madagascar [13], Tanzania [91], Zambia [91] and Benin [92]. Electrification investment scenarios also feature for 56 countries in open access web-based platforms [87,91].

Other geospatial web-based applications are also available. The Off-grid Market Opportunities Tool uses geospatial information (such as population density, proximity to transmission and road network and others) to help private companies, governments, academia and civil society to develop a high-level view of where markets for off-grid electrification may exist to better inform decision-making [93]. The Nigeria Rural Electrification Plans (NESP) [94] web platform provides least-cost geo-spatial electrification plans for five Nigerian States including detailed standalone and mini-grid assessments together with grid extension modelling [95,96]. Myanmar off-grid analytics [97] is a web tool that maps village location in Myanmar and based on available GIS data (local resources and nearby infrastructure) provides information for potential investment in off-grid electrification technologies. Ghana Energy Access Toolkit (GhEA) [98] and ECOWREX GIS [99] are mapping tools used to monitor and evaluate renewable energy resources and energy access progress in the country using geospatial datasets.

Finally, there are few noteworthy methodologies that utilize geospatial information to inform electrification plans. They have not led to functional tools however they may be replicable. Tiba et al. [100] proposed—and applied in the case study of northeast Brazil—a GIS-based methodology that supports rural electrification. Kaijuka et al. [101] used GIS information to identify patterns of electricity demand in Uganda and suggest priority areas for energy investment in the country. Teske et al. [102] developed a comprehensive multi-sectoral approach aiming to provide universal access in Tanzania only though renewable energy based technologies. They used open access data and maps in order to visualize and analyze key parameters for the analysis of Tanzania's future energy situation. These included solar and wind resources, population density, access to electricity via the central power grid or mini grids, the distribution of wealth or the economic development projections as well as energy demand projection for each settlement.

2.3. The Role of Open Access Data and Modelling Frameworks in Electrification Planning

Naturally, data and tools are designed in different contexts and may serve specific purposes. Even though their capabilities and objectives may vary per case, we find that most electrification efforts follow a conceptual framework as illustrated in Figure 1.

Figure 1. Conceptual flowchart of GIS -electrification modelling frameworks.

The flowchart in Figure 1 is far from exhaustive; it captures however the main components that we feel are crucial in GIS based electrification modelling. These include data collection, data transformation, model selection and configuration, result analysis and dissemination. Each component can be useful for policy making towards SDG 7; and here is where, we believe, open access can have the highest impact.

If open, such frameworks can enable the replicability and reproducibility of embedded processes as well as reusability of input/output data. In this way they can yield rapid techno-economic screening analyses—usually at low cost—in order to delineate the high level spatial contours of immediate (or intermediate) investment plans for electrification. They can also provide a test bed for cumbersome, long-term implementation roadmaps; support the decision making process; facilitate investment mobilization and speed up the implementation process. Finally, if transparently designed they can by audited by third parties; this is critical for assuring quality, control and demonstrating due diligence in administering public funds. With this in mind, we try to answer a set of questions commonly encountered in SDG 7 related planning and policy development activities. These may involve, among others, the following:

(1) Where is the population located?

   a. What is the population density and how are settlements distributed in the country?
   b. What are the settlements' characteristics?

(2) Which areas are currently electrified?

   a. What is the level of access and use?
   b. What is the expected/targeted electricity demand for different locations or types of settlements?

(3) What is the optimal technology mix in order to achieve SDG7?

   a. What equipment capacity is required?
   b. What is the potential role of different types of electricity supply technology?

(4) What is geospatial extent of the rollout electrification plan?

   a. Where can the national grid reach?
   b. Where do off-grid systems step up to provide access?
   c. Which areas may get access to electricity first?

(5) What is the cost of electrification?

   a. What is the total investment required to achieve full access by 2030?
   b. Where is investment most needed and in what form?
   c. Where can households afford electricity and where should subsidization be considered?

The case of Malawi is selected since it is one of the countries with the lowest electrification rate in Sub-Saharan Africa. Following the conceptual flowchart presented in Figure 1 we set up an electrification investment scenario (EIS) using an updated version of the OnSSET modelling framework. We use entirely open access data, software and methods. It is to be noted that our findings are illustrative only. The aim is primarily to highlight the power of open access information and the positive impact they might have in supporting sustainable electrification policies.

3. Electrification Policy Insights for Malawi

3.1. Data Collection and Transformation

3.1.1. Question 1 on Population Distribution & Characteristics

Malawi is a south eastern African country with population of about 18.62 million people [103]. The population growth is 2.83%, leading to an estimated population of 26.03 million in 2030 [104] and the urbanization rate 4.41% [104] per year. The average estimated household size is 4.3 and 4.5 people for urban and rural settlements respectively [105]. Identifying the location and type of settlements is a very important first step in the geospatial electrification analysis, as it can help denote several other characteristics.

Population-based datasets exist mostly in the form of grids. A grid comprises a number of spatially identical cells. The size of the cell determines the spatial resolution or else the area it represents. Each cell is used to represent a settlement and usually comes along with an attribute that specifies either total number of people or population density. It is often the case that gridded population datasets use interpolation or extrapolation techniques in order to fill data gaps [106]. This can cause false positives/negatives—areas that seem to be populated but in reality are not or reverse—and skew the electrification results. In reality, human settlements have various geometries. In a perfect modelling world, human settlements would be spatially represented by delineated vector polygons (referred to hereafter as population clusters) with full description of the settlement's characteristics (e.g., acreage, population, number and size of households). However, datasets of this nature are available only for limited locations. To overcome this, we introduce a new methodology aiming to delineated and attribute population clusters. This is achieved by using existing gridded population datasets and a set of open source geospatial processing tools. A step by step description of the methodology is presented in Appendix B. The methodology was tested upon the case study of Malawi. The derivative dataset yielded ~198,900 population clusters of various geometries and size as shown in Figure 2.

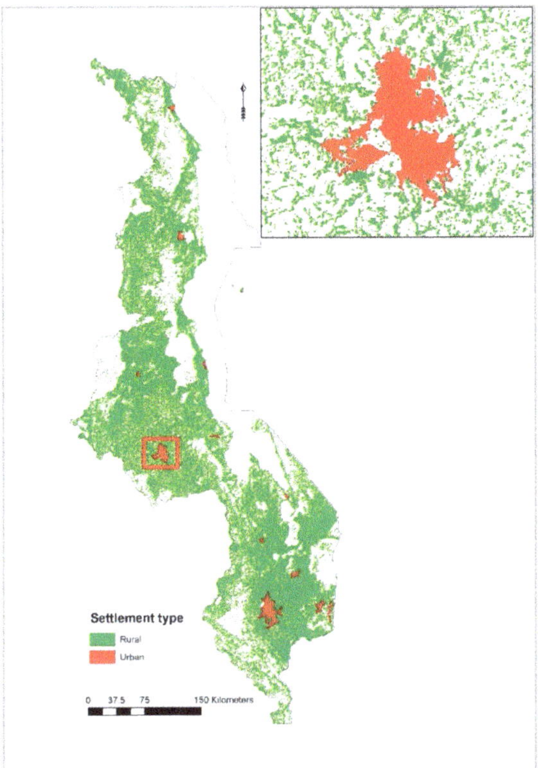

Figure 2. Characterization and spatial distribution of population clusters in Malawi as identified by the OnSSET model.

The aggregated population in the clusters was estimated as 17.19 million people, 7.65% lower than national statistics provide. The difference can likely be attributed to compounding uncertainty in geospatial processing and was mitigated through a calibration process. Once calibrated, each cluster was then characterized as either urban or rural based on information available at the GHS (S-MOD) layer. The layer provides a standardized distinction between: (a) urban centers, (b) urban clusters (peri-urban) and (c) rural settlements. For simplification, both (b) and (c) were considered as rural in this study. The process yielded 16 big urban clusters with an aggregated population of about 3 million people (in line with national statistics). The rest were identified as rural clusters.

Urban population settlements are often located closer to the existing grid network, they show higher population density, increased economic activity and (usually) higher electricity access rates and demand; the opposite applies to rural settlements [43]. In order to capture this dynamic, poverty and GDP data [55] were extracted to each cluster as shown in Figure 3a,b, accordingly. The poverty map indicates the headcount poverty rate in each cluster; the GDP map indicates the estimated total gross domestic product in each cluster. Information regarding settlements' socio-economic characteristics can be an important indicator for the selection of an "appropriate" electrification technology that will assure long-term sustainability of this solution.

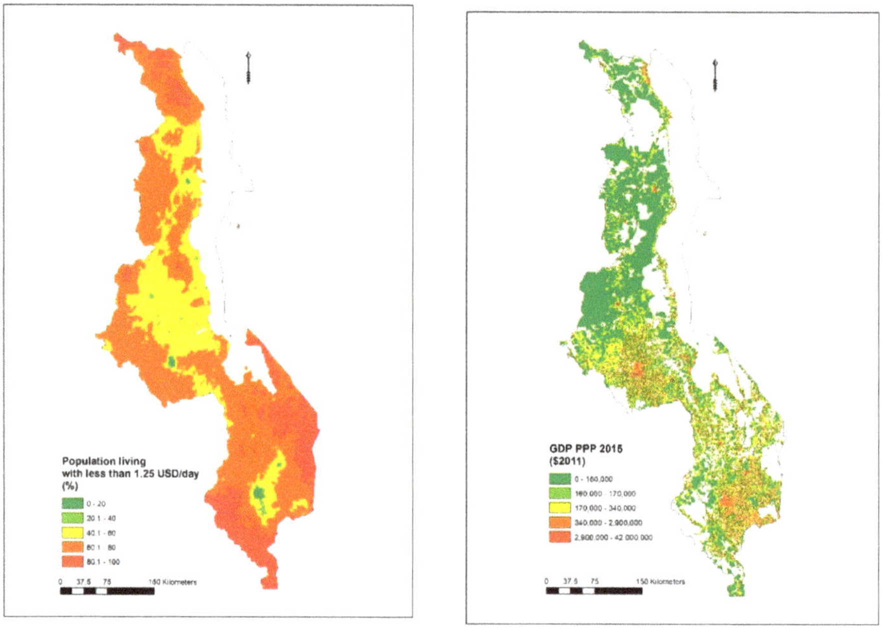

(**a**) (**b**)

Figure 3. Poverty rates (**a**) and estimated purchasing power parity Gross Domestic Product (GDP-PPP) in 2011 USD values (**b**) as distributed over population clusters in Malawi.

3.1.2. Question 2 on Current Electrification Status

The electrification rate in Malawi is among the lowest in the continent; it is estimated that about 49.2% of population living in urban areas has access to electricity while the rate is merely 3.2% in rural areas [107]. With the urban ratio in Malawi being roughly 17% [97], the national electrification rate stands at ~11%. Knowing where currently electrified clusters are located is an initial step needed for the electrification analysis. With OnSSET, already electrified clusters are used as anchor points for the electrification model. Once identified, they are, together with the known existing and planned grid lines, considered as starting points from which the grid network can be further extended. The location of electrified clusters and the access rate within those, is information often not easily accessed. Thus, in order to identify already electrified settlements rapidly a heuristic is added to OnSSET. That heuristic relies on a GIS-based multi-criteria evaluation. Note that this can easily be updated with actual figures when—and if—available (if can be the case, that with informal connections national statistics may be unhelpful in determining the extent of the electrification. National statistics may count only formal connections). The evaluation is based on five spatial attributes for each one of which a default threshold (the suggested values were reflective for Malawi; threshold values may vary per country) is defined as shown below:

(A) Distance to service transformers (initial threshold, <1 km)
(B) Distance to MV lines (initial threshold, <1 km)

(C) Distance to HV lines (initial threshold, <5 km)
(D) Nigh-time light intensity (initial threshold, >0)
(E) Population (initial threshold, >300 people)

Priority factors can be assigned according to data availability and the level of confidence on the quality of the datasets. Independently, (A) serves as a priority proxy for identifying electrified locations. In case (A) is insufficient or not available, (B) is a considered a useful alternative. Finally, (C) might be used if none of the above is available. Yet, the use of (A), (B) or (C) alone might cause the selection of locations that are close to a line or transformer but not necessarily electrified; therefore, these layers shall be used in combination with (D) and/or (E). Note that in absence of both (A)–(C), the combination of solely (D) and (E) can yield alternative proxies. In fact, for the case of Malawi, 86.7% of all clusters with night-time light greater than zero are located within 1 km from a service transformer, and 96.7% are located within two km. Ideally as detailed surveys and measurement become available the validity of (and even the need for this) heuristic might be assessed.

While the authors had access to (A) and (B), these datasets were not openly available at the time of writing. For consistency with the narrative of this paper, we relied only on the use of (D) and (E) and identified 814 electrified population clusters. Then, for each one of these clusters, we calculated the ratio between lit and non-lit area (using NTL) and provided an estimate of the electrification rate within the cluster. Finally, we used an iterative routine in Python where the electrification rate in each cluster was calibrated so that the aggregated electrified urban and rural population matches the values indicated by national statistics. Results for Malawi are illustrated in Figure 4.

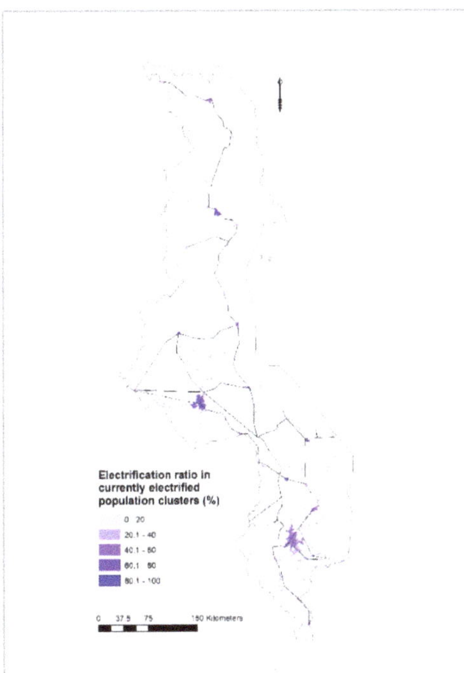

Figure 4. Distribution of settlements that indicate current access to electricity in Malawi. The multi-criteria evaluation yielded 16 urban and 798 rural electrified settlements with average electrification rates of 46.3% and 21.2% respectively.

According to the country's SE4ALL Action Agenda [108], the government in Malawi envisions that it will provide affordable and sustainable electricity services to all households at a level at least equivalent to Tier 1 (~38.7 kWh/household/year [109]) by 2030. Stimulated by this target and building upon the previous geospatial information, we prepare a map indicating targeted electricity levels per settlement as expected in Malawi by 2030. It should be noted that the current average household electricity consumption in Malawi is approximately 1072 kWh/year [108]. That is, all currently electrified settlements in Malawi were assigned a demand target equivalent to Tier 4 as in [109]. As illustrated in Figure 5, average electricity demand is expected to be higher in big urban clusters (Lilongwe, Blantyre, Zomba, Mzuzu); for the urban clusters identified in this analysis the median value of electricity demand was estimated at 32.4 GWh/year. For rural clusters the average electricity demand was estimated at 763 kWh/year.

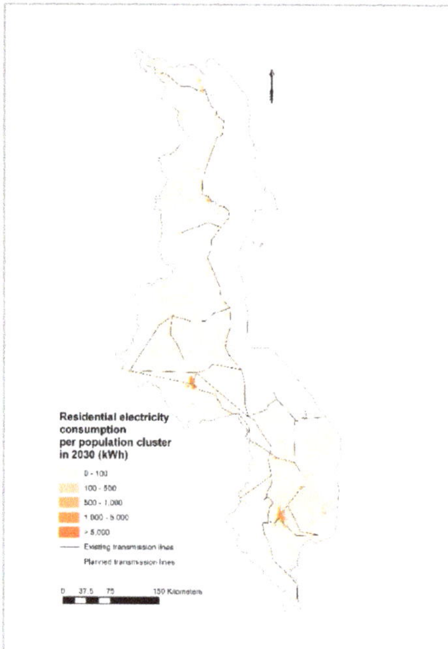

Figure 5. Distribution of the expected residential electricity demand per population cluster based on specified access targets (Tier 4 for urban and Tier 1 for rural clusters) in Malawi.

3.1.3. Additional Background Information

Malawi's current power system has a total installed generation capacity of about 361 MW with import capacity estimated less than 30 MW [110], whereas the country's current (actual and latent) demand is estimated to be as much as 700 MW leading to supply deficits [110]. According to the Master Plan and the rural electrification plan (MAREP), grid generation capacity will gradually increase to 1500 MW in 2020, 1859 MW in 2025 and 2519 MW in 2030 [108]. Grid extension currently plans to provide electricity to 31.6% of rural population by 2030 [108]. Beyond grid expansion, the government plans to electrify approximately 29.3% of rural population through solar home systems; and provide pico-solar systems to all the remainder (~39%) rural households by 2030. Other mini-grids are expected to electrify less than 0.1% of rural population [108]. Ramping up electricity access is a capital intensive process, especially in the rate under which this is expected to take place in Malawi. According to the SE4ALL Action Agenda, the cost of the suggested interventions for Malawi is estimated at $5.3 billion [108]. It shall be noted that similar electrification targets have been established in in many developing countries nowadays [111]. Also, in a historical parallel, the electrification of 1.7 million farms in 1930s in the USA came at a cost of $321 million [112] (or ~$5.7 billion in 2018 values). The development of a cost effective and sustainable rollout plan for Malawi is therefore essential in order to avoid unnecessary sunk costs and sub-optimal investment portfolios.

3.2. Geospatial Modelling Framework Configuration

The electrification investment scenario was developed so as to reflect the background information presented in the previous section. Therefore, we assumed that urban settlements target achieving Tier 4 by 2030 while rural settlements aim at Tier 1. We assumed that all currently electrified settlements are grid-connected; in these settlements full access is achieved through grid intensification only. In contrast, the un-electrified settlements are assessed for electrification using all electrification technologies. The selection of electrification technology is based on the lowest cost required to meet the specified Tier in each cluster.

It was assumed that the electrification progress is gradual. That is, the national access rate was set to reach 50% in 2023 and 100% in 2030 [108]. This was achieved with the introduction of a time step function in OnSSET that allowed the definition of explicit access targets per time interval. In this case we selected two time intervals in the means of representing the first five-year investment perspective (2018–2023) and the overall target up to 2030. The time step function relies on a prioritization algorithm developed to first pick "Low hanging Fruit" sites. That is, the algorithm prioritizes grid intensification first; then it continues electrifying other settlements based on the lowest (to highest) investment cost per capita achieved (either grid or off-grid).

From a techno-economic standpoint the following assumptions were made. For the centralized grid generation, the average investment cost was assumed as 1874 $/kW based on the expected generation mix (this might include: large hydro at 1471.5 MW (58.4%), small hydro at 103.4 MW (4.1%), solar at 550 MW (21.8%), biomass (bagasse) at 46 MW (1.8%), coal at 300 MW (12%) and diesel at 48 MW (1.9%) [108]). in the country by 2030. Similarly, the grid generating cost of electricity was assumed as 0.076 $/kWh. It should be noted that this value does not reflect the customer tariff but the estimated cost of producing 1 kWh of electricity. (It is assumed that taxes and subsidies are applied ex-ante. Indeed, this needs to be the case in order to rationalise the level of subsidy required for electrification.) Other costs related to grid extension (T&D costs, losses and connection costs) were also considered. Techno-economic parameters for the off-grid electrification technologies included (a) investment cost ($/kW of installed capacity, including batteries), (b) operation and maintenance cost (% of investment cost per year), (c) capacity factor and (d) expected technology lifetime. Further, efficiency values and fuel costs were included for the diesel-based technologies. Finally, the discount ratio was set at 8%. A more detailed description of all assumptions is available in Appendix C. It should be noted that here for the purpose of this paper—introducing a cluster based approach to geospatial electrification—several parts of OnSSET were modified considerably. One of these, refers to the modelling of essential power components (e.g., type and size of substations, transformers, conductors). A more elaborate explanation of these modifications is available in Appendix D.

3.3. Output, Analysis and Sensitivity

The following section provides a brief analysis and visualization of key findings from the electrification investment scenario in relation to the policy questions posted in Section 2.3.

3.3.1. Question 3 on Optimal Technology Mix

The model suggests that national grid may electrify 32.6% of population in 2030. More specifically, grid extension may provide electricity to 6.3 million people by 2023 increasing to 8.5 million by 2030. It is also noteworthy that all new grid connections derive mainly from intensification, or else ramping up connections in already electrified locations. Extension of the grid network was only observed in a limited number of areas due to the low access target levels set in this scenario. In contrast, off-grid technologies do play a very important role in this scenario. As indicated, the majority (67.4%) of the population in Malawi is expected to get access to electricity by off-grid stand-alone PV systems (Figure 6a).

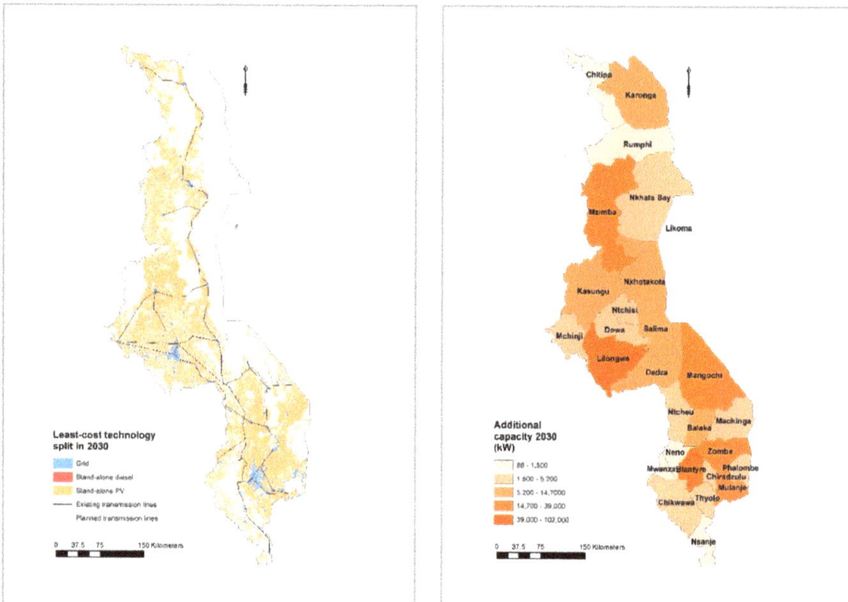

(a) (b)

Figure 6. Least cost technology split (**a**) and additional capacity (**b**) required per province to reach universal access to electricity in Malawi by 2030.

A few (<0.03%) stand-alone diesel systems were identified with no mini-grids (PV, wind, hydro or diesel) being included in the electrification mix in this scenario. In total, the country will need to increase the generating capacity by 351.8 MW by 2030 (168.1 by 2023 and 183.7 between 2023–2030) in order to meet the increased residential demand indicated by this scenario. From these, 23.9% shall derive from the deployment of stand-alone PV systems. That said and by assuming that grid generating capacity mix will be as described in Section 3.1, it is estimated that renewable technologies in Malawi can account for up to 89.5% of the additional generating capacity needed to achieve universal access goals by 2030.

3.3.2. Question 4 on Electrification Rollout Plan

From a geospatial perspective, national grid coverage is expected to cover 8817 km^2 or else about 14.4% of the populated land in Malawi. The majority of these areas are in close proximity to the existing network; in particular, 99.5% of the grid electrified population in 2030 (as per this scenario) is located within 5 km from the current grid network. Stand-alone PV systems have been identified as least cost electrification options in the rest of the country. In the districts of Nkhata Bay, Dedza, Ntcheu and Neno the electrified population by off-grid PV systems is expected to surpass 95%. Based on the time step function presented in Section 3.2 it was estimated that in the first five years of the analysis electricity service will reach about 8.76 newly electrified million people (Figure 7); from those about 51.5% will get access via off-grid systems while the rest through new grid connections (Figure 8). That is, grid connections will need to increase by a rate of 198,000 households per year until 2023 and slow down to 72,000 households per year between 2023–3030.

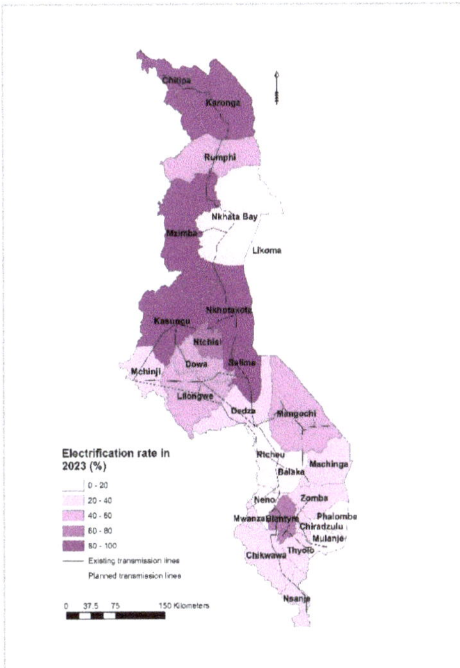

Figure 7. Percentage (%) of electrified population per province as in 2023.

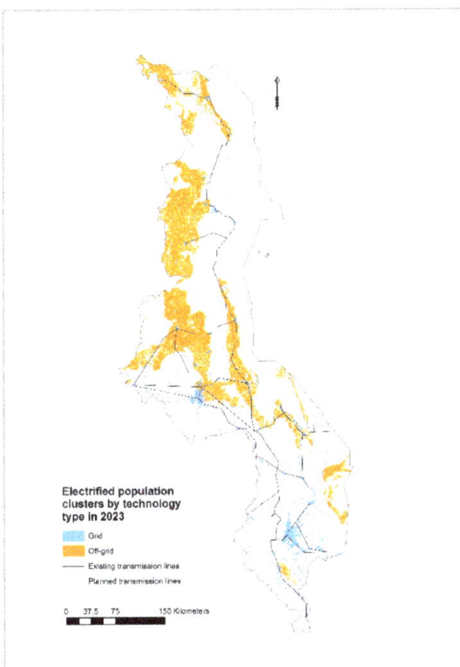

Figure 8. Grid vs. Off-grid split as per 2023 rollout plan estimated to electrify 50% of Malawians.

3.3.3. Question 5 on Cost of Electrification

The total investment required to achieve full electrification in Malawi by 2030, is \$1.83 billion. New grid connections will require \$1.48 billion. The investment cost per household varies depending on the distance to the transmission lines as well as the population in each settlement (Figure 9). The average cost of connecting to the grid amounted to \$228.2 per person or else about \$981 per household. It should be noted that these costs reflect mainly intensification of network; grid extension to new settlements even though slightly observed in this scenario might induce higher connection costs. Investment for decentralized technologies (stand-alone PV systems) is estimated to reach \$351.9 million. The average connection cost for stand-alone PV systems was estimated at \$26.3 per person or about \$118 per household. Finally, for the few stand-alone diesel systems identified the average connection cost was estimated at \$28.5 per person or about \$128 per household. The distribution of required investment over Malawi is presented in Figure 10.

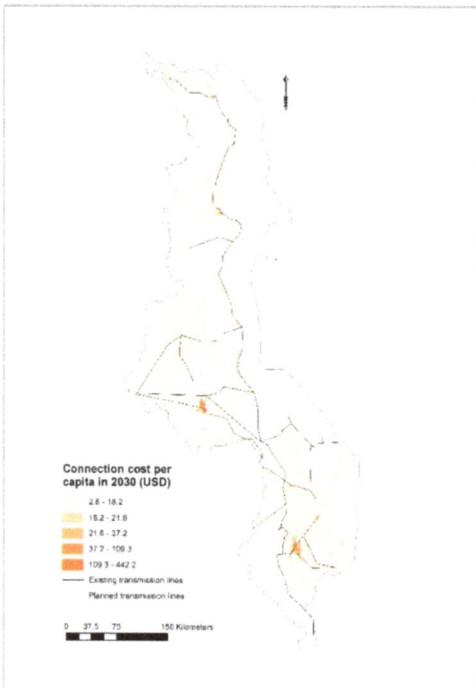

Figure 9. Connection cost per capita based on the least cost option identified in the selected electrification scenario.

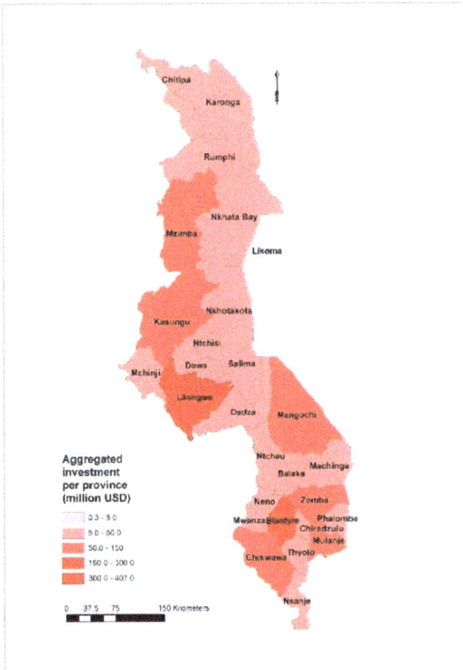

Figure 10. Investment requirements for the achievement of universal access as defined in the selected electrification scenario for Malawi by 2030. Results are aggregated per province.

3.3.4. Synthesis and Sensitivity Analysis

Comparing the results of this analysis with the government's estimates on achieving universal access in Malawi (Section 3.1), some noteworthy observations stand out. In both cases, grid connection is expected to provide electricity to approximately one third (31–33%) of the population. Also, the role of off-grid PV systems is crucial; solar home and pico-solar systems are expected to provide electricity to two thirds (67–69%) of the population by 2030. The additional capacity needed to achieved universal access based on the above statements is 267.5 MW for the grid and 84.2 MW for off-grid systems. However, government generation expansion plans will reflect general expansion vision that includes electricity demand not only for the residential sector. That also explains the disparity detected in terms of the total investment requirements. The electrification model estimated that $1.83 billion are needed to achieve the access target specified about a third of the government's estimates (~$5.3 billion). Lack of more detailed information on the rollout, investment plan from the government of Malawi limits the possibility of a more in depth comparison. Otherwise, based on these results the least cost electrification plan seem to be in alignment.

It is important at this point to highlight that any electrification analysis is subject to certain assumptions on the decision parameters. In this study we have selected to run a sensitivity analysis for six input parameters including population growth, electricity demand target, electrification rate in 2023, grid generation cost of electricity, PV cost and diesel cost. In total, ninety-six scenarios were generated and analysed indicating that the total investment requirements to achieve universal access to electricity in Malawi ranges between $1.65–7.78 billion. We find that the electricity demand target is the strongest determinant of both electrification investment and grid penetration in the total mix in comparison to the rest of parameters studied. A more detailed description of the findings is available in Appendix E.

4. Discussion

Open GIS data and modelling tools are increasingly being used in project development and planning in the energy sector. Their adoption and use can bring considerable advantages. It can provide a fast and cost effective way to map information that has a strong geospatial nature such as grid infrastructure, energy resources and settlement patterns. This can consequently, empower governments to effectively monitor progress, rationalize policy making and better inform strategic decisions in the energy field.

Electrification planning is no exception. The achievement of universal access to electricity is a crucial yet challenging task. It requires the motivation of significant financial resources in a timely and well-coordinated manner. This, given the rapid socio-economic changes and development particularly in currently unelectrified areas, makes the availability of good, up-to-date and consistent energy related information very important. This paper attempted to map existing data, tools and methods that have been commonly used to support SDG7 implementation efforts. It was observed that their number and importance has been progressively increasing over the past few years. Upon this, we provided key additions to the OnSSET methodology to form OnSSET 2018. Specifically, with this paper we have added an updated grid extension algorithm, a time step functionality and a new prioritization algorithm that allows the development of dynamic roll out plans for electrification. In addition, we introduce a restructured code basis that allows for a vector based approach of population settlements and the integration of new or upcoming geospatial datasets (MV lines, service transformers, poverty data, electricity demand for residential—e.g., Appendix F—as well as other productive activities). Despite that, limitations still exist and should be highlighted in the context of this analysis.

For geospatial data, the level of granularity is a key concern. Usually, open access data are available at low spatial or temporal resolution. Higher granularities are available either at a premium or under special agreement with the provider. Take for example T&D infrastructure; while at high level (e.g., HV lines) data have long been open and available for public consumption, at lower level (e.g., MV or LV lines) openly available data are scattered and inconsistent. This often leads to generalized assumptions, which in turn increase uncertainty in geospatial analysis. Reliability is another common concern. Open access geospatial data can be of unknown origin, questionable quality, poorly maintained, lack proper metadata or in some cases purposefully false. This makes quality assessment processes necessary for most practitioners before use, which can be time and resource consuming. Furthermore, despite progress, many socio-economic datasets potentially useful to electrification planning e.g., energy demand data, income level and distribution, energy expenditure, location of schools, health clinics and other productive nodes as well as mobile phone coverage are still limited or un-available in an open, geo-spatial format.

Similarly, the available GIS-based planning tools and methods, have one or more limitations: they are partially or fully proprietary; they focus only on rural areas and do not provide an overall electrification expansion indication for an entire country; they deploy a limited number of electrification technologies; they have restricted representation of demand; they lack a grid expansion algorithm or they do not account for a dynamic change of the bulk grid electricity supply. In OnSSET for example, the electricity demand is exogenous (layers imported from external calculations) and provide only an educated estimate. In addition, demand currently reflects only residential electrification targets. The model considers a set of static end-states (myopic optimization) thus, it does not use perfect foresight. Load profile is also represented on the basis of peak-to-average demand; that is reliability is incorporated but not optimized for. Despite its limitations, the basic OnSSET model is simple and open, allowing for a more tailored analysis to suite needs as needed—including improving all the above. Looking forward, we identify a clear need for synergies between the existing initiatives in the geospatial electrification field. The development of a single tool that incorporates all dimensions mentioned above could be theoretically feasible. Yet, we suggest that the development of a collaborative, open-source environment including interoperable data and tools with different characteristics might be more desirable. It could conceivably cover a wider range of applications and solutions—as well as harness a greater volume of analysts and communities. This consequently would require that both data and tools should be democratized so that electrification analytics become accessible to more actors. Thus, global partnerships that promote collaboration between stakeholders who collect, create, manage or use geospatial data are particularly needed. A notable effort in this regard is the multi-country, multi-agency 'round-table' effort championed by DFID, as well as the fledgling Open Tools, Integrated Modelling and Upskilling for Sustainable-development (OpTIMUS) community of practice. These, might be enhanced through an inclusive, open and scalable platform that allows universal access global data layers as well as customizable modelling solutions. Such a platform would help build spatial literacy in the field of energy access and enable better decision making to deliver SDG7.

5. Conclusions and Final Remarks

As elements in a growing energy planning ecosystem, open access geospatial data and models have started a paradigm shift; a shift that constitutes a significant improvement over conventional planning efforts. Their availability and accessibility can help policy makers, government agencies, investors and project developers to overcome paucity of information and better inform decision making mechanisms. Despite its limitations, we hope that this study will help setting up new ground in the field of geospatial electrification planning and accelerate progress against the achievement of SDG7. Thus, the code basis of the updated electrification toolkit (OnSSET 2018) as well as input/output files for all electrification scenarios included in this paper are publicly available in [85] and open to review, update and/or reproduction.

Author Contributions: Conceptualization, A.K. and M.H.; Methodology, A.K., A.S. and B.K.; Software, A.K., B.K., A.S. and C.A.; Validation, M.H.; Formal Analysis, A.K.; Investigation, A.K.; Resources, A.K. and B.K.; Data Curation, A.K., B.K. and A.S.; Writing—Original Draft Preparation, A.K.; Writing—Review & Editing, A.K., M.H., C.A.; Visualization, A.K. and B.K.; Supervision, M.H.; Project Administration, M.H.; Funding Acquisition, M.H. and A.K.

Funding: This research was funded by the World Bank under the contract number 7185716.

Conflicts of Interest: The authors declare no conflict of interest.

Appendix A. Listing and Gaps of GIS Data in Geospatial Electrification Modelling

Table A1. GIS data gap analysis in electrification modelling.

#	Dataset	Type	Description	Status
			Infrastructure	
1	High Voltage (HV) lines	Line vector	Spatial distribution of (Existing & Planned) the transmission network. HV capacity definition depends on the country but usually refers to lines above 69 kV.	Publicly Available
2	Medium Voltage (MV) lines	Line vector	Spatial distribution of the medium voltage transmission network. What is defined as medium voltage depends on the country but usually refers to lines between 11–69 kV.	Not publicly available
3	Substations	Point vector	The location of currently available substations. Capacity and type should be provided as attributes.	Publicly Available
4	Transformers (primary or service)	Point vector	The location of currently available transformers. Capacity and type should be provided as attributes.	Not publicly available
5	Road Network	Line vector	Existing & planned road infrastructure. The road network may include major roads such as highways, primary and secondary roads. Detail should go as low on the road scale as can accommodate a pickup/truck.	Publicly Available
6	Power Plants (Existing & Planned)	Point vector	The locations of existing and planned power plants. It is important that the dataset includes attributes regarding each plant's minimum capacity.	Publicly Available

Table A1. *Cont.*

#	Dataset	Type	Description	Status
			Energy Resources	
7	Global Horizontal Irradiation (GHI)	Raster	Provide information about the Global Horizontal Irradiation (kWh/m^2/year) over an area.	Publicly Available
8	Small scale Hydropower potential	Point vector	Points showing potential mini/small hydropower potential. The layer shall include information regarding the location of potential sites, power output (kW), head (m) and the discharge (m^3/year).	Publicly Available
9	Wind speed or Power Density	Raster	Provide information about the wind velocity (m/sec) over an area. This layer may be substituted by wind power density maps (W/m^2).	Publicly Available
10	Biomass	Raster	Current and potentially productive agricultural activity as an indicator of agricultural residues.	Publicly Available
			Socio-economic	
11	Population density and distribution	Raster or vector	Spatial quantification of the population for a selected area of interest (usually country or continent).	Publicly Available
12	Administrative Boundaries	Polygon vector	Includes information (e.g., name) of the country(s) to be modelled and delineates the boundaries of the analysis.	Publicly Available
13	Residential demand	Raster	Layer that indicates electricity demand for residential sector	Not publicly available
14	Poverty maps	Raster or vector	Poverty maps stating the headcount rate (%) for the population below the poverty line. The poverty line used should be clearly stated.	Publicly Available (to some extent)
15	Income level or expenditure indicators	Vector or Raster	The income level or energy expenditure in an area ($/km^2). Map can be either in raster format or vector data on the basis of administrative areas.	Not publicly available
16	Gross Domestic product (GDP)	Raster	GDP map showing the purchasing power parity over an area. Map can be either in raster format or vector data on the basis of administrative areas.	Publicly Available
17	Human Development Index (HDI)	Raster	Providing information regarding the Human Development Index in an area of interest. Map can be either in raster format or vector data on the basis of administrative areas.	Publicly Available
18	Productive uses—Education facilities	Point vector or raster	Locations of schools as vector with relevant attributes (e.g., size of school, no of students, electricity needs/consumption).	Not publicly available
19	Productive uses—Health facilities	Point vector or raster	Locations of health clinics/hospitals as vector with relevant attributes (e.g., type or size of clinic, electricity needs/consumption)	Not publicly available
20	Productive uses—Commercial	Point vector or raster	Locations of commercial units (mines, businesses et.) as vector with relevant attributes (e.g., type or size, electricity needs/consumption.).	Not publicly available
21	Productive uses—Agricultural demand	Raster	Electricity demand layer (e.g., raster) indicating per capita (kWh/pp/year) or per settlement values (kWh/settlement/year) and is related to agriculture (e.g., pumping irrigation, post-harvesting).	Not publicly available
			Other	
22	Travel time	Raster	Visualizes spatially the travel time required to reach from any individual cell to the closest urban centre. The unit shall be in minutes/hours.	Publicly Available
23	Elevation	Raster	Filled Digital Elevation Model (DEM) maps.	Publicly Available
24	Land cover	Raster	Land cover classification. Currently OnSSET uses 17 classes as described in [32].	Publicly Available
25	Slope	Raster	A sub product of DEM. The slope map visualizes the terrain slope in degrees. Any slope map that is to be used has to provide the slope in degrees.	Publicly Available
26	Night-time Lights	Raster	Night-time light maps showing light pollution. The map has a relative scale for the intensity of light.	Publicly Available

Appendix B. Methodology to Generate Population Clusters Using the High Resolution Settlement Layer and GIS Processing

The following methodology (Figure A1) has been developed in order to create population clusters based on open access population datasets from the HRSL and a series of processes developed in QGIS, an open source desktop geographic information system application.

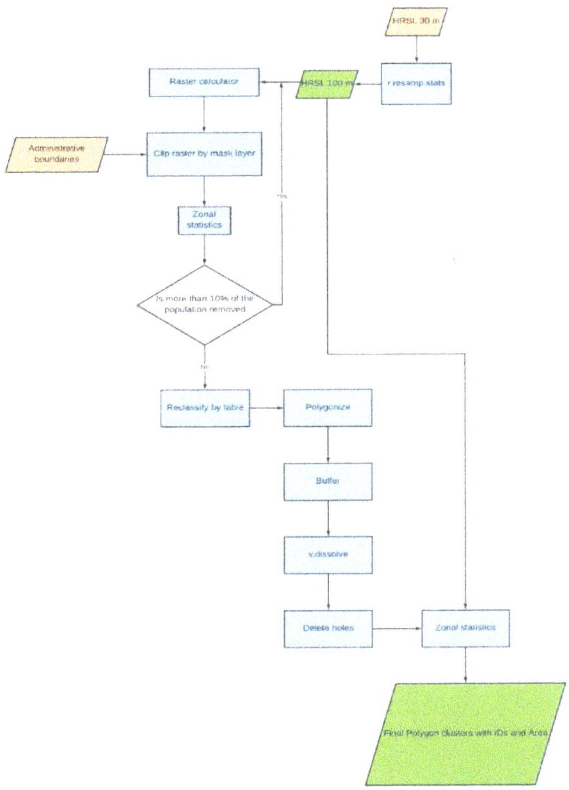

Figure A1. Methodological flowchart of creating population clusters using the High Resolution Settlements Layer and Geographic Information Systems processing.

Appendix B.1. Resampling Population Layer

The original spatial resolution of HRSL is 900 m^2. In the case of Malawi this translates to 3.2 million grid cells that have to be processed in the GIS environment. This is problematic due to (a) computational limitations of the GIS software used (QGIS) and (b) memory/running time complications of the electrification model used (OnSSET). Therefore, reducing the spatial resolution (resampling) of HRSL, is a sensible—and highly suggested—first step in the process. A final resolution of 0.1 km^2/10,000 m^2 is a good compromise as it will significantly reduce computational limitations while maintaining a good level of granularity. Lower resolution than 0.1 km^2 (or 10,000 m^2) will cause undesirable distortion of the layer's values and therefore is not considered as a viable option.

Suggested tool in QGIS: "r.resamp.stats".

Notes/Comments: This tool is part of GRASS GIS and enables the user to resample raster datasets. As of the time of writing, this is the only tool included in QGIS 3.2 that allows for increasing the cell size while automatically aggregating the raster values.

Appendix B.2. Removing Redundant Cells

HRSL population density values derive from interpolating recent census data [113]. This creates grid cell "neighborhoods" in the raster that have the exact same value to the 16th digit. These grid cells are considered false positives and thus shall be removed. In order to eliminate falsely populated grid cells, a threshold value is defined through an iterative process described below.

Step 1. Calculate the total population in the area of interest.

Suggested tool in QGIS: "Zonal statistics" from the QGIS package.

Step 2. Initialize the threshold value; the initial value can be anything within the density range in the area of interest.

Notes/Comments: The threshold value can be determined by examining the distribution of pixel values for the raster dataset. Also, removing low populated grid cells increases the share of coinciding built-up areas in comparison to Google map tiles.

Step 3. Zero out all grid cells with raster value below the threshold.

*Suggested tool in QGIS: "Raster calculator" e.g., (HRSL > 6) * HRSL removes all values below 6.*

Notes/Comments: The "Raster calculator" rounds the coordinates for the raster to the first six digits. Therefore, there might be a slight offset between the datasets after using the tool. Since this raster is the base of the clusters the raster calculator should be used twice; once to multiply by one and once to carry out the operation described above. This way there will not be any offset between the different datasets used in the analysis.

Step 4. Re-calculate the total population in the area of interest. If this loss is larger than 10% repeat again from Step 2 using a lower threshold value. Repeat until loss is acceptable.

Appendix B.3. Reclassify HRSL

The re-classification of the HRSL is necessary for the population clusters to be formed uniformly during the next step. This process creates the conditions for all adjacent cells to become part of the same cluster (Figure A2-left). If not re-classified, the clusters will be comprised by multi-part polygons as shown in Figure A2-right.

Suggested tool in QGIS: "Reclassify by table" from the QGIS package.

Notes/Comments: There is a number of tools that can be used in order to reclassify a raster layer in QGIS. This specific tool is from the same package as the raster calculator and therefore it does not create any further distortion or offset.

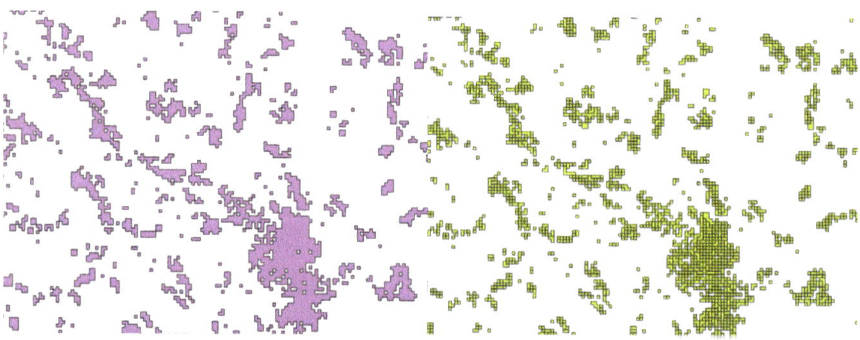

Figure A2. Uniform (**left**) and multi-part (**right**) population clusters created from HRSL.

Appendix B.4. Convert the HRSL Raster to Vector Polygons

In this process QGIS is used in order to convert the format of the processed HRSL from raster to vector polygons.

Suggested tool in QGIS: "Polygonize" from the GDAL package.

Appendix B.5. Buffering Polygons

A buffer of 10 m is applied to the polygons. This is due to QGIS treating polygons with one common corner as separate even in cases in which they touch. By applying a small buffer, it is ensured that these polygons are overlapping.

Suggested tool in QGIS: "Buffering vectors" from the GDAL package.

Appendix B.6. Dissolving Polygons

Dissolving the polygons ensures that overlapping polygons from the previous step are all merged.

Suggested tool in QGIS: "v.dissolve" from the GRASS package.

Appendix B.7. Remove Gaps and/or Slivers inside Polygons

Converting a raster layer to vector polygons as in previous step, can generate gaps and slivers to some of the polygons due to holes in the raster layer and due to the buffering process. These need to be removed/dissolved so that uniform population clusters are created.

Suggested tool in QGIS: "Delete holes" from the QGIS package.

Notes/Comments: It is important to only cover holes and slivers caused by the clustering process and not holes naturally occurring holes (e.g., lakes, forests etc.). Therefore, a maximum area is specified in the tool and all holes smaller are deleted.

Appendix B.8. Assigning Population Values to Clusters

Due to the population being reclassified when generating the clusters there is no population value connected to the clusters. In order to assign population values the raster values are aggregated for every cluster.

Suggested tool in QGIS: "Zonal statistics" from the QGIS package.

Appendix C. Techno-Economic Input Parameters in OnSSET

Table A2. Techno-economic parameters for off-grid technologies included in the electrification analysis.

Plant Type	Indicative Capacity (kW)	Investment Cost ($/kW)	O&M Costs (% of Inv. Cost/Year)	Efficiency	Capacity Factor *	Life (Years)
Mini-grid diesel	100	721	10%	33%	0.7	15
Mini-grid hydro	1000	5000	2%	-	0.5	30
Mini-grid PV	100	4300	2%	-	Obtained by model	20
Mini-grid wind	100	2500	2%	-	Obtained by model	20
Stand-alone diesel	1	938	10%	28%	0.5	10
Stand-alone PV	0.3	5,500	2%	-	Obtained by model	15
Diesel pump price	1.2 **	$/litre				
Connection cost Mini-grid	125	$/household				
Connection cost Stand-alone	0	$/household				
Discount rate	8	%				

* An indicative capacity factor was specified externally for diesel based technologies and hydro; capacity factor values for solar and wind were estimated by the model based on natural resource availability at each location; ** The diesel pump price was assumed at ~1.2 $/liter (900 MWK) [114,115]; exchange ratio used as $1 to 714.3 MWK.

Table A3. Techno-economic parameters related to the operation of the centralized grid and its extension process.

Parameter	Value *	Unit
HV cost (69 kV)	28,000	\$/km
MV cost (33 kV)	13,000	\$/km
MV amperage limit	8	Ampere
LV cost (0.2 kV)	10,000	\$/km
Max LV line length	0.5	km
Load moment	9643 [116]	For 50 mm aluminum conductor under 5% voltage drop (kW m)
Service transformer (50 kVA)	3500	\$
Max nodes per transformer	300	nodes
MV to MV substation (400 kVA)	10,000	\$
HV to MV substation (1000 kVA)	25,000	\$
MV max reach	50	km
Base to peak ratio	0.5	-
Connection cost per household	150	\$
T&D losses	10% [117]	of capital cost/year
O&M costs of distribution	2% [117]	of capital cost/year
Grid extension cost ratio	10	%
Power factor	0.9	-
System life	30	years
Discount rate	8	%

* The cost of grid components adopted in this study was primarily based on reference values provided by [118,119]; the selected values reflect authors' best estimate for the case of Malawi and they are only indicative.

Table A4. Expected generation mix, investment costs and generating costs for the expected centralized grid technologies in Malawi in 2030.

Technology Type	Expected Capacity (MW) in 2030 [110]	Share (%)	Investment Cost * (\$/kWe)	Generating Cost ** (\$/kWh)
Hydro (large)	1471.5	58.4%	1929	0.05
Hydro (medium/small)	103.4	4.1%	5025	0.08
Solar (utility)	550	21.8%	935	0.15
Coal	300	11.9%	2080	0.08
Diesel	48	1.9%	708	0.23
Biomass	46	1.8%	4105	0.07
Average weighted	2518.9	100%	1874	0.076

* Estimated overnight capital costs were retrieved from [117,120]; ** Estimates for Hydro, Solar and Biomass were retrieved from [121]; estimates for coal from [120] and for diesel from [122].

Appendix D. Updated Grid Extension Algorithm

The following paragraphs describe the modifications induced on the grid extension algorithm in OnSSET 2018. As of the previous version of the tool, the grid extension algorithm was based on the square geometry of a grid mesh with equal sized grid cells being adjacent to each other [123]. The integration of population clusters in the analysis, required the modification of the algorithm so that is it able to process vector data (polygons) of various geometry, size and spatial orientation. We describe the updated process in five distinctive steps.

Step 1. Sizing transmission lines (HV or MV)

As a first step, the algorithm decides the type of extension line (HV or MV) to be used to connect a settlement; the decision is based on two parameters as presented in (A1):

$$transmission\_line\_type = \begin{cases} \text{MV}, & grid\_distance \leq \text{MV}_{\text{max}\_reach} \; || \; peak\_load \leq \text{max\_MV\_}load \\ \text{HV}, & \text{otherwise} \end{cases} \quad \text{(A1)}$$

where:

$$peak\_load = \frac{\frac{Cluster\_electricity\_demand \div (1 - \text{T\&D losses})}{8760}}{Base\ to\ peak\ load\ ratio} \tag{A2}$$

$$\text{max\_MV\_}load = \text{MV}_{type} \times \text{MV}_{amp\_limit} \times \frac{\text{HV}cost}{\text{MV}cost} \tag{A3}$$

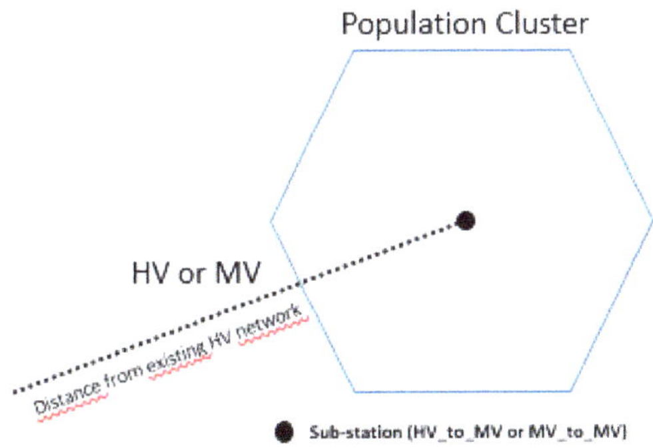

Figure A3. Estimating transmission line length from existing grid network.

Then, the mileage of additional transmission lines required to reach the cluster is estimated using (A4)–(A7) as follows:

$$transmission\_line_{km} = grid\_distance \times No\_of\_transmission\_lines \tag{A4}$$

where:

$$No\_of\_transmission\_lines = \frac{peak\_load}{line_{amperage} \times line\_type} \tag{A5}$$

$$line_{amperage} = \frac{substation\_type}{transmission\_line\_type} \tag{A6}$$

$$substation\_type = \begin{cases} \text{MV to MV}, & line\_type : \text{MV} \\ \text{HV to MV}, & line\_type : \text{HV} \end{cases} \tag{A7}$$

Step 2. Sizing transformers and connection to sub-station

Then, the algorithm estimates the number of service transformers required to provide full coverage of the population cluster:

$$No\_of\_service\_transformers = max\left\{\frac{S_{max}}{service\_transformer\_type}, \frac{total\_nodes}{nodes\_per\_transformer_{max}}, \frac{cluster's\ area}{transformer\_area\_coverage_{max}}\right\} \tag{A8}$$

where:

$$S_{max} = \frac{peak\_load}{power\_factor} \tag{A9}$$

$$transformer\_area\_coverage_{max} = \pi \times \text{LV\_}line\_length_{max}{}^{2} \tag{A10}$$

$$total\_nodes = \frac{cluster\_population}{No\_of\_people\_per\_household} + productive\ nodes \tag{A11}$$

$$No\_of\_people\_per\_household = \begin{cases} 4.5, & cluster\ type : Urban \\ 4.3, & cluster\ type : Rural \end{cases} \tag{A12}$$

The transformer load is the sum of the load of all households connected to a single transformer:

$$transformer\ load = \frac{peak\_load}{No\_of\_service\_transformers} \tag{A13}$$

It should be noted that the transformers are assumed to be evenly spaced within a cluster, thus the average distance from the service transformer to the substation is 2/3 of the cluster's radius, and the average distance between two service transformers is twice the transformer radius:

$$transformer\_distance_{average} = \frac{2}{3} \times cluster\_radius \tag{A14}$$

$$cluster\_radius = \sqrt{\frac{cluster\_area}{\pi}} \tag{A15}$$

$$transformer\_\ radius = \sqrt{\frac{\frac{cluster\_area}{No\_of\_service\_transformers}}{\pi}} \tag{A16}$$

If the estimated load moment is larger than 9643 (see Appendix C) an MV line is used to connect the service transformer to the substation; if not, a LV line is used. If connected by LV lines, each service transformer is assumed to have its own connection to the substation. With MV lines, multiple transformers may be connected in series:

$$load\_moment = transformer\_distance_{average} \times transformer\_load \tag{A17}$$

$$connection\_line_{km} = \begin{cases} \frac{2}{3} \times cluster\_radius \times No\_of\_service\_transformers, & load\_moment \leq 9643\ (LV) \\ 2 \times transformer\_radius \times No\_of\_service\_transformers, & load\_moment > 9643\ (MV) \end{cases} \tag{A18}$$

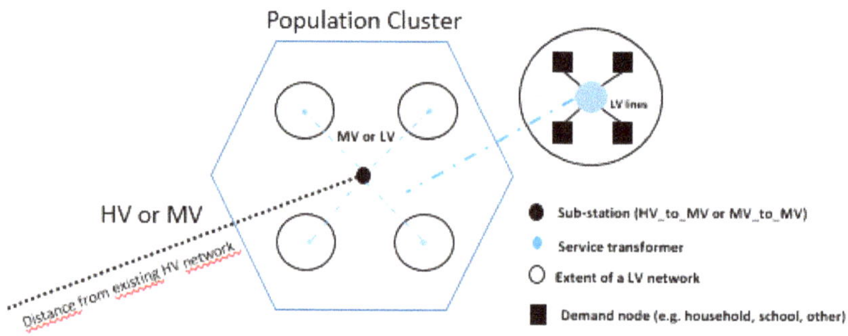

Figure A4. Estimating the size of transformers and their connection to sub-station.

Step 3. Sizing distribution lines (LV)

The area of each service transformer is then divided into a number of smaller circles (Figure A5) each one representing a demand node, assumed to be equally spaced within the larger circle. The distance between two demand nodes is defined as twice the radius of one of the smaller circles. The calculations do not consider the routing of LV lines from the transformer.

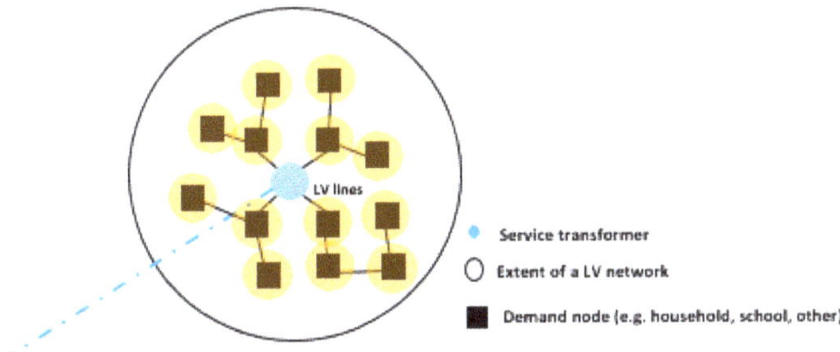

Figure A5. Sizing the LV network for each transformer in the population cluster.

The total length of LV lines per transformer is defined as described in (A19):

$$LV\_km\_per\_transformer = 2 \times r_{demand\ node} \times total\_nodes \tag{A19}$$

where:

$$r_{demand\ node} = \sqrt{\frac{demand\ node\ area}{\pi}} \tag{A20}$$

$$demand\ node\ area = \frac{transformer\_area\_coverage_{max}}{total\_nodes} \tag{A21}$$

Finally, the total number of distribution (LV) lines per cluster is estimated by (A22):

$$distribution\_line_{km} = LV\_km\_per\_transformer \times No\_of\_service\_transformers \tag{A22}$$

Step 4. Estimating the total investment cost for grid extension per cluster

In the last step, the total cost of grid extension per cluster is estimated by taking into account all partial costs as described in (A23):

$$\begin{aligned} grid\ extension\ cost_{per\ cluster} &= (transmission\_line_{km} \times transmision\_line\_cost) \\ &+ (connection_{linekm} \times connection\_line\_cost) \\ &+ (distribution_{linekm} \times distribution\_line\_cost \\ &+ (substation\_type \times substation\_cost) \\ &+ (No\_of\_service\_transformers \times service\_transformer\_cost) \\ &+ (total\_nodes \times node\_connection\_cost) \end{aligned} \tag{A23}$$

Appendix E. Detailed Results of Sensitivity Analysis

The sensitivity analysis in this study was conducted in order to identify which are the most critical parameters and how they affect the least cost electrification mix and investment requirements. Six parameters were selected as shown in Table A5. Option 1 (or Baseline) includes the values as presented in previous paragraphs and used in the analysis so far. Option 2 includes modification of these values; for parameters 1–3 modifications intend to a more aggressive electrification strategy; for parameters 4–6 modifications suggest a cost increase in selected technologies. Finally, option 3 suggest an alternative approach to electricity demand targeted for each population cluster. The latter adopted an approach based on available poverty and GDP data (elaborate description in Appendix F). In total, ninety-six scenarios were generated and analysed.

Table A5. List of parameters used in the sensitivity analysis and their selected available options.

#	Parameters	Option 1—Baseline	Option 2	Option 3
1	Population growth (PG)	2.83%	3.10% *	-
2	Electricity demand target (EDT)	Urban—Tier 4 Rural—Tier 1	Urban—Tier 5 Rural—Tier 3	Custom Residential Electricity Demand Indicative Target Layer (CREDIT)
3	Electrification rate in 2023 (ER23)	50%	80%	-
4	Grid generating cost of electricity (GGC)	0.076 $/kWh	+25%	-
5	PV cost factor (PVC)	0%	+25%	-
6	Diesel cost (DC)	1.2 $/liter	1.5 $/liter	-

* Based on the highest variant of population growth as in [104].

Between all scenarios, the total investment requirements to achieve universal access to electricity in Malawi ranged between $1.65–7.78 billion. As seen in Figure A6, parameter 2 shows very low variance in all options studied. That is, parameter 2 is a quite strong determinant of electrification investment in comparison to the rest of parameters studied. Higher level of targeted electricity demand in population clusters rises significantly the total cost of electrification. Parameters 1, 3 and 6 do have a noticeable—yet not as strong—impact on the total investment; option 2 of these parameters indicates higher median value. For parameter 1 this is naturally explained by higher population growth, which also causes the min/max values to shift upwards. The second option for parameter 3 mandates the electrification of bigger part of population in the first five years; this results in higher penetration of off-grid systems which in turn are more capital intensive in terms of per unit capacity ($/kW). Higher diesel price leads to lower penetration of diesel based systems which are replaced either by other off-grid systems or grid connection; both alternatives have higher cost per capacity unit, explaining the variation observed in parameter 6. Finally, minor changes in total investment were observed by the variation of parameters 4 and 5.

The share of grid connected population ranges between 32.6–80.1%. Parameter 2 is the strongest determinant of grid penetration in the total mix, defining therefore the above limits. Parameters 3, 4 & 5 can induce a maximum of 1.3%, 1.2% and 3% increase in grid share respectively between options 1 and 2. No effect on grid share was observed by parameter 6. The share of stand-alone systems varies reversely with their share ranging between 10.2–64.7%. Mini-grids share ranges between 0–0.7% with the upper limit observed only when parameters 1, 2, 4 & 5 are set to option 2. The interplay between decentralized technologies is notably affected by parameter 5. Higher PV costs allow the penetration of other renewable off-grid technologies in the optimal mix; the cost of diesel affects the optimal mix only when parameter 5 is set at option 2, otherwise its impact is negligible.

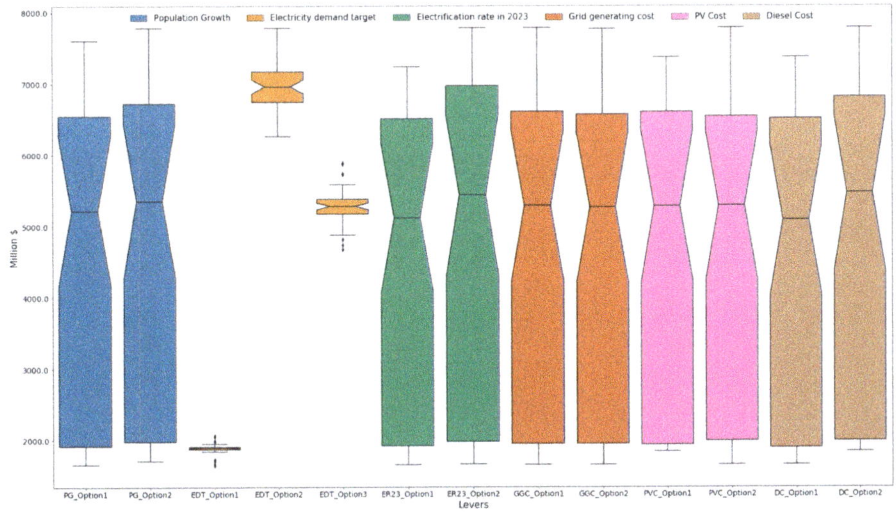

Figure A6. Investment variation for the achievement of universal electrification in Malawi as retrieved by the 96 scenarios developed in this study. The scenarios reflect the modification of six selected parameters and represent the impact of each one on the total investment.

Appendix F. The Custom Residential Electricity Demand Indicative Target (CREDIT) Layer

A customized raster layer indicating residential electricity demand target over Malawi has been developed by using open access poverty and GDP maps as described in Section 2.1. First, an equal interval classification technique using five classes was applied on the poverty map; the breaking values indicated intervals between 0–100% of headcount poverty rate. The GDP map was classified based on geometric intervals since this technique is particularly useful for datasets that are not normally distributed; it creates a balance between highlighting changes in the middle values and the extreme values; therefore, a good fit for the GDP data available in this case. Then, the two layers were reclassified as shown in Table A6 and added under equal weighted factors (0.5) using raster calculation.

Table A6. Re-classification of GDP and poverty layers into five classes. I_{1-5} are the geometric intervals of the classification process.

Initial Poverty Layer	Poverty Classification	Initial GDP Layer	GDP Classification
$0 \leq$ poverty < 0.2	5	$0 <$ GDP $< I_1$	1
$0.2 \leq$ poverty < 0.4	4	$I_2 \leq$ GDP $< I_3$	2
$0.4 \leq$ poverty < 0.6	3	$I_3 \leq$ GDP $< I_4$	3
$0.6 \leq$ poverty < 0.8	2	$I_4 \leq$ GDP $< I_5$	4
poverty ≥ 0.8	1	GDP $\geq I_5$	5

The output provided an indicative demand target index ranging from 0 to 5; 0 indicating the lowest potential target and 5 the highest. Finally, using 1-D linear interpolation the above target index was translated into kWh/capita/year as shown in Figure A7. The interpolation was based on the multi-tier framework for energy access adapted to reflect the situation in Malawi; that is, the lowest and highest values were set at 8.8 and 680.2 kWh/capita/year for Malawi.

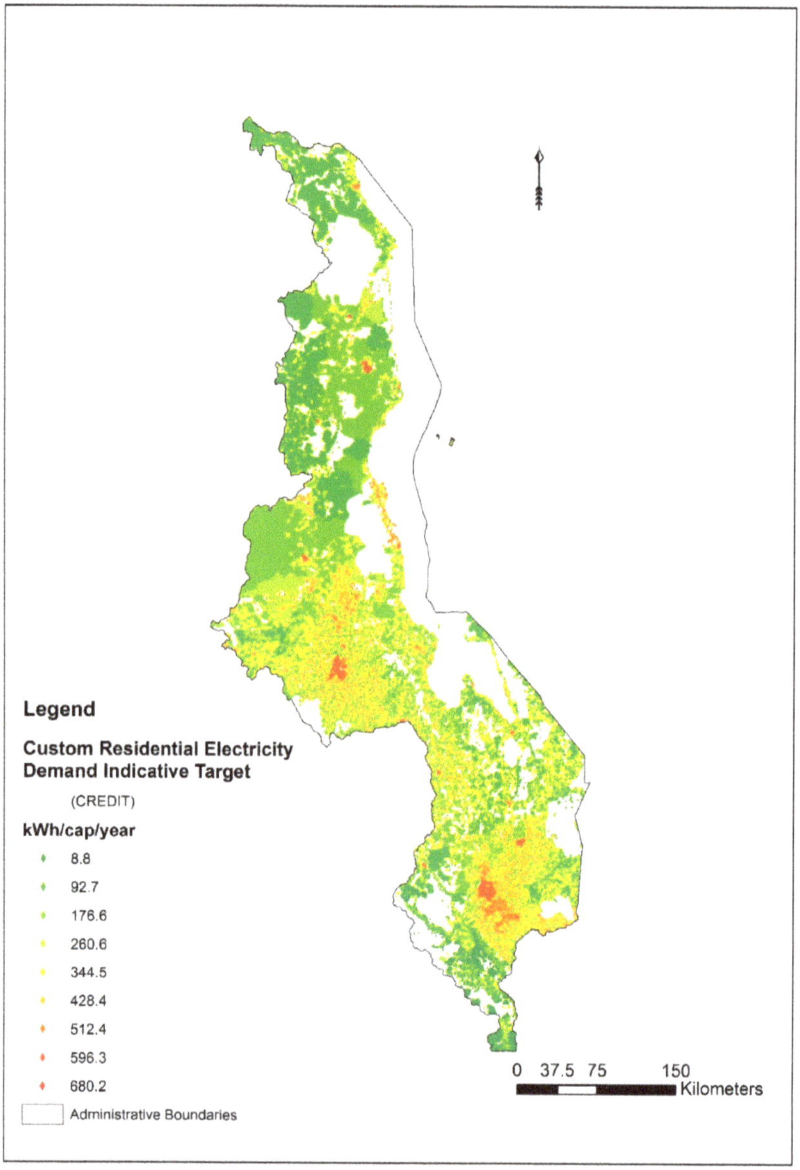

Figure A7. Customized layer indicating electricity demand target levels (in kWh/capita/year) over Malawi, based on openly available poverty and GDP maps.

References

1. United Nations: Division of Economic and Social Affairs, Sustainable Development-Knowledge Platform. 2015. Available online: https://sustainabledevelopment.un.org/sdg7 (accessed on 16 January 2019).
2. The International Energy Agency (IEA). *Energy Access Outlook 2017. From Poverty to Prosperity*; International Energy Agency: Paris, France, 2017; Available online: https://www.iea.org/publications/freepublications/publication/WEO2017SpecialReport_EnergyAccessOutlook.pdf (accessed on 16 January 2019).

3. Bijker, W.E.; Hughes, T.P.; Pinch, T.J. *The Social Construction of Technological Systems*; The MIT Press: Cambridge, MA, USA, 1987. [CrossRef]
4. Gurin, J.; Manley, L.; Ariss, A. Sustainable Development Goals and Open Data, World Bank Information Communications Development Blogs. 2015. Available online: http://blogs.worldbank.org/ic4d/sustainable-development-goals-and-open-data (accessed on 16 January 2019).
5. United Nations. Big Data for Sustainable Development, Glob. Issues. 2018. Available online: http://www.un.org/en/sections/issues-depth/big-data-sustainable-development/index.html (accessed on 16 January 2019).
6. UN Global Pulse. Big Data for Development: Challenges & Opportunities. 2012. Available online: http://www.unglobalpulse.org/sites/default/files/BigDataforDevelopment-UNGlobalPulseMay2012.pdf (accessed on 16 January 2019).
7. Sustainable Energy for All. Global Tracking Framework Chapter 2: Universal Access, Vienna, Austria. 2013. Available online: https://www.seforall.org/sites/default/files/l/2013/09/7-gtf_ch2.pdf (accessed on 16 January 2019).
8. Mentis, D.; Howells, M.; Rogner, H.; Korkovelos, A.; Arderne, C.; Zepeda, E.; Siyal, S.; Taliotis, C.; Bazilian, M.; De Roo, A.; et al. Lighting the World: The first application of an open source, spatial electrification tool (OnSSET) on Sub-Saharan Africa. *Environ. Res. Lett.* **2017**, *12*. [CrossRef]
9. Moner-Girona, M.; Puig, D.; Mulugetta, Y.; Kougias, I.; AbdulRahman, J.; Szabó, S. Next generation interactive tool as a backbone for universal access to electricity, Wiley Interdiscip. *Rev. Energy Environ.* **2018**, *7*, e305. [CrossRef]
10. Innovation Energie Développement (iED), United Republic of Tanzania: National Electrification Program Prospectus. 2014. Available online: http://www.tzdpg.or.tz/fileadmin/documents/dpg_internal/dpg_working_groups_clusters/cluster_1/Energy_and_Minerals/Key_Documents/Strategy/PROSPECTUS_-_Report_v4.pdf (accessed on 16 January 2019).
11. Korkovelos, A.; Bazilian, M.; Mentis, D.; Howells, M. A GIS Approach to Planning Electrification in Afghanistan, Washington, DC, USA. 2017. Available online: https://energypedia.info/wiki/File:A_GIS_approach_to_electrification_planning_in_Afghanistan.pdf (accessed on 16 January 2019).
12. Deloitte Consulting LLP, Zambia Electrification Geospatial Model: Executive Summary. 2018. Available online: https://dec.usaid.gov/dec/content/Detail_Presto.aspx?vID=47&ctID=ODVhZjk4NWQtM2YyMi00YjRmLTkxNjktZTcxMjM2NDBmY2Uy&rID=NTA2MTEw (accessed on 16 January 2019).
13. Kappen, J.F. Project Information Document-Integrated Safeguards Data Sheet-Madagascar-Least-Cost Electricity Access Development Project-LEAD-P163870. 2019. Available online: http://documents.worldbank.org/curated/en/281861547039951916/Project-Information-Document-Integrated-Safeguards-Data-Sheet-Madagascar-Least-Cost-Electricity-Access-Development-Project-LEAD-P163870 (accessed on 16 January 2019).
14. World Resources Institute (WRI). Global Power Plant Database. 2018. Available online: http://datasets.wri.org/dataset/globalpowerplantdatabase (accessed on 16 January 2019).
15. Center for International Earth Science Information Network-CIESIN-Columbia University and Information Technology Outreach Services-ITOS-University of Georgia. *Global Roads Open Access Data Set*; Version 1 (gROADSv1); Columbia University: New York, NY, USA, 2013. [CrossRef]
16. Ibisch, P.L.; Hoffmann, M.T.; Kreft, S.; Pe'er, G.; Kati, V.; Biber-Freudenberger, L.; DellaSala, D.A.; Vale, M.M.; Hobson, P.R.; Selva, N. A global map of roadless areas and their conservation status. *Science* **2016**, *354*, 1423–1427. [CrossRef] [PubMed]
17. OpenStreetMap Contributors. Planet OSM: Complete OSM Data. 2015. Available online: https://planet.openstreetmap.org/ (accessed on 16 January 2019).
18. The World Bank. Africa Electricity Grids Explorer. 2019. Available online: http://africagrid.energydata.info/ (accessed on 16 January 2019).
19. Arderne, C. Africa-Electricity Transmission and Distribution Grid Map. 2019. Available online: https://energydata.info/dataset/africa-electricity-transmission-and-distribution-2017 (accessed on 16 January 2019).

20. ECOWAS Centre for Renewable Energy and Energy Efficiency. Transmission Grid-ECOWAS. 2017. Available online: http://www.ecowrex.org:8080/geonetwork/srv/eng/catalog.search#/home (accessed on 16 January 2019).
21. Gershenson, D.; Rohrer, B.; Lerner, A. Predictive model for accurate electrical grid mapping. *Connect. Netw. Traffic.* **2019**. Available online: https://code.fb.com/connectivity/electrical-grid-mapping/ (accessed on 16 January 2019).
22. Arderne, C. Gridfinder. 2019. Available online: https://github.com/carderne/gridfinder (accessed on 16 January 2019).
23. Seed, D. Mapping the Electric Grid. 2018. Available online: https://devseed.com/ml-grid-docs/ (accessed on 16 January 2019).
24. Matikainen, L.; Lehtomäki, M.; Ahokas, E.; Hyyppä, J.; Karjalainen, M.; Jaakkola, A.; Kukko, A.; Heinonen, T. Remote sensing methods for power line corridor surveys. *ISPRS J. Photogramm. Remote Sens.* **2016**, *119*, 10–31. [CrossRef]
25. Duke University. Energy Data Analytics Lab: Energy Infrastructure Map of the World through Satellite Data (2018–2019). 2018. Available online: https://bassconnections.duke.edu/project-teams/energy-data-analytics-lab-energy-infrastructure-map-world-through-satellite-data-2018 (accessed on 16 January 2019).
26. DLR Institute for Networked Energy Systems. Open Source Reference Model of European Transmission Networks for Scientific Analysis (SciGRID). 2017. Available online: https://www.power.scigrid.de/pages/general-information.html (accessed on 16 January 2019).
27. Szabó, S.; Bódis, K.; Huld, T.; Moner-Girona, M. Sustainable energy planning: Leapfrogging the energy poverty gap in Africa. *Renew. Sustain. Energy Rev.* **2013**, *28*, 500–509. [CrossRef]
28. Solargis. Global Solar Atlas 1.0. 2017. Available online: https://globalsolaratlas.info/ (accessed on 16 January 2019).
29. Technical University of Denmark (DTU). Global Wind Atlas 2.0. Available online: https://globalwindatlas.info/ (accessed on 16 January 2019).
30. Goddard Earth Sciences Data and Information Services Center (GES DISC). *Global Modeling and Assimilation Office (GMAO), MERRA-2 tavgU_2d_flx_Nx: 2d, diurnal, Time-Averaged, Single-Level, Assimilation, Surface Flux Diagnostics V5.12.4*; Goddard Earth Sciences Data and Information Services Center: Washington, DC, USA, 2015. [CrossRef]
31. Jarvis, A.; Guevara, E.; Reuter, H.I.; Nelson, A.D. Hole-Filled SRTM for the Globe Version 4, Available from the CGIAR-CSI SRTM 90m Database, CGIAR-CSI, Cali, Colombia. 2008. Available online: http://srtm.csi.cgiar.org/ (accessed on 23 October 2018).
32. Friedl, M.A.; Sulla-Menashe, D.; Tan, B.; Schneider, A.; Ramankutty, N.; Sibley, A.; Huang, X. MODIS Collection 5 Global Land Cover: Algorithm Refinements and Characterization of New Datasets, 2001–2012, Collection 5.1 IGBP Land Cover. 2010. Available online: http://glcf.umd.edu/data/lc/ (accessed on 16 January 2019).
33. USGS/Earth Resources Observation and Science (EROS) Center. Land Cover Type Yearly L3 Global 0.05Deg CMG, MCD12C1 Courtesy of the NASA Land Processes Distributed Active Archive Center (LP DAAC), Sioux Falls, South Dakota. 2014. Available online: https://lpdaac.usgs.gov/dataset_discovery/modis/modis_products_table/mcd12c1 (accessed on 23 October 2018).
34. European Space Agency (ESA). Cci Land Cover-S2 Prototype Land Cover 20M Map of Africa. 2016. Available online: http://2016africalandcover20m.esrin.esa.int/ (accessed on 16 January 2019).
35. European Space Agency (ESA). GlobCover. 2010. Available online: http://due.esrin.esa.int/page_globcover.php (accessed on 16 January 2019).
36. Lehner, B.; Verdin, K.; Jarvis, A. New Global Hydrography Derived From Spaceborne Elevation Data. *Eos Trans. Am. Geophys. Union* **2008**, *89*, 93. [CrossRef]
37. Lehner, B.; Grill, G. Global river hydrography and network routing: Baseline data and new approaches to study the world's large river systems. *Hydrol. Process.* **2013**, *27*, 2171–2186. [CrossRef]
38. Beck, H.E.; de Roo, A.; van Dijk, A.I.J.M. Global Maps of Streamflow Characteristics Based on Observations from Several Thousand Catchments. *J. Hydrometeorol.* **2015**, *16*, 1478–1501. [CrossRef]
39. Barbarossa, V.; Huijbregts, M.A.J.; Beusen, A.H.W.; Beck, H.E.; King, H.; Schipper, A.M. FLO1K, global maps of mean, maximum and minimum annual streamflow at 1 km resolution from 1960 through 2015. *Sci. Data* **2018**, *5*, 180052. [CrossRef] [PubMed]

40. Mentis, D.; Siyal, S.H.; Korkovelos, A.; Howells, M. Estimating the spatially explicit wind generated electricity cost in Africa-A GIS based analysis. *Energy Strateg. Rev.* **2017**, *17*. [CrossRef]
41. Korkovelos, A.; Mentis, D.; Siyal, S.; Arderne, C.; Rogner, H.; Bazilian, M.; Howells, M.; Beck, H.; De Roo, A.; Korkovelos, A.; et al. A Geospatial Assessment of Small-Scale Hydropower Potential in Sub-Saharan Africa. *Energies* **2018**, *11*, 3100. [CrossRef]
42. Brass, J.N.; Carley, S.; MacLean, L.M.; Baldwin, E. Power for Development: A Review of Distributed Generation Projects in the Developing World. *Annu. Rev. Environ. Resour.* **2012**, *37*, 107–136. [CrossRef]
43. Nerini, F.F.; Broad, O.; Mentis, D.; Welsch, M.; Bazilian, M.; Howells, M. A cost comparison of technology approaches for improving access to electricity services. *Energy* **2016**, *95*, 255–265. [CrossRef]
44. Linard, C.; Gilbert, M.; Snow, R.W.; Noor, A.M.; Tatem, A.J. Population Distribution, Settlement Patterns and Accessibility across Africa in 2010. *PLoS ONE* **2012**, *7*, e31743. [CrossRef]
45. Worldpop. Africa Continental Population Datasets (2000–2020). 2016. Available online: http://www.worldpop.org.uk/ (accessed on 16 January 2019).
46. European Commission Joint Research Centre (JRC), Columbia University Center for International Earth Science Information Network-CIESIN, GHS Population Grid, Derived from GPW4, Multitemporal (1975, 1990, 2000, 2015). 2015. Available online: http://data.europa.eu/89h/jrc-ghsl-ghs_pop_gpw4_globe_r2015a (accessed on 17 January 2019).
47. Esch, T.; Schenk, A.; Ullmann, T.; Thiel, M.; Roth, A.; Dech, S. Characterization of Land Cover Types in TerraSAR-X Images by Combined Analysis of Speckle Statistics and Intensity Information. *IEEE Trans. Geosci. Remote Sens.* **2011**, *49*, 1911–1925. [CrossRef]
48. Esch, T.; Heldens, W.; Hirner, A.; Keil, M.; Marconcini, M.; Roth, A.; Zeidler, J.; Dech, S.; Strano, E. Breaking new ground in mapping human settlements from space–The Global Urban Footprint. *ISPRS J. Photogramm. Remote Sens.* **2017**, *134*, 30–42. [CrossRef]
49. Esch, T.; Bachofer, F.; Heldens, W.; Hirner, A.; Marconcini, M.; Palacios-Lopez, D.; Roth, A.; Üreyen, S.; Zeidler, J.; Dech, S.; et al. Where We Live—A Summary of the Achievements and Planned Evolution of the Global Urban Footprint. *Remote Sens.* **2018**, *10*, 895. [CrossRef]
50. Facebook Connectivity Lab and Center for International Earth Science Information Network-CIESIN-Columbia University, High Resolution Settlement Layer (HRSL). 2016. Available online: https://www.ciesin.columbia.edu/data/hrsl/ (accessed on 16 January 2019).
51. Cader, C.; Pelz, S.; Radu, A.; Blechinger, P. Overcoming data scarcity for energy access planning with open data-The example of Tanzania. *Int. Arch. Photogramm. Remote Sens. Spat. Inf. Sci.* **2018**. [CrossRef]
52. NOAA. Version 1 VIIRS Day/Night Band Nighttime Lights. 2018. Available online: https://ngdc.noaa.gov/eog/download.html (accessed on 16 January 2019).
53. NOAA. Version 4 DMSP-OLS Nighttime Lights Time Series. 2013. Available online: https://ngdc.noaa.gov/eog/dmsp/downloadV4composites.html (accessed on 16 January 2019).
54. NASA Earth Observatory. Earth at Night. 2012. Available online: https://earthobservatory.nasa.gov/features/NightLights (accessed on 16 January 2019).
55. Kummu, M.; Taka, M.; Guillaume, J.H.A. Gridded global datasets for Gross Domestic Product and Human Development Index over 1990–2015. *Sci. Data* **2018**, *5*, 180004. [CrossRef]
56. Energydata.info. 2018. Available online: https://energydata.info/dataset?q=afghanistan (accessed on 16 January 2019).
57. Gorelick, N.; Hancher, M.; Dixon, M.; Ilyushchenko, S.; Thau, D.; Moore, R. Google Earth Engine: Planetary-scale geospatial analysis for everyone. *Remote Sens. Environ.* **2017**, *202*, 18–27. [CrossRef]
58. The International Renewable Energy Agency (IRENA). Global Atlas Version 3.0. 2018. Available online: https://www.irena.org/globalatlas (accessed on 16 January 2019).
59. World Resources Institute (WRI). Datasets. 2019. Available online: http://datasets.wri.org/dataset (accessed on 16 January 2019).
60. United Nations (UN). Biodiversity Lab. 2018. Available online: https://www.unbiodiversitylab.org/ (accessed on 16 January 2019).
61. National Renewable Energy Laboratory (NREL). Geospatial Data Science. 2018. Available online: https://www.nrel.gov/gis/data.html (accessed on 16 January 2019).
62. NREL. OpenEI. 2018. Available online: https://openei.org/wiki/Data (accessed on 16 January 2019).

63. Earth Observing System Data and Information System (EOSDIS). EarthDATA. Available online: https://earthdata.nasa.gov/ (accessed on 16 January 2019).
64. Infraestructura de Datos Espaciales del Estado Plurinacional de Bolivia (IDE-EPB). GeoBolivia. 2014. Available online: http://geo.gob.bo/portal/#map (accessed on 16 January 2019).
65. Instituto Brasileiro de Geografia e Estatística (IBGE). Geosciences-Products. 2011. Available online: https://ww2.ibge.gov.br/english/geociencias/default_prod.shtm#REC_NAT (accessed on 16 January 2019).
66. Ministry of Information Communications and Technology. Kenya Open Data. 2018. Available online: http://www.opendata.go.ke/ (accessed on 16 January 2019).
67. National Spatial Data Center-Department of Surveys, Malawi Spatial Data Platform (MASDAP). 2019. Available online: http://www.masdap.mw/ (accessed on 16 January 2019).
68. Africam Center for Media Excellence, DATA.Ug: Open Data in Uganda. 2018. Available online: http://acme-ug.org/ (accessed on 16 January 2019).
69. Ministry of Environment and Tourism, Digital Atlas of Namibia. 2002. Available online: http://www.uni-koeln.de/sfb389/e/e1/download/atlas_namibia/main_namibia_atlas.html (accessed on 16 January 2019).
70. Innovation Energie Développement (iED), IMPROVES-RE. 2009. Available online: https://www.improves-re.com/sig/ (accessed on 16 January 2019).
71. Sustainable Engineering Lab, Network Planner. 2015. Available online: https://github.com/SEL-Columbia. (accessed on 16 January 2019).
72. Cader, C.; Blechinger, P.; Bertheau, P. Electrification Planning with Focus on Hybrid Mini-grids—A Comprehensive Modelling Approach for the Global South. *Energy Procedia* **2016**, *99*, 269–276. [CrossRef]
73. Team, E.P.; Modi, V.; Adkins, E.; Carbajal, J.; Sherpa, S. Liberia Power Sector Capacity Building and Energy Master Planning Final Report, Phase 4: National Electrification Master Plan. 2013; pp. 1–52. Available online: https://qsel.columbia.edu/assets/uploads/blog/2013/09/LiberiaEnergySectorReform_Phase4Report-Final_2013-08.pdf (accessed on 16 January 2019).
74. Kemausuor, F.; Adkins, E.; Adu-Poku, I.; Brew-Hammond, A.; Modi, V. Electrification planning using Network Planner tool: The case of Ghana. *Energy Sustain. Dev.* **2014**, *19*, 92–101. [CrossRef]
75. Parshall, L.; Pillai, D.; Mohan, S.; Sanoh, A.; Modi, V. National electricity planning in settings with low pre-existing grid coverage: Development of a spatial model and case study of Kenya. *Energy Policy* **2009**, *37*, 2395–2410. [CrossRef]
76. Sanoh, A.; Parshall, L.; Sarr, O.F.; Kum, S.; Modi, V. Local and national electricity planning in Senegal: Scenarios and policies. *Energy Sustain. Dev.* **2012**, *16*, 13–25. [CrossRef]
77. European Commission Joint Research Centre (JRC)-Renewable Energy Mapping and Monitoring in Europe and Africa (REMEA), Renewable Energies for Rural Electrification of Africa (RE2nAF). 2014. Available online: https://iet.jrc.ec.europa.eu/remea/re2naf (accessed on 16 January 2019).
78. Szabó, S.; Bódis, K.; Huld, T.; Moner-Girona, M. Energy solutions in rural Africa: mapping electrification costs of distributed solar and diesel generation versus grid extension. *Environ. Res. Lett.* **2011**, *6*, 34002. [CrossRef]
79. Moner-Girona, M.; Bódis, K.; Huld, T.; Kougias, I.; Szabó, S. Universal access to electricity in Burkina Faso: Scaling-up renewable energy technologies. *Environ. Res. Lett.* **2016**, *11*, 84010. [CrossRef]
80. Borofsky, Y.; Perez-Arriaga, I.; Stoner, R. A model for better electrification planning. *ABB Rev.* **2017**, 23–27. Available online: http://search-ext.abb.com/library/Download.aspx?DocumentID=9AKK107045A1041&LanguageCode=en&DocumentPartId=&Action=Launch (accessed on 16 January 2019).
81. Ellman, D. The Reference Electrification Model: A Computer Model for Planning Rural Electricity Access, Massachusetts Institute of Technology. 2015. Available online: https://dspace.mit.edu/bitstream/handle/1721.1/98551/920674644-MIT.pdf?sequence=1 (accessed on 16 January 2019).
82. Borofsky, Y. Towards a Transdisciplinary Approach to Rural Electrification Planning for Universal Access in India, Massachusetts Institute of Technology. 2015. Available online: https://dspace.mit.edu/bitstream/handle/1721.1/98731/920874583-MIT.pdf?sequence=1 (accessed on 16 January 2019).
83. Cotterman, T. Enhanced Techniques to Plan Rural Electrical Networks Using the Reference Electrification Model, Massachusetts Institute of Technology. 2017. Available online: https://dspace.mit.edu/bitstream/handle/1721.1/111229/1003284003-MIT.pdf?sequence=1 (accessed on 16 January 2019).
84. O'Neil, K.M. Going off Grid: Tata Researchers Tackle Rural Electrification, MIT News. 2016. Available online: http://news.mit.edu/2016/tata-researchers-tackle-rural-electrification-0121 (accessed on 16 January 2019).

85. KTH dESA, OnSSET 2018. 2019. Available online: https://github.com/KTH-dESA/OnSSET-2018 (accessed on 16 January 2019).
86. International Energy Agency. *World Energy Outlook*; International Energy Agency: Paris, France, 2014. [CrossRef]
87. UNDESA/UNDP. Modelling Tools for Sustainable Development. 2016. Available online: https://un-modelling.github.io/electrification-paths-presentation/ (accessed on 10 November 2018).
88. Mentis, D.; Andersson, M.; Howells, M.; Rogner, H.; Siyal, S.; Broad, O.; Korkovelos, A.; Bazilian, M. The benefits of geospatial planning in energy access-A case study on Ethiopia. *Appl. Geogr.* **2016**, *72*. [CrossRef]
89. Mentis, D.; Welsch, M.; Fuso Nerini, F.; Broad, O.; Howells, M.; Bazilian, M.; Rogner, H. A GIS-based approach for electrification planning—A case study on Nigeria. *Energy Sustain. Dev.* **2015**, *29*, 142–150. [CrossRef]
90. Moksnes, N.; Korkovelos, A.; Mentis, D.; Howells, M. Electrification pathways for Kenya-linking spatial electrification analysis and medium to long term energy planning. *Environ. Res. Lett.* **2017**, *12*. [CrossRef]
91. The World Bank. Electrification Pathways Web Application. Available online: http://electrification.energydata.info/presentation/ (accessed on 10 November 2018).
92. Sustainable Energy for All. KTH division of Energy Systems Analysis & SNV. Electrification pathways for Benin-A spatial electrification analysis based on the Open Source Spatial Electrification Tool (OnSSET), Stockholm, Sweden. 2018. Available online: http://www.snv.org/update/mini-grids-and-stand-alone-pv-systems-serve-millions-benin-quest-universal-electricity-access (accessed on 16 January 2019).
93. Development Seed, Offgrid Market Opportunity Tool. 2016. Available online: http://offgrid.energydata.info/#/?_k=0exsuj (accessed on 16 January 2019).
94. INTEGRATION Environment & Energy GmbH and Reiner Lemoine Institut gGmbH, Nigeria Rural Electrification Plans. 2017. Available online: http://rrep-nigeria.integration.org/ (accessed on 16 January 2019).
95. Bertheau, P.; Cader, C.; Blechinger, P. Electrification Modelling for Nigeria. *Energy Procedia* **2016**, *93*, 108–112. [CrossRef]
96. Bertheau, P.; Oyewo, A.; Cader, C.; Breyer, C.; Blechinger, P.; Bertheau, P.; Oyewo, A.S.; Cader, C.; Breyer, C.; Blechinger, P. Visualizing National Electrification Scenarios for Sub-Saharan African Countries. *Energies* **2017**, *10*, 1899. [CrossRef]
97. INTEGRATION Environment & Energy GmbH and Reiner Lemoine Institut gGmbH, Myanmar Off-grid Analytics. 2017. Available online: http://adb-myanmar.integration.org/ (accessed on 16 January 2019).
98. Ghana Energy Commission. Ghana Energy Access Toolkit (GhEA). 2018. Available online: http://167.114.144.200/Home/Project (accessed on 16 January 2019).
99. ECOWAS Regional Centre for Renewable Energy and Energy Efficiency (ECREEE), ECOWREX GIS. 2013. Available online: http://www.ecowrex.org/mapView/ (accessed on 16 January 2019).
100. Tíba, C.; Candeias, A.L.B.; Fraidenraich, N.; Barbosa, E.D.S.; de Carvalho Neto, P.B.; de Melo Filho, J.B. A GIS-based decision support tool for renewable energy management and planning in semi-arid rural environments of northeast of Brazil. *Renew. Energy* **2010**, *35*, 2921–2932. [CrossRef]
101. Kaijuka, E. GIS and rural electricity planning in Uganda. *J. Clean. Prod.* **2007**, *15*, 203–217. [CrossRef]
102. Teske, S.; Morris, T.; Nagrath, K. 100% Renewable Energy for Tanzania—Access to Renewable and Affordable Energy for All Within One Generation, Sydney, Australia. 2017. Available online: https://www.worldfuturecouncil.org/wp-content/uploads/2017/11/Tanzania-Report-8_Oct-2017-BfdW_FINAL.pdf (accessed on 16 January 2019).
103. The World Bank-Data Catalog, Total Population. 2018. Available online: https://data.worldbank.org/indicator/SP.POP.TOTL?locations=MW (accessed on 18 January 2019).
104. United Nations | DESA Population Division. World Population Prospects-Population Division-United Nations. Available online: https://esa.un.org/unpd/wpp/ (accessed on 18 January 2019).
105. National Statistical Office (NSO) of Malawi and ICF, Malawi Demographic and Health Survey 2015–16, Zomba, Malawi, and Rockville, Maryland, USA. 2017. Available online: http://www.nsomalawi.mw/images/stories/data_on_line/demography/mdhs2015_16/MDHS2015-16FinalReport.pdf (accessed on 18 January 2019).

106. Tobler, W.; Deichmann, U.; Gottsegen, J.; Maloy, K. World population in a grid of spherical quadrilaterals. *Int. J. Popul. Geogr.* **1997**, *3*, 203–225. [CrossRef]
107. The International Energy Agency (IEA), Energy Access Database. 2017. Available online: https://www.iea.org/energyaccess/database/ (accessed on 16 January 2019).
108. Government of Malawi (GoM)-Department of Energy Affairs, Malawi SEforALL Action Agenda, Lilongwe, Malawi. 2017. Available online: https://energy.gov.mw/index.php/resource-centre/documents/policies-strategies (accessed on 16 January 2019).
109. Energy Sector Management Assistance Program (ESMAP), Beyond Connections-Energy Access Redefined, Washington, DC, USA. 2015. Available online: https://openknowledge.worldbank.org/bitstream/handle/10986/24368/Beyond0connect0d000technical0report.pdf?sequence=1&isAllowed=y (accessed on 16 January 2019).
110. The Government of Malawi (GoM), Malawi National Energy Policy, Lilongwe, Malawi. 2018. Available online: https://energy.gov.mw/index.php/resource-centre/documents/policies-strategies (accessed on 16 January 2019).
111. The World Bank. Regulatory Indicators for Sustainable Energy (RISE). 2017. Available online: http://rise.esmap.org/ (accessed on 21 January 2019).
112. Slattery, H. Rural America Lights Up, National Home Library Foundation, Washington, DC. 1940. Available online: http://hdl.handle.net/2027/coo.31924073970919 (accessed on 16 June 2017).
113. Tiecke, T. Open population datasets and open challenges. *Connect. Netw. Traffic.* **2016**. Available online: https://code.fb.com/core-data/open-population-datasets-and-open-challenges/ (accessed on 16 January 2019).
114. Malawi Energy Regularoty Authority. Energy Prices in Malawi. 2019. Available online: https://www.meramalawi.mw/ (accessed on 16 January 2019).
115. Global Petrol Prices, Malawi Diesel Prices. 2019. Available online: https://www.globalpetrolprices.com/Malawi/diesel_prices/ (accessed on 16 January 2019).
116. Lenz, V. Generation of Realistic Distribution Grid Topologies Based on Spatial Load Maps, Swiss Federal Institute of Technology (ETH) Zurich. 2015. Available online: https://www.ethz.ch/content/dam/ethz/special-interest/itet/institute-eeh/power-systems-dam/documents/SAMA/2015/Lenz-SA-2015.pdf (accessed on 16 January 2019).
117. Pappis, I. Electrified Africa–Associated Investments and Costs. 2016. Available online: http://kth.diva-portal.org/smash/record.jsf?pid=diva2%3A1070760&dswid=5999 (accessed on 6 February 2019).
118. The World Bank. Reducing the Cost of Grid Extension for Rural Electrification, Washington, DC, USA. 2000. Available online: http://documents.worldbank.org/curated/en/209121468740401066/Reducing-the-cost-of-grid-extension-for-rural-electrification (accessed on 7 March 2018).
119. Energy Sector Management Assistance Program (ESMAP). Model for Electricity Technology Assessment (META). 2014. Available online: http://esmap.org/node/3051 (accessed on 16 January 2019).
120. Nuclear Energy Agency (NEA) International Energy Agency (IEA) Organization for Economic Co-operation and development (OECD), Projected Costs of Generating Electricity 2015 Edition, Paris, France. 2015. Available online: https://www.oecd-nea.org/ndd/pubs/2015/7057-proj-costs-electricity-2015.pdf (accessed on 6 February 2019).
121. The International Renewable Energy Agency. *Renewable Power Generation Costs in 2017*; IRENA: Abu Dhabi, UAE, 2018; ISBN 978-92-9260-040-2.
122. Lazard, N. Lazards Levelized Cost of Energy Analysis-Version 11.0. 2017. Available online: https://www.lazard.com/media/450337/lazard-levelized-cost-of-energy-version-110.pdf (accessed on 16 January 2019).
123. van Ruijven, B.J.; Schers, J.; van Vuuren, D.P. Model-based scenarios for rural electrification in developing countries. *Energy* **2012**, *38*, 386–397. [CrossRef]

energies

MDPI

Article

Evaluation of Energy Distribution Using Network Data Envelopment Analysis and Kohonen Self Organizing Maps

Thiago Gomes Leal Ganhadeiro [1], Eliane da Silva Christo [1,*], Lidia Angulo Meza [2], Kelly Alonso Costa [3] and Danilo Pinto Moreira de Souza [1]

1 Postgraduate Program in Computational Modeling in Science and Technology, Fluminense Federal University, Volta Redonda 27255-125, Brazil; thiago_ganhadeiro@hotmail.com (T.G.L.G.); danilop20@gmail.com (D.P.M.d.S.)

2 Postgraduate Program in Production Engineering, Fluminense Federal University, Niterói 24220-900, Brazil; lidiaangulomeza@id.uff.br

3 Postgraduate Program in Production Engineering, Fluminense Federal University, Volta Redonda 27255-125, Brazil; kellyalonso@id.uff.br

* Correspondence: elianechristo@id.uff.br; Tel.: +55-24-2107-3510

Received: 1 August 2018; Accepted: 5 October 2018; Published: 9 October 2018

Abstract: This article presents an alternative way of evaluating the efficiency of the electric distribution companies in Brazil. This assessment is currently performed and designed by the National Electric Energy Agency (ANEEL), a Brazilian regulatory agency, to regulate energy prices. This involves calculating the *X*-factor, which represents the efficiency evolution in the price-cap regulation model. The proposed model aims to use a network Data Envelopment Analysis (DEA) model with the network dimension as an intermediate variable and to use Kohonen Self-Organizing Maps (SOM) to correct the difficulties presented by environmental variables. In order to find which environmental variables influence the efficiency, factor analysis was used to reduce the dimensionality of the model. The analysis still uses multiple regression with the previous efficiency as the dependent variable and the four factors extracted from factor analysis as independent variables. The SOM generated four clusters based on the environment and the efficiency for each distributor in each group. This allows for a better evaluation of the correction in the *X*-factor, since it can be conducted inside each cluster with a maintained margin for comparison. It is expected that the use of this model will reduce the margin of questioning by distributors about the evaluation.

Keywords: data envelopment analysis; Kohonen self-organizing maps; factor analysis; multiple regression; energy efficiency

1. Introduction

In Brazil, the electricity sector remains treated as a natural monopoly, which is regulated by the government through its own regulatory agency, the National Electric Energy Agency (ANEEL) to in order to prevent possible abuses of market power.

In order to stimulate the search for efficiency [1,2], ANEEL currently uses the price-cap model, which provides periodic price corrections based on several factors, like the service quality and to the productivity of the energy distributor.

The efficiency in the price-cap regulation model is calculated by the *X*-factor, which is related to increases in productivity. This leads to changes in the price that distributors can charge the consumer. As it affects the profitability of the regulated entities, it is important to calculate the *X* factor to be grounded in a consistent explanation in order to convince the society and industry that the calculated value is fair to everyone. Inconsistencies can serve as a basis for discussions and may generate points for modifying the calculation method used [3].

The model used by ANEEL allows an objective evaluation of the efficiencies of each company, since there is a comparison with the others, and considers the influence of the environment where the distributors operate, including factors that can affect the efficiencies of the same. However, there are some points in ANEEL's analysis that need attention.

First, in the DEA model used, it is considered that operational expenditure (OPEX) are a process input, while network extension, number of consumers and consumption are outputs from the process. However, the network extension variable is peculiar: depending on the analysis, it can behave as much as input, since it is used to generate consumption and to serve the consumers, as well as output, because in order to maintain the operational cost demand.

Another point is the fact of using multiple regression as a way of correcting the efficiencies found. It can be argued that when multiple regression is used, it is believed that only environmental factors influence the efficiency of the distributor, so that the intrinsic factors related to the operation and management of the distributors themselves would not be essential for efficiency, which runs counter to the very purpose of the price-cap model.

The evaluation of efficiency in electric energy distributors in Brazil is addressed in several articles. The difference between them is mainly in the combination of the techniques used. The common goal is always to improve the results with the reality of the country.

In this work, some of the most relevant articles are highlighted. The article [4] deals with an evaluation of the electric power distributors using Kohonen self-organizing maps (SOM) and DEA. The article [5] studies the use of undesirable outputs in DEA with application in the electric power sector. The article [6] studies the application of a DEA network model with shared inputs to analyze the efficiency of Brazilian energy distributors. Article [7] used game theory applied to the DEA and later used mode clustering to evaluate the Brazilian energy distributors in the year. In addition, [8] finally reviews the main applications of DEA in the energy field.

This work aims to propose an alternative way to that currently used by ANEEL to evaluate the efficiency of energy distributors. In the first phase of the proposal, it is intended to modify the DEA model, failing to use a non-decreasing composite model by an input and three outputs, and using a non-decreasing DEA network model, taking the network as an intermediate variable, maintaining the OPEX as input and the consumption and number of consumers as outputs. Such an approach would avoid questioning the use of the network as input or output of the process.

In the second phase, as some environmental variables have correlations between them, a factor analysis is done to avoid multicollinearity. After that, a multiple regression is performed with the environmental factors to verify which affect in the efficiency.

In the third phase, the Kohonen maps are used to group the distributors based on the results of phase 2. Efficiency for each group found is calculated with a non-decreasing additive DEA model. In addition, in the final phase, these results are normalized.

2. Materials and Methods

2.1. Price-Cap Model

The price-cap model is a regulation model that intends to support the search for efficiency, while also stopping monopoly-associated practices, in special overpricing. As stated in a previous reference [3], the price-cap model assumes that the price charged must pay the total costs and contain a margin that generates an attractive internal rate of return to the investor. This is done by setting an initial price and correcting this price in prefixed time periods, by analyzing some factors. The general formula of the price-cap model is given by Equation (1):

$$P_t = P_{t-1} + \pi \pm X \pm Q \tag{1}$$

where P_t is the price in period t; P_{t-1} is the price in period of $t-1$; π is the inflation of the period; X is the factor related to the productivity of the company; Q is the factor related to the quality of services provided.

The X factor is related to advances in methods used by the company to increase its productivity, which would lead to decreases in prices due to the competitive market. Since there is basically no competition in the regulated sector, there is the need to adjust prices by adding this variable in the model. The Q factor is related to the quality of the service in a way that better service allows higher prices to be charged.

2.2. DEA BCC Model and Non-Decreasing Returns to Scale

A concept of DEA was given by a previous reference [9] as follows: "DEA evaluates the relative efficiencies of a homogeneous set of decision making units (DMUs) having multiple inputs and outputs."

This means that DEA is an approach used to evaluate efficiency by comparing Decision Making Units (DMUs) in a way that each DMU tries to maximize its own efficiency, but with the restriction that no DMU can be more than 100% efficient. A DMU is an individual unit that performs a process that is similar in its entries (inputs) and exits (outputs) to other DMUs. In this work, an example of a DMU is an energy distributor in the year of 2012.

One of the benefits of using DEA is that "it provides a non-parametric estimate of the efficiency of each DMU compared to the best practice frontier constructed by the best-performing DMUs" [10].

Furthermore, one study [11] states that "DEA showed great promise to be a good evaluative tool for future analysis on energy efficiency" as a conclusion. One of the reasons for this great promise is the facility of multiple inputs and multiple outputs of the DEA model.

The BCC (Banker, Charnes and Cooper) model [12] is a DEA model that uses variable returns to scale (VRS). This means that a linear proportion between the inputs and outputs is not constant. The BCC model considers a DMU to be efficient if it uses the smallest input produces the maximum value of an output. If the returns to scale were constant, this would not necessarily be true. This DMU is considered efficient in constant returns to scale (CRS) only if a relation output/input is maximized. The CRS model is also called the CCR model, which is the first DEA model to be introduced [13].

The BCC model is described in Equations (2)–(6):

$$\text{Max } Eff_o = \sum_{i=1}^{s} u_i y_{jo} + \eta^* \tag{2}$$

$$\text{subject to } \textstyle\sum_{i=1}^{r} v_i x_{ik} = 1 \tag{3}$$

$$\sum_{j=1}^{s} u_j y_{jk} - \sum_{i=1}^{r} v_i x_{ik} + \eta^* \leq 0,\ \forall\, k \tag{4}$$

$$u_j, v_i \geq 0,\ \forall j,\ i \tag{5}$$

$$\eta^* \text{ free} \tag{6}$$

where u are the weights associated with the outputs y are the outputs; j is the index of the output; v are the weights associated with the inputs; x are the inputs; i is the indicator index of the input; s is the number of outputs; r is the number of inputs; Eff_o is the efficiency of DMU_0 k is the DMU identifier; and η^* is a variable indicating the type of return scale.

With a modification in the restriction of η^*, it is possible to adapt non-decreasing returns of scale. This is done by restricting η^* values to be only positive values, so Equations (5) and (6) will transform into Equation (7):

$$u_j, v_i,\ \eta^* \geq 0,\ \forall j,\ i \tag{7}$$

Using Equation (7) in the model implies that some of the DMUs will be evaluated using variable returns to scale (specifically those that have a smaller input variable) while others will be evaluated using constant returns to scale.

Since the objective of the price-cap model is to adjust distributor incomes in order to reflect the increase in efficiency given by the proper usage of resources, it is only natural that the input of the model represents the costs associated with the service. In that sense, operational expenses (OPEX) are used as an input in this present work. The outputs should represent the amount of service being delivered, related to the OPEX. Therefore, the extension of the network, that represents the extension of land covered by the distributor, consumption, that is a direct output of the process, and one of the major sources of company's variable costs, and quantity of consumers, that represents the final clients of the process, could be used as outputs of the process. However, there is discussion about the nature of the extension of the network, since it is not one of the final outputs of the process, but it is a means of achieving the other two outputs. Even more, it is shown in Technical Note nº 101/2011-SRE/ANEEL that the non-decreasing returns to scale hypothesis cannot be rejected. Therefore, distributors that use less OPEX should have greater benefits under the efficiency evaluation.

The sources of the variables are the Public Audience 23/2014 from ANEEL, for the network extension, the number of consumers, and for OPEX. For consumption, data was obtained from the Associação Brasileira de Distribuidores de Energia Elétrica (in English, Brazilian Association of Electric Energy Distributors, known by its Portuguese acronym ABRADEE).

2.3. Network DEA Models

A problem identified in the classic DEA models is the fact that there is no clarification about what happens within the process. A previous study [14] proposed the first DEA network model to solve this question.

These models divide the process into two parts: the first with the objective of transforming the inputs into intermediate variables, which will be used in the process, and the second one with the objective of transforming these intermediate variables into outputs of the process. As stated by a previous study [15], "the division of the production process makes it easier to identify the sources of inefficiency in the process as a whole".

One network DEA model of high importance for this work is the additive Network DEA model, which was first proposed by reference [16]. The major relevance of this model to the present work is the fact that it incorporates variable returns to scale and the characteristics of the Network DEA models, which makes it possible to incorporate the network model in the non-decreasing returns to scale model used by ANEEL. To calculate the overall efficiency of the DMU, this model uses a weighted sum of the stages' efficiency. This model is shown in Equations (8)–(13):

$$\text{Max } Eff_o = \sum_{i=1}^{s} u_j y_{jo} + \sum_{t=1}^{T} w_t z_{to} + \eta_1 + \eta_2 \tag{8}$$

$$\text{subject to} \sum\nolimits_{t=1}^{T} w_t z_{rk} + \sum\nolimits_{i=1}^{r} v_i x_{ik} = 1 \tag{9}$$

$$\sum_{j=1}^{s} u_j y_{jk} - \sum_{i=1}^{T} w_t z_{tk} + \eta_2 \leq 0,\ \forall\, k \tag{10}$$

$$\sum_{i=1}^{T} w_t z_{tk} - \sum_{i=1}^{r} v_i x_{ik} + \eta_1 \leq 0,\ \forall\, k \tag{11}$$

$$u_{j,} v_{i,} w_r \geq 0,\ \forall j,\ i, r \tag{12}$$

$$\eta_1,\ \eta_2 \in R \tag{13}$$

In this model, the objective function in Equation (8) is generated by the weighted sum of the stages. There is a new variable of z, which represents the intermediate variable, and a new weight of w, which is associated to this variable. Equation (10) ensures that the second stage efficiency is less or equal to 1, while Equation 11 ensures this condition for the first stage. As discussed previously, for this model, the OPEX is used as input, while the consumers and consumption are used as outputs. Since the extension of network can be understood as an output for the OPEX and as an input for consumption and number of consumers, it is used as an intermediate variable. For the DEA model proposed in this work, Equations (15)–(17) are added to the additive model, while Equation (6) becomes Equation (14) to ensure non-decreasing returns to scale:

$$\eta_1, \eta_2 \geq 0 \tag{14}$$

$$\frac{v_{consumers}}{u} \geq 30 \tag{15}$$

$$\frac{v_{network}}{u} \geq 580 \tag{16}$$

$$\frac{v_{consumption}}{u} \geq 1 \tag{17}$$

Equations (15)–(17) are restrictions to the weights and are added because they are also used by the current model used by ANEEL [17].

2.4. *Neural Networks and Kohonen Self-Organizing Maps (SOMs)*

Neural networks can be understood as a mathematical way of trying to simulate the physical functioning of the brain. To do so, they are composed of computational units called neurons, which are responsible for acting on the received data.

Kohonen self-organizing maps (SOMs) are neural networks whose main purpose is to find similarities in a group of elements. In this model, the input data is distributed to all neurons, but only one neuron is assigned to each element. SOMs have been used in several applications, such as the modeling of hippocampal dynamics [18]. A SOM's algorithm is made by the iterative steps [19]:

1. Initialization;
2. Competition;
3. Cooperation;
4. Synaptic Adaptation.

The competition process finds the nearest neuron. After this, the cooperation process finds the neighborhood of the winning neuron. The synaptic adaptation adjusts the weights of the winning neuron, based on the neighborhood function and the number of iterations, to make the neuron closer to the input element [20].

By the end of the iterative process, each element is related to a single neuron in a way that similar elements tend to be in the same neuron. In this way, each neuron is attached to a cluster of elements.

In the scope of this work, the SOM is used to group energy distributors by environmental similarity. Since Brazil has a large territorial extension, the environment changes greatly within the country. As so, there are significant differences in the way each company can use its resources, or have its efficiency affected by the environment. Therefore, it could be unfair to compare all the distributors without accounting for these effects. In that sense, one way of keeping the isonomy of treatment between distributors is to group them based on their environmental area of activity similarities. By doing so, it is possible to obtain a cluster efficiency and to adjust the previously found efficiency by seeing how close the distributor is to the other distributors in the cluster. The data for environmental variables were extracted from Public Audience 23/2014 from ANEEL.

2.5. Multiple Regression

According to reference [21], the "main application [of multiple regression], after finding the mathematical relationship, is to produce values for the dependent variable when the independent variables are present."

As in the simple regression analysis, the objective of this analysis is to verify if there is a correlation between the dependent variable and the independent variables in such a way that a change in a value in the independent variables may cause a proportional change in the dependent variable. If such correlation exists, the objective is to find a linear expression that defines the dependent variable in terms of the independent ones.

In this work, this technique is used to verify which environmental variables actually influence the efficiency of a DMU. The general form of the multiple regression is given by Equation (18):

$$Y = b_0 + \sum_{i=1}^{k} b_i X_i + e \tag{18}$$

where Y is the independent variable; b_0 is the bias of the equation, which is the value that the dependent variable would assume if there were no errors in the analysis and all the independent variables were assumed to have the null value; i is the indicator index of the independent variable; k is the number of independent variables; b_i is the change in the dependent variable relative to the independent variable X_i; X_i is the independent variable i; and e is the associated error, which is also called the residue.

We used the minimum squares method to discover the values of b_i. It is also possible to get the t value associated with the b_i, which can be used for statistical purposes in order to verify if the independent variable is in fact correlated with the dependent variable.

According to a previous reference [22], "residuals in a regression are obtained from the difference between the observed value of the response variable and that forecast by the regression model". Therefore, in order to obtain adequate results from a regression and possibly to use it to forecast the results of new elements, it is important that the residues are as small as possible.

2.6. Factorial Analysis

According to the concept given by a previous study [23], factorial analysis aims to explain the correlations between a large set of variables in terms of a set of few unobservable random variables, which are called factors.

One of the advantages of factor analysis and the main reason for using this technique in this work is its ability to reduce the number of variables and eliminate multicollinearity, which is able to maintain the explanatory power of these variables to an adequate extent.

The technique of factor analysis consists of finding the values of the coefficients such that they reproduce the variables from the factors with the highest degree of confidence, according to Equation (19):

$$Y_i = \sum_{j} a_{ij} F_j + a_i U_i \tag{19}$$

where Y is the input variable; F is the factor; i is the indicator in the variable in the input element; j is the factor indicator; U is the specific factor associated; and a is the factor load.

In this work, analysis of the principal components was conducted, which involves a model that aims to find factors that have little errors or unique variance, but explains most of the variance of the original variables.

The Varimax rotation was also used, which is an approach that aims to make factor loads close to 0 or 1 in order to facilitate interpretation.

3. Methodology

In this work, the data from the year of 2012 and previous years were used. However, the focus was to find the efficiency of the distributors in 2012. Previous data was used for historical analysis and to enhance DEA's results. Distributors that did not have data for a certain year were excluded from the analysis in the given year. For the year of 2012, the distributors analyzed were distributed by states in the following way:

- Bahia: Coelba;
- Ceará: Coelce;
- Distrito Federal: Ceb;
- Espírito Santo: Escelsa;
- Goiás: Celg;
- Maranhão: Cemar;
- Mato Grosso: Cemat;
- Mato Grosso Do Sul: Enersul;
- Minas Gerais: Dme-Poços De Caldas; Cemig;
- Pará: Celpa;
- Paraíba: Ene. Paraíba;
- Paraná: Copel;
- Pernambuco: Celpe;
- Rio De Janeiro: Ampla; Light;
- Rio Grande Do Norte: Cosern;
- Rio Grande Do Sul: Aes Sul; Rge; Santa Maria; Ceee
- Santa Catarina: Celesc;
- São Paulo: Bandeirante; Bragantina; Elektro; Eletropaulo; Cpfl Paulista; Piratininga; Nacional;
- Sergipe: Sulgipe; Ene. Sergipe;
- Tocantins: Celtins.

In order to understand the model proposed, a diagram is presented in Figure 1. The first stage is to calculate the non-decreasing returns to scale DEA model with the restrictions presented in Section 2.3, using the OPEX as input and the extension of the distribution network, the number of consumers and the consumed amount of electricity as outputs. This model is called the Retorno Contábil Médio (RCM, or in English, Average Accounting Return) model, which is the model used in the 2-phase analysis by ANEEL. The second phase of ANEEL's analysis is a multiple regression using the environmental variables as independent variables and the efficiencies as the dependent variable.

The RCM model was used for every distributor with data from 2012 and previous years in order to avoid false correlations between environmental variables and efficiency. It is important to note that due to practical issues, the effects of inflation were not considered in this step, which may cause some variations in the final result if such consideration is made in the future.

The second stage of Figure 1 was needed because the environmental variables presented had multicollinearity, which could negatively impact the analysis. The factor analysis reduced the number of variables to a number of factors that would still represent the environment properly, but would not affect the regression.

The third stage of Figure 1 involves a regression analysis with the previously found factors as the independent variables and the RCM efficiency as the dependent variable. This allows us to discover which factors actually influenced the efficiency of the distributors. Following this, these factors were used as the inputs for the SOM in stage 4.

The fourth stage of the model involves the execution of SOM with the factors that influence efficiency in order to cluster the distributors by environment. This was conducted only with the distributors in 2012 in order to avoid clustering one distributor with itself in previous years. Since the objective is to find the efficiencies in 2012, it is more logical to use only the distributors in this year.

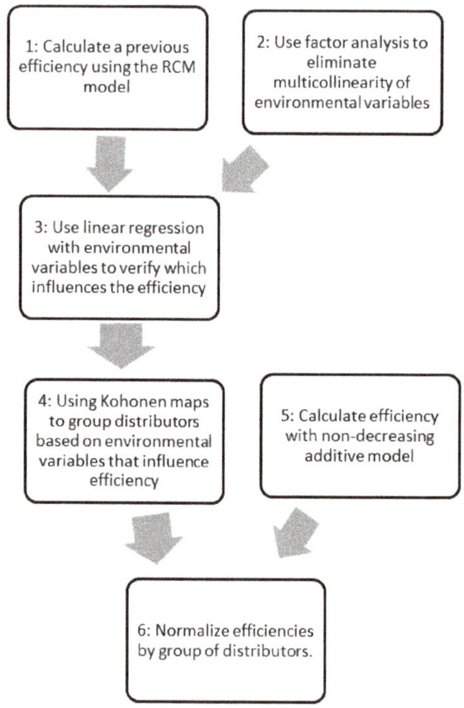

Figure 1. Diagram of proposed model.

The fifth stage calculates the non-decreasing additive model, with the restrictions presented in Section 2.3. Here, since the final result will be normalized by clustering using only data from 2012, all the DMUs were used with one DMU being the data of one distributor in a given year. The OPEX was used as the input, the extension of the distribution network was used as an intermediate variable, while the consumed amount and number of consumers were used as outputs. In order to understand the model proposed in this stage, a diagram is presented in Figure 2.

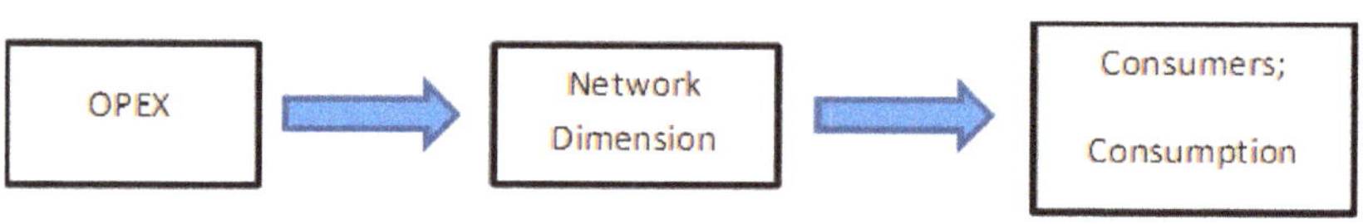

Figure 2. Diagram of DEA model.

The assumption of non-decreasing returns to scale is based on Technical Note n° 101/2011–SRE/ANEEL, that performs the Banker test [23] in several configurations of inputs and outputs. The results show that, for the configuration used in this work, the non-decreasing returns to scale hypothesis cannot be rejected. Finally, the sixth stage normalizes the efficiency from stage 5 within the clusters of stage 4. The data used was obtained from the ANEEL's public audience no. 023/2014 and data from ABRADEE.

4. Results

For the sake of space and relevancy, 411 DMUs were considered in stages 1–3 of Figure 1, with some specific data from these analyzes having been omitted. These include data from inputs and outputs used in the DEA model as well as environmental variables and the efficiency found in the

DEA model. However, this data is used primarily for statistical purposes in stage 3, with no other purpose in this article.

In stage 2 of the analysis, two variables were excluded for the purpose of fixing the Kaiser-Meyer-Olkin (KMO) test value. These factors were area and low vegetation. Subsequently, four factors were found in the analysis. The participation of each environmental variable in each factor is shown in Figure 3.

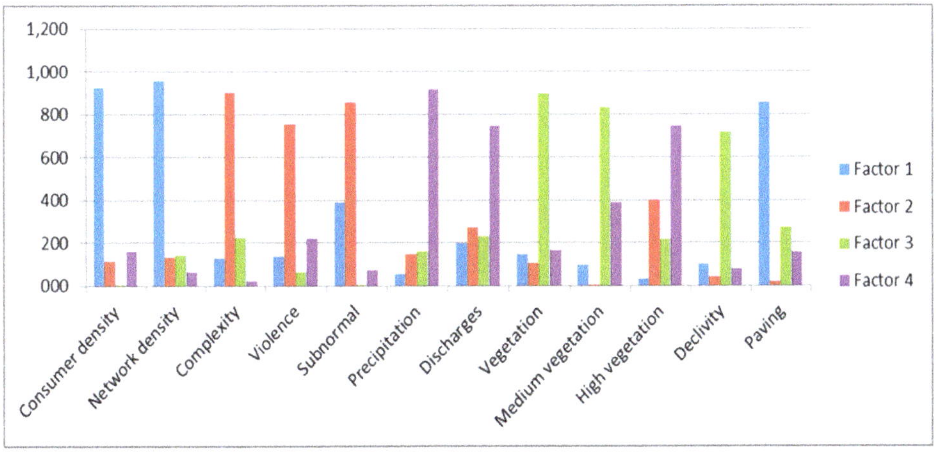

Figure 3. Participation of variables in each factor.

From Figure 3, it is possible to observe that:

- Factor 1 is composed primarily of the following variables: consumer density, network density and paving;
- Factor 2 is composed primarily of complexity, violence and subnormal;
- Factor 3 is composed primarily of vegetation, vegetation and declivity;
- Factor 4 is composed primarily of precipitation, discharges and high vegetation.

In stage 3, R^2 was found to be 0.356. This means that there is some correlation between the factors and the RCM efficiency, but this relationship is not enough for the factors to solely explain the efficiency of the distributors. The significance levels of factors really influencing the efficiency must be further analyzed, which is provided in Table 1.

Table 1. Significance levels for factors in regression.

Variable	*t*	Level of Significance
(Constant)	65.279	0
Factor 1	−11.593	0
Factor 2	2.375	0.018
Factor 3	4.992	0
Factor 4	−6.124	0

From the analysis of Table 1, it can be seen that all factors are significant. Therefore, all four factors should be included for the creation of SOM in stage 4. In order to find the number of clusters needed, the technique of observing the U matrix and the weighted maps was used. This strategy is explained in further detail in reference [24]. For the purposes of this work, a larger than required map is created and the distances between the neurons are plotted with a color scale. The brighter, connected regions are usually linked to the same cluster. This technique displays a natural number of clusters formed by the network. It is mostly a visual technique, but provides a good estimate of the number of clusters that should be used.

The U matrix (unified distances matrix) for this work is shown in Figure 4 and the weight distances are plotted in Figure 5. Both were generated with help of MATLAB® (version R2017b).

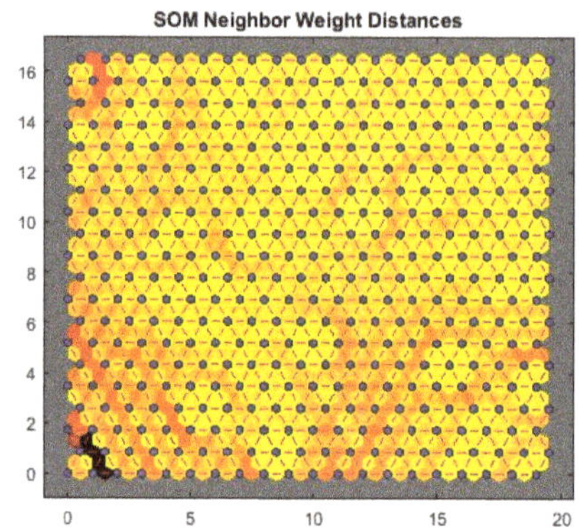

Figure 4. U-Matrix (unified distances matrix).

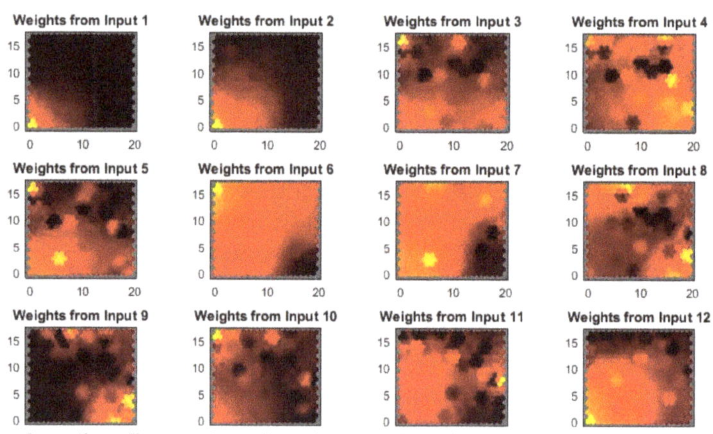

Figure 5. Weight distances.

From the U-matrix in Figure 4, it is not possible to obtain enough data. However, the input weights in Figure 5 provides some clues. It is possible to observe a cluster in superior left part of the map, while there are two others in the lower left and right extremes of the map. Further, there is a clearer area in the top central area that can be interpreted as another cluster.

Maps ranging from 3 to 6 clusters were created. The maps with six clusters presented a cluster without any elements, so it was discarded. The map with three clusters did not represent well the ecological and economical diversity of the area in analysis. Both the four clusters and the five clusters maps could explain reasonably well the similarities of the environments. In synthesis, some distributors that belong in clusters 1 and 4 when used the four clusters map created a new cluster in the 5 clusters map. In this work, the four clusters map was used for the analysis, since it provides a greater number of distributors by clusters, therefore improving efficiency comparability in DEA.

In such way, a map with a size of 2 rows × 2 columns was used for clustering, with the results shown in Table 2. For better visualization, these results are shown graphically in Figure 6.

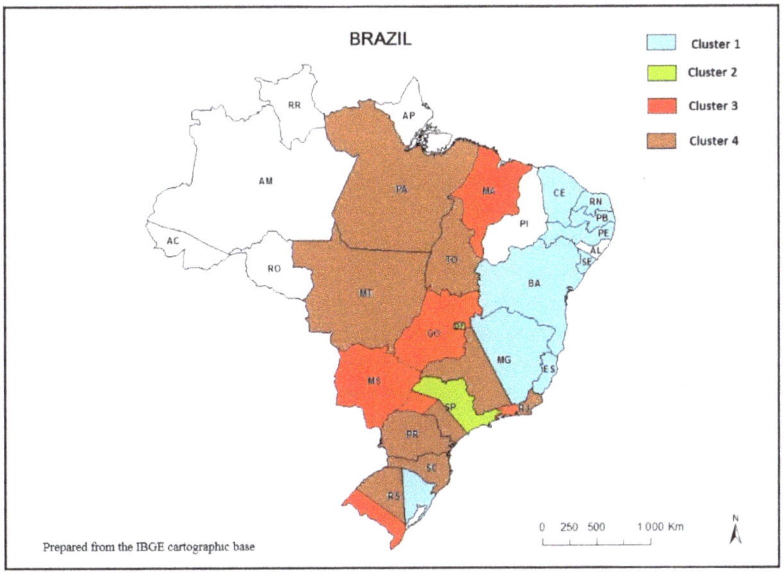

Figure 6. Cluster representation by area.

In Figure 6, there are areas of blank spaces. This is caused by the absence of data relative to the electric distributor that is located in these areas. Furthermore, some states are divided in color. This is due to there being more than one distributor serving these states, with these distributors being allocated in different clusters. However, it is possible to see that there is at least some consistency to the clustering. The final results and comparison with the ANEEL results are given in Table 2.

From Table 2, it is possible to say that there is no direct relationship between the model results and the ANEEL model. Some distributors gain efficiency, while others lose it. In fact, the correlation between the models is 24%, with a R^2 of only 0.06. Such divergence is due to the great changes made, especially in relation to the network DEA model.

Figure 7 presents efficiencies of each stage. Generally, distributors are more efficient in Stage 1. It is observed in Figure 7 that, seven distributors, such as Energy Company of Brasília (CEB) and Eletropaulo for example, are more efficient in Stage 2.

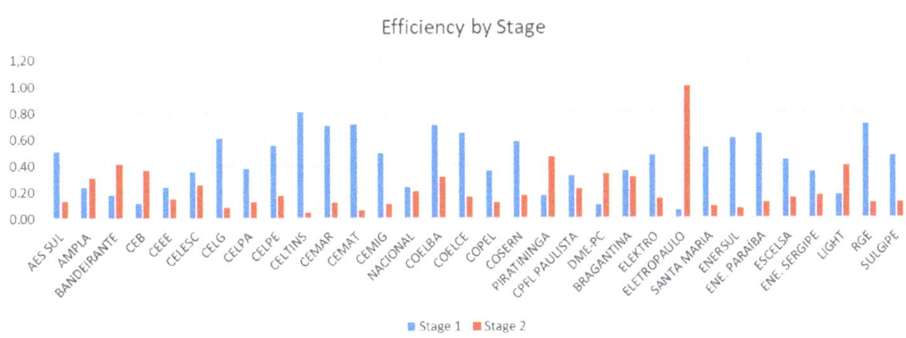

Figure 7. Efficiency of each concessionaire by stage.

Table 2. Significance levels for factors in regression.

Distributor	Efficiency of Proposed Model	Cluster	Normalized Efficiency	Efficiency of ANEEL Model	Difference
AES SUL	0.36	4	0.82	0.80	−0.02
AMPLA	0.23	3	0.54	0.70	0.16
BANDEIRANTE	0.19	4	0.44	0.77	0.33
BRAGANTINA	0.12	3	0.30	0.74	0.44
CEB	0.20	2	0.70	0.49	−0.21
CEEE	0.27	3	0.65	0.39	−0.26
CELESC	0.38	4	0.89	0.59	−0.30
CELG	0.29	3	0.69	0.66	−0.03
CELPA	0.39	4	0.89	0.50	−0.39
CELPE	0.49	1	1.00	0.79	−0.21
CELTINS	0.43	4	1.00	1.00	0.00
CEMAR	0.41	3	1.00	0.82	−0.18
CEMAT	0.34	4	0.79	0.71	−0.08
CEMIG	0.21	1	0.43	0.64	0.21
COELBA	0.44	1	0.90	0.89	−0.01
COELCE	0.43	1	0.87	0.95	0.08
COPEL	0.27	4	0.62	0.60	−0.02
COSERN	0.40	1	0.82	0.89	0.07
CPFL PAULISTA	0.19	3	0.46	0.84	0.38
DME-PC	0.25	4	0.58	0.47	−0.11
ELEKTRO	0.11	3	0.26	0.76	0.50
ELETROPAULO	0.28	4	0.64	0.67	0.03
ENE. PARAÍBA	0.34	1	0.70	0.75	0.05
ENE. SERGIPE	0.10	1	0.20	0.56	0.36
ENERSUL	0.35	3	0.85	0.69	−0.16
ESCELSA	0.38	1	0.77	0.78	0.01
LIGHT	0.41	4	0.94	0.69	−0.25
NACIONAL	0.33	3	0.79	0.65	−0.14
PIRATININGA	0.28	2	1.00	0.88	−0.12
RGE	0.19	4	0.43	1.00	0.57
SANTA MARIA	0.44	1	0.88	0.90	0.02
SULGIPE	0.33	1	0.68	0.66	−0.02

5. Conclusions

A different model was proposed for the evaluation of the efficiency of the different Brazilian energy distributors, which aims to reduce the margin of potential doubts by using the environmental variables in the regression and extending the distribution network in the DEA model.

It must be considered that the distributors were divided into clusters for the calculation of the *X* factor. Therefore, it becomes possible to perform the comparison within each cluster in the final calculation of the *X* factor. Despite this division, there was comparison between the distributors, so that some proportionality can be achieved in the corrections of the defined energy tariffs that should be charged, which is a positive point of the presented model.

The proposed model presented differences compared to the model currently used by ANEEL, with some distributors having high efficiency, while others had their efficiency reduced. This indicates the need for adjustments to the data used. Although the proposed model presents results with high sensitivity to variable specifications.

It is reasonable to assume that there may be distortions due to the limitations presented in this paper, such as the lack of use of shared inputs or the absence of correction of inflationary effects.

This work can be used in countries of great territorial extension that deals with at least some influence of private sector in energy distribution, or any other regulated service. It allows objective evaluation of efficiency, which is relevant when the discussion involves economical rights, such as the present case. Moreover, the resulting efficiency accounts for environmental aspects, but is not defined by them.

Author Contributions: T.G.L.G. carried out the computational implementations and the writing of the article. E.d.S.C. and L.A.M. contributed to the analysis of DEA, SOM, factor analysis and Multiple Regression. K.A.C. and D.P.M.d.S. revised of the text and the orientation of the construction of the work.

Funding: This research was funded by Coordenação de Aperfeiçoamento de Pessoal de Nível Superior and Conselho Nacional de Desenvolvimento Científico e Tecnológico.

Acknowledgments: The authors thank the Federal University Fluminense—UFF for the teachers. In addition, to the Coordination for the Improvement of Higher Education Personnel—CAPES for the financial assistance during the course of Post-Graduation.

Conflicts of Interest: The authors declare no conflict of interest.

References

1. Silva, R.D.S.; Oliveira, R.C.; Tostes, M.E.L. Analysis of the Brazilian Energy Efficiency Program for Electricity Distribution Systems. *Energies* **2017**, *10*, 1391. [CrossRef]
2. Liu, J.-P.; Yang, Q.-R.; He, L. Total-Factor Energy Efficiency (TFEE) Evaluation on Thermal Power Industry with DEA, Malmquist and Multiple Regression Techniques. *Energies* **2017**, *10*, 1039. [CrossRef]
3. Pires, J.C.L.; Piccinini, M.S. Modelos de regulação tarifária do setor elétrico. *Rev. do BNDES* **1998**, *5*, 147–168.
4. Andrade, G.N.; Alves, L.A.; Silva, C.E.R. F.; De Mello, J.C.C.B.S. Evaluating Electricity Distributors Efficiency Using Self-Organizing Map and Data Envelopment Analysis. *IEEE Lat. Am. Trans.* **2014**, *12*, 1464–1472. [CrossRef]
5. Tschaffon, P.B.; Angulo-Meza, L. Um estudo de outputs indesejáveis em dea com aplicação no setor de distribuição de energia elétrica. In Proceedings of the XLIII Simpósio Brasileiro de Pesquisa Operacional—SBPO, São Paulo, Brasil, 15–18 August 2011.
6. Moreno, P.; Andrade, G.; Angulo-Meza, L.; De Mello, J.C.B. S. Evaluation of brazilian electricity distributors using a network dea model with shared inputs. *IEEE Lat. Am. Trans.* **2015**, *13*, 2209–2216. [CrossRef]
7. Machado, L.G.; de Mello, J.C.C.B.S.; Roboredo, M.C. E-ciency evaluation of brazilian electrical distributors using data envelopment analysis game and cluster analysis. *IEEE Lat. Am. Trans.* **2016**, *14*, 4499–4505. [CrossRef]
8. Mardani, A.; Zavadskas, E.K.; Streimikiene, D.; Jusoh, A.; Khoshnoudi, M. A comprehensive review of data envelopment analysis (dea) approach in energy e-ciency. *Renew. Sustain. Energy Rev.* **2017**, *70*, 1298–1322. [CrossRef]
9. Yu, P.; Lee, J.H. A hybrid approach using two-level SOM and combined AHP rating and AHP/DEA-AR method for selecting optimal promising emerging technology. *Expert Syst. Appl.* **2013**, *40*, 300–314. [CrossRef]
10. Jebali, E.; Essid, H.; Khraief, N. The analysis of energy efficiency of the Mediterranean countries: A two-stage double bootstrap DEA approach. *Energy* **2017**, *134*, 991–1000. [CrossRef]
11. Mardani, A.; Zavadskas, E.K.; Streimikiene, D.; Jusoh, A.; Koshnoudi, M. Data Envelopment Analysis in Energy and Environmental Economics: An Overview of the State-of-the-Art and Recent Development Trends. *Energies* **2018**, *11*, 2002. [CrossRef]
12. Banker, R.D.; Charnes, A.; Cooper, W. Some models for estimating technical and scale inefficiencies in data envelopment analysis. *Manag. Sci.* **1984**, *30*, 1078–1092. [CrossRef]
13. Charnes, A.; Cooper, W.; Rhodes, E. Measuring the efficiency of decision making units. *Eur. J. Oper. Res.* **1978**, *2*, 429–444. [CrossRef]
14. Färe, R.; Grosskopf, S. Network DEA. *Socio-Econ. Plan. Sci.* **2000**, *34*, 35–49. [CrossRef]
15. Júnior, S.F.G.; Beltrán, P.M.; de Mello, J.C.C.B.S.; Angulo-Meza, L. Utilização de modelo network dea na avaliação de cursos de pós-graduação stricto sensu em engenharia. In Proceedings of the XVII Simpósio de Pesquisa Operacional e Logística da Marinha—SPOLM, Rio de Janeiro, Brasil, 6–9 August 2014; pp. 99–111. [CrossRef]
16. Cook, W.D.; Zhu, J.; Bi, G.; Yang, F. Network dea: Additive efficiency decomposition. *Eur. J. Oper. Res.* **2010**, *207*, 1122–1129. [CrossRef]
17. Superintendência de Regulação Econômica SRE. *Análise de Eficiência dos Custos Operacionais das Distribuidoras de Energia Elétrica*; Relatório Técnico; ANEEL: Brasília, Brazil, 2014.

18. Vitral, R.W.; de Araújo, G.F.; de Oliveira, F.C.; Martins, D.M.; Christo, E.D.S.; Vitral, C.M.; Abramov, D.M. Neurobiological data sustaining opponent processing operations on self-organizing networks as tools for the modeling of hippocampal dynamics. In Proceedings of the 2005 IEEE International Joint Conference on Neural Networks, Montreal, QC, Canada, 31 July–4 August 2005; pp. 2135–2138. [CrossRef]
19. Haykin, S. *Neural Networks: A Comprehensive Foundation*, 1st ed.; Prentice Hall PTR: Upper Saddle River, NJ, USA, 1994; ISBN 978-0132733502.
20. Kohonen, T. The self-organizing map. *Proc. IEEE* **1990**, *78*, 1464–1480. [CrossRef]
21. Sassi, C.P.; Perez, F.G.; Myazato, L.; Ye, X.; Ferreira-Silva, P.H.; Louzada, F. *Modelos de Regressão Linear Múltipla Utilizando os Softwares e Estatística: Uma Aplicação a Dados de Conservação de Frutas*; ICMC, USP, CP668, n° 377; ICMC-USP: São Carlos, Brazil, 2012.
22. De Silva, R.O.; da Christo, E.S.; Costa, K.A. Analysis of Residual Autocorrelation in Forecasting Energy Consumption through a Java Program. *Adv. Mater. Res.* **2014**, *962*, 1753–1756. [CrossRef]
23. Baptistella, M.; Steiner, M.T.A.; Neto, A.C. O Uso de Redes Neurais e Regressão Linear Múltipla na Engenharia de Avaliações: Determinação dos Valores Venais de Imóveis Urbanos. Available online: http://www.din.uem.br/sbpo/sbpo2006/pdf/arq0172.pdf (accessed on 15 June 2017).
24. Silva, M.A.S. da. Mapas Auto-Organizáveis na Análise Exploratória de Dados Geoespaciais Multivariados. Master's Thesis, INPE, São José dos Campos, Brasil, 2005.

energies

MDPI

Article

Automatic Processing of User-Generated Content for the Description of Energy-Consuming Activities at Individual and Group Level

Roos de Kok, Andrea Mauri * and Alessandro Bozzon

Faculty of Electrical Engineering, Mathematics and Computer Science, Delft University of Technology, 2628 XE Delft, The Netherlands; R.E.deKok@student.tudelft.nl (R.d.K.); a.bozzon@tudelft.nl (A.B.)

* Correspondence: a.mauri@tudelft.nl

Received: 31 October 2018; Accepted: 18 December 2018; Published: 21 December 2018

Abstract: Understanding and improving the energy consumption behavior of individuals is considered a powerful approach to improve energy conservation and stimulate energy efficiency. To motivate people to change their energy consumption behavior, we need to have a thorough understanding of which energy-consuming activities they perform and how these are performed. Traditional sources of information about energy consumption, such as smart sensor devices and surveys, can be costly to set up, may lack contextual information, have infrequent updates, or are not publicly accessible. In this paper, we propose to use social media as a complementary source of information for understanding energy-consuming activities. A huge amount of social media posts are generated by hundreds of millions of people every day, they are publicly available, and provide real-time data often tagged to space and time. We design an ontology to get a better understanding of the energy-consuming activities domain and develop a text and image processing pipeline to extract from social media the description of energy-consuming activities. We run a case study on Istanbul and Amsterdam. We highlight the strength and weakness of our approach, showing that social media data has the potential to be a complementary source of information for describing energy-consuming activities.

Keywords: social media; energy-consuming activities; energy consumption; machine learning; ontology

1. Introduction

Europe's 2030 Energy Strategy targets a 40% cut in greenhouse gas emissions compared to 1990 levels, at least a 27% share of renewable energy consumption and at least 27% energy savings compared with the business-as-usual scenario (https://ec.europa.eu/energy/en/topics/energy-strategy-and-energy-union/2030-energy-strategy). To meet this target, energy policies and programs should be formed and individuals should be motivated to change their energy consumption behavior [1], both in terms of energy conservation and energy efficiency. Energy efficiency involves using less energy to provide the same service; for instance, replacing a single-pane window in the house with an energy-efficient one. On the other hand, energy conservation involves saving energy by reducing or omitting an activity; for instance, turning a light off or reducing the time one watches television.

Multiple studies have examined how energy efficiency and conservation could be motivated among policy makers and citizens. In [2] the author explains how comparative feedback on energy usage with others can generate feelings of competition, social comparison, or social pressure, which appears to be more effective in motivating energy conservation than temporal self-comparisons. The author of [3] endorses this in his Social Electricity case study, which "allows people to compare their energy footprint with other online peers or with the consumption at their neighborhood, village or town, to perceive if their own consumption is low, average or high". Multiple energy saving applications [4] have been developed, using visualized consumption feedback and gamified social interactions to motivate people to adopt energy-efficient lifestyles.

Before we can motivate individuals to change their energy consumption behavior, we need a thorough understanding of why and how they consume energy. To do so, insights into the individual's activities behind the energy consumption should be gathered at a high-granular level.

Multiple data sources are used to provide insights into energy-consuming activities (i.e., an activity that have a direct or indirect impact on energy consumption). Smart meters and smart plugs give insights into domestic energy consumption by providing aggregated energy consumption data. Techniques have been developed to isolate the signal of each appliance by looking at the total power consumed, the different current waveform and the voltage signature [5–7]. Surveys and interviews are used to break down the energy consumption into different end-uses through several questions (e.g., how much time you watch TV at home? How often do you use public transportation?) [8–10]. While being the most reliable source of quantitative data and qualitative information, the aforementioned sources come with drawbacks: surveys are costly to perform, they do not scale and are done infrequently; while smart sensors and smart plugs are costly, the data obtained lack of contextual information and is often not accessible. Moreover, smart sensor devices neglect indirect energy usage [11] (i.e., related to the production, transportation, and disposal of a variety of consumer goods and services [12]) and the disaggregation process is far from perfect [5].

On the other hand, hundreds of millions of people frequently use social media to share, communicate, connect, and interact. Although being noisy and biased (i.e., used by a subset of the population), they are publicly available and provide real-time and semantically rich data.

For these reasons, social media has proven to be a good source for human activity recognition [13–15], including, but not limited to, travel behavior [16–18], mode of transportation [16] and nutrition patterns [19–21].

This work puts the following intuition at test: since social media posts relate to different aspects of daily activities, they may either directly refer to energy-consuming activities, or contain relevant information about energy-consuming activities in their semantic signature. Therefore, by processing the content of social media posts, we aim at extracting information about the energy-consuming activity it refers to.

Hence, we aim to answer the following research question:

RQ How can we automatically process user-generated content to describe energy-consuming activities at individual and group level?

We focus on four categories of energy-consuming activities: dwelling, mobility, food consumption, and leisure. Based on the literature [22–24], they cover a considerable spectrum of the activities impacting on the energy footprint of an individual's lifestyle.

Dwelling refers to the consumption of energy due to the usage of home appliances (e.g., washing machine, gaming console), mobility includes the energy required for moving from one place to another, food consumption refers to the use of resources associated with the preparation and processing of food and leisure indicate the energy required for performing recreational activities (e.g., watching TV, playing video-games, partying). Activities related to industry—e.g., the individual being at work—are not taken into account.

Figure 1 illustrates the intuition behind this work, the message (*Great dinner at Hotel de Goudfazart [...]*) suggests that the picture is taken by the user during dinner. In addition, in the image we can indeed identify some kind of cooked fish and vegetables. Furthermore, the hash tags and the location where the user has checked in indicate that the dinner took place in the Hotel de Goudfazant. By looking at the place properties, we discover that the restaurant is located in Amsterdam, the Netherlands. Moreover, we can suppose that the person travelled to the restaurant using either a car or by public transportation. To conclude, this post discloses information about food (i.e., the dinner was cooked), leisure (i.e., the activity takes place in) and mobility (i.e., the individual had to travel to get at the venue) energy-consuming activities.

Figure 1. Example of social media post on Instagram.

Contribution: The objective of this work is to automatically extract information about energy-consuming activities from social media posts. To do so, we (1) create an *ontology* of the domain to identify relevant and important concepts and how these are interrelated. It provides terms for describing our knowledge about the energy consumption domain in a structured manner and it facilitates to draw the link between the social media post and the activity performed in the physical world. Then (2), we design a *data processing pipeline* that extract the characteristics of energy-consuming activities from the social media data. This pipeline includes multiple components: (i) the data collection (and pre-processing) from the social media data sources; (ii) different steps of data enrichment; (iii) a dictionary and rule-based classification model that outputs to which categories of energy-consuming activities social media posts are classified; and (iv) a linked data publisher that use the information gathered by the previous modules to create instances of the ontology and output them using the JSON-LD format (https://json-ld.org/).

The pipeline is evaluated through a case study performed on the social media activity in the cities of Amsterdam and Istanbul.

2. Materials and Methods

2.1. The Social Smart Meter Ontology

In this section, we present the Social Smart Meter ontology (SSMO). We create this ontology with two objectives in mind: (i) understand the domain of energy-consuming activities and (ii) identify relevant and important concepts and how these are interrelated, by providing terms for describing and representing our knowledge about this domain in a structured manner [25].

In addition, the ontology allows for an unambiguous conceptual description of the targeted domain and can be also used to enable better interaction among different fields of studies concerned with energy consumption.

Since social media data refer to individual's daily activities [15], we include social media concepts in the definition as well, by linking them to the relevant concepts of energy-consuming activities. Adding meaning to a user's social media data help us understand to what extent these data sources reflect the individual's energy-consuming activities.

The design of the ontology has been performed according to the *Methontology* guidelines [26]. We follow the methodological guidelines for specifying ontology requirements presented in [27] to compose a set of functional requirements for the SSMO ontology, which are presented in Table A1 in Appendix A.

2.1.1. The Ontology Definition

As depicted in Figure 2, an *Individual* consumes energy by performing an *Activity* at a certain *Location*, at a certain time, and for a certain period of time. That activity can be of multiple types: *Dwelling*, *Mobility*, *Food Consumption*, and/or *Leisure*.

A *Location* can either be a *Path* or *Place*. A *Place* can be a geographical location (e.g., a town or country) or a venue (e.g., a restaurant or airport) and is characterized by its corresponding coordinates and a category. A *Path* is composed of multiple (at least two) places, among which the origin and destination.

In case of a domestic activity, generally, one or more *Appliances* are used. Among appliances, *Brown Goods* (small household electrical entertainment appliances) and *White Goods* (major household appliances) are distinguished [28].

In food consumption-related activities (having breakfast or lunch, dining, cooking, etc.), the *Food* product itself and its *Ingredients*, the *Tableware* used for consumption, the food *Source*, and the (cooking) *Process* are relevant entities. Among processes, cooking and *Modification* are distinguished. Modification involves a technique used to modify raw food into food that is ready for cooking.

In leisure, several subcategories can be distinguished, among which: culture, event, gastronomy, playful, relaxation, social interaction, etc. In general, leisure activities require the use of one or more *Artifacts*, for instance, an appliance.

An activity that involves mobility is characterized by the transportation along a path. People travel by a certain *Mode of transport*, for which the type indicates whether the mode of transport is public or private.

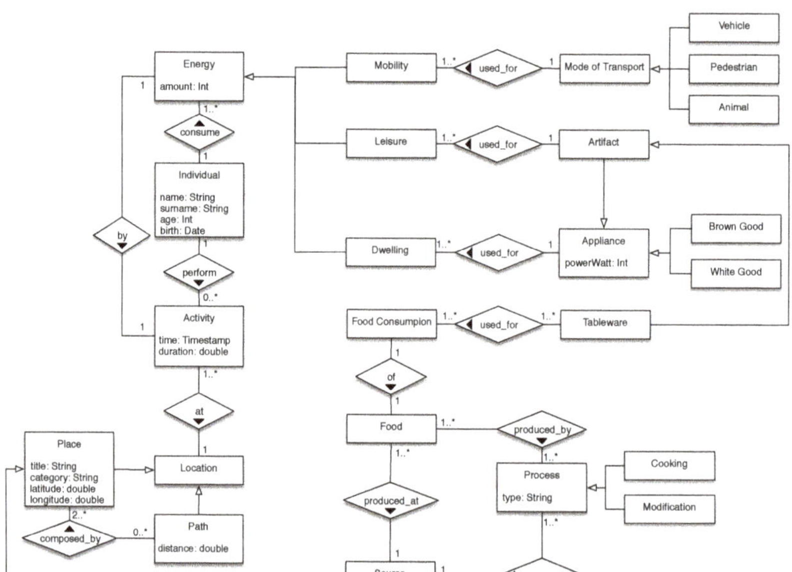

Figure 2. Conceptual data model of energy-consuming activities.

For our ontology it is also important to include social media data. Therefore, based on the existing ontologies and studies [29,30], we created a conceptual data model, depicted in Figure 3, including the following elements:

- A *User* has a social media user account, including a user *Profile*, containing information such as name, gender, age, etc.
- A *User* can create one or more social media *Posts*, which can be placed at a timeline or newsfeed to share those with other social media users.
- A *Post* contains one or more *Items*, which can be of type image, video, link, etc.
- Within a *Post*, a *User* can *Mention* a concept, such as another *User* or a *Location*. This mention provides a link to this concerning concept. Often, more information about the location is available, such as the corresponding coordinates or the location category.

Then the two parts are linked by the following relations: a *User* is an *Individual* and *Post* may reflect an *Activity*.

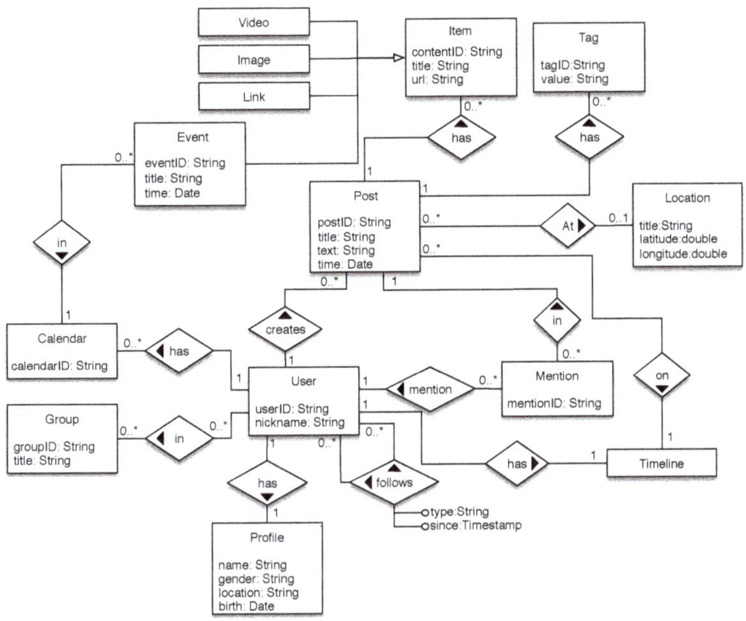

Figure 3. Conceptualization of social media activity.

2.1.2. Implementation of the Ontology

To prevent a proliferation of ontologies covering the same entities and relationships, it is important to determine which existing ontologies can be integrated and extended to develop ours. For this reason, we looked at existing ontologies about energy consumption, travel, food, and social media.

The Suggested Upper Merged Ontology (SUMO) [31] has been designed as a foundation ontology and is the largest formal public ontology today, used for research and applications in search, linguistics, and reasoning (in computer information processing systems). Since it covers most of the concepts of our conceptual data model of energy-consuming activities, it is used as the foundation to be extended for our SSMO ontology.

The Semantic Tools for Carbon Reduction (SEMANCO) Energy Model [32] focuses on terms and attributes describing energy consumption and CO_2 emission indicators for regions, cities, neighborhoods, and buildings, along with climate and socioeconomic factors affecting energy consumption. We include it to model the energy consumption part of our ontology.

The EnergyUse (EU) platform [33] is built upon the PowerOnt [28] ontology that provides information of energy consumption for numerous household appliances and extends the DogOnt [34] ontology, which aims to model intelligent domotic environments. We integrate this ontology to cover the concepts related to appliances.

The Food Ontology (FO) [35] encompasses information about recipes, their ingredients, along with suitable diets, menus, seasons, courses, and occasions. Also, entities about food chain (i.e., methods and techniques used to process the food) are promising for the integration in the SSMO ontology. FO does not cover the tableware entities; yet, this is not problematic since the SUMO ontology covers them. Finally, the Travel Ontology (TO) by Stevens [36], covers most of the relevant entities within the mobility concept, except for the actual mobility activity itself.

In Table 1 for each ontology is indicated to what extent the entities within the high-level concepts (energy activity, location, dwelling, food consumption, leisure, and mobility) are covered. A "+" indicates the entity occurs in the ontology, a "+/−" indicates the entity is covered to some extent, and a "−" indicates the ontology does not include the entity.

Table 1. Overview of the current state-of-the-art related ontologies with a focus on the previously distinguished domains of energy-consuming activities (+: included; +/−: covered to some extent; −: not included).

	SUMO [31]	SEMANCO [32]	EU [33]	FO [35]	TO [36]
Energy activity					
- Energy units	+	+	+	−	−
- Consumption	+/−	+	+	−	−
- Individual	+	+	+	+	−
Location					
- Location	+	+	+	−	+
- Path	+	−	−	−	+
Dwelling					
- Activity	+	+	−	−	−
- Appliance	+	+	+	−	−
Food consumption					
- Activity	+/−	−	−	+	−
- Food	+	−	−	+	−
- Food chain	−	−	−	+	−
- Tableware	+	−	−	−	−
Leisure					
- Activity	+	+	−	−	+
- Artifact	+	−	−	−	
Mobility					
- Activity	+	+	−	−	−
- Mode of transport	+	−	−	−	+

Regarding the social media activity, we reuse the Friend of a Friend (FOAF) [30] and the Semantically-Interlinked Online Communities (SIOC) [29] ontologies. In general, both cover the concepts of user account, post, and item; but the *mention* entity only recurs in the SIOC ontology, whereas the location entity can only be found in the FOAF ontology.

To a great extent, the SSMO ontology can be built upon existing ontologies, as can be deduced from the overview in Table 1; many classes can be reused. Table 2 summarizes the classes that are reused from existing ontologies.

On the other hand, the existing ontologies serve other purposes than identifying and describing energy-consuming activities, so even though some concepts are already covered (e.g., the mobility activity by the *SUMO:Motion* class), the exact semantic of the class is slightly different. For those cases, we create new entities for those classes and we draw the equivalence relationship between them (e.g., our *ssmo:MobilityActivity* class and the *SUMO:Motion* class). Table 3 summarizes the entities created in this way.

In addition, not all entities from the conceptual data models can be covered by existing ontologies. The new entities that had to be created for the SSMO are listed in Table 4.

The ontology was then implemented using the Web Ontology Language (OWL) [37] with Protégé (https://protege.stanford.edu), Stanford University's free, open-source ontology editor.

Finally, the ontology is available on the companion website (http://social-glass.tudelft.nl/social-smart-meter/#ontology).

Table 2. Overview of the entities in the SSMO ontology reused from existing ontologies.

	Ontology	Prefixed Class Name
Energy activity		
- Energy	SEMANCO	SEMANCO:Energy_Quantity_And_Emission
- Individual	SUMO; SEMANCO	SUMO:Human; SEMANCO:Household_Member
Location		
- Place	SUMO	geo:SpatialThing
- Path	TO; SUMO	upper.owl#Pattern; SUMO:TransitRoute
Dwelling		
- Activity	SUMO	SUMO:Cooking
- Appliance	EU	DogOnt:Appliances
Food consumption		
- Activity	SUMO	SUMO:Cooking
- Food	FO	fo/Food
- Ingredient	FO	fo/Ingredient
- Modification	FO	fo/Technique
Leisure		
- Artifact	SUMO	SUMO:Artifact
Mobility		
- Activity	SUMO	SUMO:Motion
- Mode of transport	TO	travel.owl#ModeOfTransport
- Vehicle	SUMO; TO	SUMO:Vehicle; travel.owl#VehicleTransport
Social Media		
- User account	FOAF	foaf:OnlineAccount
- Post	FOAF; SIOC	foaf:Document; ns1:Post
- Mention	SIOC	sioc:link
- Location	FOAF	foaf:based_near

Table 3. Overview of the new entities equivalent to reused entities in the SSMO ontology.

	Ontology	Prefixed Class Name
Energy activity		
- Energy	SEMANCO	ssmo:Energy ≡ SEMANCO:Energy_Quantity_And_Emission
- Individual	SUMO	ssmo:Individual ≡ SUMO:Human
Location		
- Place	SUMO	ssmo:Place ≡ geo:SpatialThing
- Path	TO	ssmo:Path ≡ upper.owl#Pattern
Food consumption		
- Modification	FO	ssmo:Modification ≡ fo/Technique
Mobility		
- Mobility	SUMO	ssmo:MobilityActivity ≡ SUMO:Motion

Table 4. Overview of the new entities in the SSMO ontology.

	Ontology	Prefixed Class Name
Location		
- Location	SSMO	ssmo:Location
Dwelling		
- Activity	SSMO	ssmo:DwellingActivity
Food consumption		
- Activity	SSMO	ssmo:FoodConsumption
- Process	SSMO	ssmo:Process
- Tableware	SSMO	ssmo:Tableware
Leisure		
- Activity	SSMO	ssmo:LeisureActivity
- Artifact	SSMO	ssmo:Artifact

2.2. Data Processing Pipeline

The data processing pipeline, shown in Figure 4 is composed of four modules: *Data Collection*, *Data Enrichment*, *Classifier* and *Linked Data Publisher*.

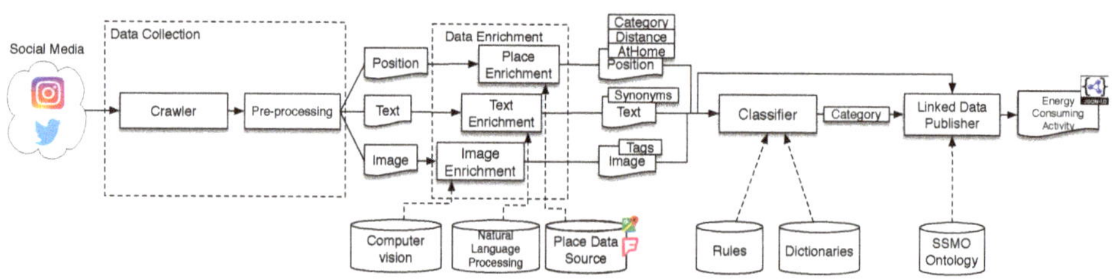

Figure 4. Overview of the data processing pipeline.

During the first stage, the data is collecting through the APIs of the selected data sources. Both data (image, and text data) and metadata (user, time, and place data) are collected.

In the second stage, different enrichment steps are performed. First, for each social media post, computer vision and natural language processing techniques are applied to respectively the image and text. For the images, we use both object and scene recognition models to extract information regarding the items present in the picture and the context where the photo was taken, while for the text we apply state-of-the-art processing methods and word disambiguation techniques. We enrich the information about the place by looking for its category on external data sources such as Foursquare and Google Places.

Using the enriched data, the social media post is classified to one or more of the energy-consuming activity categories using a hybrid rule and dictionary-based approach.

Finally, the publisher module combines the output of the other modules and publish the information about the energy-consuming activity as linked data (http://linkeddata.org/) conforming to the Social Smart Meter ontology.

2.2.1. Data Collection and Pre-Processing

The pipeline collect data from Twitter and Instagram. Those sources were chosen because these are widely used, and provide public APIs to retrieve the data (text, images, places, time, user) we are interested in.

Since a social media post is very noisy, contains slang, hashtags or mentions, we apply text pre-processing techniques (stopword removal, removal of hashtags and other special characters, stemming,) before the tokenization (word segmentation of the message). This results in a set of tokens that might refer to an energy-consuming activity. To perform this task, we use the Python-based Natural Language Toolkit (NLTK (https://www.nltk.org/)) module.

2.2.2. Data Enrichment

In this section, we describe the enrichment steps performed by our pipeline. Each step aims at extracting additional data from the text, image, and place of the social media post.

Text Enrichment

To overcome the ambiguity of words we use the Lesk algorithm [38] for word sense disambiguation. Assuming that words in a particular text section (i.e., a message in our case) are likely to share a common topic, it compares the definitions of each term in the section to determine the more likely sense of the word. In particular, we use the Adapted Lesk algorithm [39], implemented in the NLTK library, that incorporates WordNet (https://wordnet.princeton.edu/)'s lexical database. For each term in the social media post, this phase output its WordNet sense and the list of synonyms.

Image Enrichment

In this phase, state-of-the-art image processing techniques are applied to provide annotations on objects and scenes that are recognized in the images.

We include both object and scene recognition models, because they provide complementary information. For instance, the objects recognized in the example in Figure 5a (e.g., various tableware), may indicate food consumption activity. The scene recognition in Figure 5b on the other hand, recognize a cafeteria scenario, suggesting a leisure activity.

(a) (b)

Figure 5. Differences in computer vision techniques applied to the same images; (**a**) uses an object recognition method that person, dining table, cup (2x), knife (2x), bowl (5x), while (**b**) uses a scene recognition one extracting dining hall, cafeteria, and delicatessen annotations.

For the image object recognition, we use a state-of-the-art pre-trained model based on the regional convolutional neural network Mask R-CNN [40] trained on the Microsoft Common Objects in Context (MS COCO) dataset using the `mask_rcnn_coco.h5` weights (https://github.com/matterport/Mask_RCNN/releases).

For the scene recognition, we incorporated the neural network model based on the ResNet50 backbone (https://github.com/CSAILVision/places365), which is pre-trained on the Places (http://places2.csail.mit.edu/index.html) data set.

Place Enrichment

In this phase, we extract the category of the place where the post was published, because it could be an indicator for the category of the energy-consuming activity. We compute the distance from the previous post created by the user to infer how far he has traveled to understand if the post refers also to an energy-consuming activity related to mobility.

For the first case, we look to retrieve more information by matching the location of the social media post with the venues in Google Places and Foursquare. Numerous studies have investigated place matching; [41] found that the mean great circle distance between two matched Points of Interest (POIs) was equal to 62.8 m and in [42] a buffer area with a radius of 25 m (per POI) was used to reduce geocoding errors. Based on these values, we use a radius of 50 m. If a match is found, the corresponding place details are requested to collect one or more place categories.

Moreover, once we have an overview of all the places a user has checked in, we infer the user's home location by using spatial clustering. Then, we estimate the distances between the home and other location check-ins. To estimate the home, we use the density-based spatial clustering of applications with noise (DBSCAN, [43]). It separates high-density clusters from low-density ones and marks outlier points lying alone in low-density areas (whose nearest neighbors are too far away). We assume that the location of a user's home will be a relatively small-sized, high-density area, whereas at other places fewer check-ins take place, resulting in areas of low density.

2.2.3. Classification

We apply a hybrid dictionary and rule-based classification approach to determine whether a social media post refers to one or more energy-consuming activities.

We used a custom rule/dictionary-based approach instead of a state-of-the-art classifier for mainly two reasons: first, traditional classification approaches need a large set of manually annotated data for the training; to the best of our knowledge, such dataset does not exist, and its creation is beyond the scope of this work. In addition, second, while lacking generalization, a rule-based approach performs better in a narrow domain.

We define a dictionary as a set of terms related to a specific energy-consuming activity type—e.g., ingredients or cooking utensils are associated with the food consumption category. Thus, each category of energy-consuming activities has a distinct dictionary. The basic idea is to compare the terms extracted from the message (text tokens), image (annotations), and place (categories) to the terms in the dictionary. For now, a distinct dictionary for each of these types of data is constructed. Undoubtedly, this comes with some hassle but it also rules out ambiguity to some extent—e.g., the text token "tram" might infer a mobility activity whereas the image annotation "tram" could also point at some tram in the background which might not be related to the user's activity.

For the text dictionaries, we reuse the ones created in [44], where the authors use a hybrid dictionary-similarity distant supervision with the purpose of classifying Twitter content to energy consumption-related content. We further expand the dictionaries by adding the corresponding synonym.

The image dictionary is composed by the predefined list of classes of the pre-trained models. The classes are manually classified to none, one or more of the different categories of energy-consuming activities. For instance, "television" relates to both dwelling and leisure and is part of both dictionaries, whereas "person" does not indicate any energy-consuming activity and is thereby not included in any dictionary.

Alike the image annotations, the sets of place categories are also predefined. As all place categories that could possibly be assigned to a place are known, these can be categorized in the same manner as the image annotation classes, by manually linking the place category to the energy-consuming category. (e.g., a "restaurant" place category is part of both food consumption and leisure dictionaries.)

The dictionaries are available on the companion website (http://social-glass.tudelft.nl/social-smart-meter/#dictionary).

Then, the post is classified according to the rules illustrated in Figure 6. For each term, we identify if it is evidence (i.e., it appears in one of the dictionaries) for one or more energy-consuming activities. In case a leisure or food consumption activity is performed at home, we can classify it to dwelling as well. Furthermore, if a food consumption activity is performed at some place other than home, we classify it as a leisure activity.

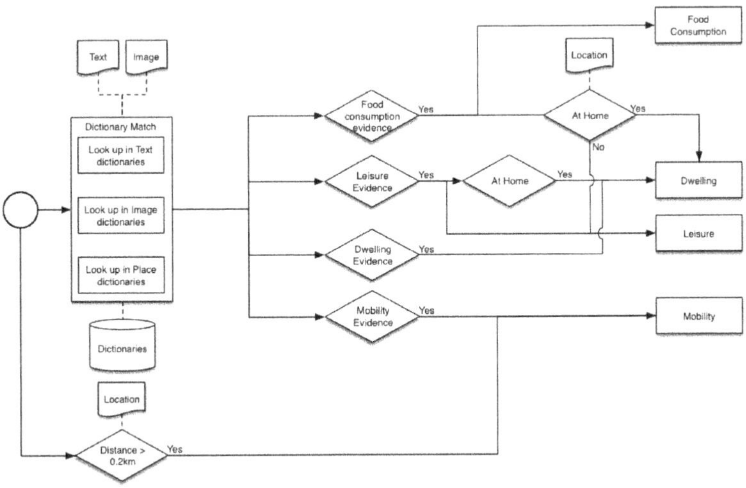

Figure 6. Illustration of the rule-based approach.

Then, we look at the user's distance to his or her previous post. If it exceeds the threshold of 0.2 km (This value was found after several test iterations of our pipeline. It seems to provide the best trade-off between precision and recall in our context), we consider it to be a mobility activity. Along with that, we analyze whether a vehicle was required to bridge this distance. If so, the mode of transport can be inferred—e.g., if the distance traveled in a day is more than 5000 km, it is very likely the individual traveled by aircraft to cover that distance.

Given the noisy nature of social media posts we tried to model the confidence of our classifier based on three parameters: (i) the ratio of relevant tokens, distinguished on type of data (text, image, place), (ii) for each term a score indicating its relevance to the category of energy-consuming activities, and (iii) a weighted factor that represents to what extent the type of data is informative for this category of energy-consuming activities. For instance, it is hard to recognize a mobility activity from an image, since individuals do not often post images of objects such as a transportation means while traveling. A check-in which is based on a mobility-related place such as an airport or train station would be far more indicative in that situation. On the contrary, if individuals perform a food consumption activity, they are more likely to post images in which food objects can be recognized.

Taking all the above into account, the calculation of our classification confidence is formulated as follow:

$$\begin{aligned}\text{confidence}_x &= \sum_y \left(\frac{N_{relevant,x,y}}{N_{relevant,y}} \cdot w_{x,y} \cdot \frac{1}{N_{relevant,x,y}} \sum_x \text{scores}_{x,y}\right) \\ &= \sum_y \left(\frac{1}{N_{relevant,y}} \cdot w_{x,y} \cdot \sum_x \text{scores}_{x,y}\right)\end{aligned} \tag{1}$$

where $N_{relevant}$ is the number of relevant terms, w is the weighted factor, x is the type of energy-consuming activity, y is the type of data (text, image, or place), and $\vec{scores}$ is the vector of the scores ($\in [0,1]$) of all relevant terms.

The relevance score of the terms ($scores_{x,y}$) are determined separately for each type of data. For a text token, the relevance is computed as the similarity between the term vectors and the word vectors included in the dictionaries obtained using Word2Vec [45] (a model used for learning vector representations of words, called "word embeddings"), whereas for an image annotation this is equal to the annotation score assigned by the object or scene recognition model.

For a place category, this score is binary (either 0 or 1), depending on whether the place category occurs in the dictionary.

To avoid possible bias due to our personal opinion, we decide to use an online survey to tune the weights ($w_{x,y}$). We showed social media posts and asked the participants to rank the data type according to their informativeness on a scale from 0 to 10 (*Not informative at all* to *Very Informative*). Figure A1a in Appendix B shows an example of question that was asked.

The users' average rankings are displayed in Table 5 and were adopted as data type weights in the classification module in the data processing pipeline for our case study. The weight values do not deviate a lot from each other. Yet, we observe that the users find images most and places least informative to describe dwelling activities. The same applies to food consumption activities.

Finally, the classifier confidence for a category x is the average of the contribution of each y data type. In future work, we will examine whether other strategies (such as taking the maximum of minimum instead of the average) provide in better results.

Table 5. The weighted factors obtained by asking the user opinions.

Category \ Data Type	Text	Image	Place
Dwelling	0.35	0.40	0.25
Food consumption	0.33	0.37	0.30
Leisure	0.35	0.32	0.33
Mobility	0.37	0.33	0.30

Hereafter, an initial threshold of 0.5 is applied to determine to which categories of energy-consuming activities the social media post is classified. This threshold value is then tuned to optimize the framework's performance.

2.2.4. Linked Data Publishing

In this final step, the label obtained by the classifier and the data extracted from the enrichment module are combined to create instances of the SSMO Ontology from the social media posts.

To do so we use Triplewave [46], an open-source, reusable and generic tool for publishing linked data streams on the web using the JSON-LD format.

Listing 1 shows an example of instance of SSMO ontology created by our pipeline. This instance was created by processing the social media post shown as example in Figure 1. Our pipeline determined that the post refers to three kind of activities (e.g., *ssmo:leisure activities*, *ssmo:food activity* and *ssmo:mobility activity*), they all take place in the venue (e.g., *ssmo:location*) of *Hotel de Godfazan*, and it involve the consumption of cooked *fish*.

Listing 1: Example of JSON-LD created with Triplewave.

```
{
"@context":{
"ssmo":"http://www.semanticweb.org/roosdekok/ontologies/2018/1/ssm",
"sioc":"http://rdfs.org/sioc/ns#",
"sem":"http://semanco02.hs-albsig.de/repository/ontology releases/eu/
semanco/ontology/SEMANCO/HEAD/SEMANCO-HEAD.owl",
"eu":"http://socsem.open.ac.uk/ontologies/eu#",
"to":"http://www.co-ode.org/roberts/travel.owl",
```

```
"foaf":"http://xmlns.com/foaf/0.1/"
},
"@id":"http://smm/i1",
"ssmo:individual":{
"@id":"http://instagram.com/userId",
"ssmo:nickname":"username"
},
"sioc:post":{
"@id":"http://instagram.com/postId",
"dcterms:created":"2018-06-24",
"sioc:content":"Great dinner at Hotel de Goudfazant in a old factory
on north side of Amsterdam...",
"sioc:hasCreator":"http://instagram.com/userId"
},
"ssmo:location":{
"ssmo:categoryOfPlace":"Restaurant",
"ssmo:address":"Aambeeldstraat 10, 1021 KB Amsterdam",
"ssmo:name":"Hotel de Godfazan",
"@id":"https://www.google.nl/maps/place/Hotel+De+Goudfazant/"
},
"ssmo:leisure activity":{
"@id":"http://ssm/lo1",
"ssmo:isOfferedAt":"https://www.google.nl/maps/place/Hotel+De+
Goudfazant/",
"ssmo:reflectedBy":"http://instagram.com/postId",
"ssmo:time":"2018-06-24"
},
"fo:food":{
"ssmo:isConsumedIn":"http://ssm/fo1",
"fo:ingridents":"fish"
},
"ssmo:food activity":{
"@id":"http://ssm/fo1",
"ssmo:isOfferedAt":"https://www.google.nl/maps/place/Hotel+De+
Goudfazant/",
"ssmo:reflectedBy":"http://instagram.com/postId",
"ssmo:time":"2018-06-24"
},
"ssmo:mobility activity":{
"@id":"http://ssm/mo1",
"ssmo:isOfferedAt":"https://www.google.nl/maps/place/Hotel+De+
Goudfazant/",
"ssmo:reflectedBy":"http://instagram.com/postId",
"ssmo:time":"2018-06-24"
}
}
```


By publishing the data as linked data we allow interoperability with other services by sharing a common understanding of the energy-consuming activities domain. In this way, others can define custom queries in a standard language (e.g., the SPARQL Protocol and RDF Query Language (https://www.w3.org/TR/rdf-sparql-query/)) and perform ad-hoc aggregations to satisfy their own research needs.

3. Evaluation

Since the behavior regarding creating social media posts might differ between cities with a different culture, for our evaluation we conducted a study on the cities of Amsterdam and Istanbul.

3.1. Dataset Collection

We collected data from 22 June until 27 June, and 27 July until 28 July 2018. At first, only social media posts created in Amsterdam were collected to provide the first round of insights and tuning of our pipeline. Hereafter, social media posts created in Istanbul were collected as well to compare the results between the two cities. An overview of the numbers of collected social media posts is provided in Table 6.

Table 6. Number of collected social media posts per day.

Date	Amsterdam		Istanbul	
	Instagram	*Twitter*	*Instagram*	*Twitter*
22/06/2018	16,099	3602	-	-
23/06/2018	15,794	3220	-	-
24/06/2018	16,365	2594	-	-
25/06/2018	15,426	3024	-	-
26/06/2018	14,985	3685	19,887	4476
27/06/2018	16,966	1929	28,346	8931
27/07/2018	17,854	1684	22,127	4818
28/07/2018	17,779	3656	21,082	11,522
Total	**131,268**	**23,394**	**91,442**	**29,747**

We observe that, in general, more social media posts are created in Istanbul than in Amsterdam. Given that Istanbul's population is more than 15 times as large as Amsterdam's population, this is expected. In both cities, Instagram yielded more posts than Twitter.

3.2. Performance Analysis

The performance of the framework was evaluated using the standard metrics of precision, recall, accuracy, and F1-score. Precision is the ratio between the posts classified correctly in one of the categories and all the classified posts, recall is the ratio between posts classified correctly in one of the categories and all the set of relevant posts. Accuracy is the fraction of posts correctly classified, taking into the account also the true negatives (i.e., the posts correctly not classified in any category). Finally, the F1-score is the harmonic average of the precision and recall.

The groundtruth was created through an online survey. We asked the participants to assess whether a social media post relates to an energy-consuming activity. We use a random sample of 100 social media posts and balanced the representation of each energy-consuming activity category. We collected 9 responses for each post and the final categories were decided with a majority vote.

Figure A1b in Appendix B shows an example of question asked in the survey.

Tables 7–9 summarize the evaluation metric values for each category of energy-consuming activities individually, as well as for the total. The evaluation metrics are calculated for different classification thresholds (from 0.3 to 0.7), to find the best-performing one. The framework's overall accuracy varies from 0.69 to 0.78. The accuracy for the classification of leisure activities is relatively low compared to the other categories due to many false negatives—i.e., social media posts that are not classified to leisure while, based on ground truth, they should be. Furthermore, the precision for dwelling activities is rather low whereas the accuracy is relatively high due to many true negatives—i.e., social media posts that (based on ground truth) do not refer to dwelling activities and are indeed not classified to this category by our classification model.

In Figure 7 the evaluation metric scores are plotted for the different threshold values. As expected, the recall scores decrease while increasing the threshold—i.e., decreasingly relevant social media posts have sufficient high confidence scores to exceed the threshold. As for the precision, we observe that the scores are fluctuating for different threshold values. Increasing the threshold results in less true positives, as well as less false positives. However, the numbers of true and false positives do not decrease proportionally. Also, there are very few social media posts with a high confidence score for dwelling. For a threshold greater than 0.4, the precision is zero for dwelling because no post was classified as such.

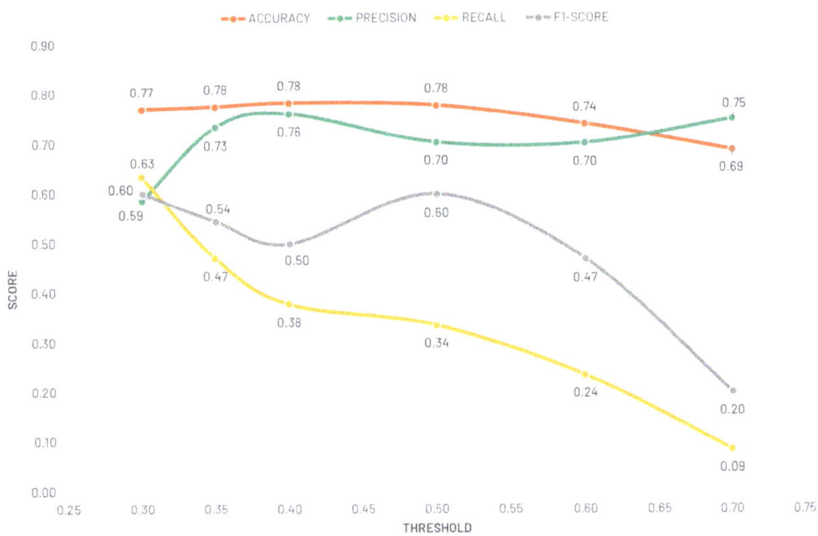

Figure 7. Evaluation metrics.

Table 7. Accuracy of the pipeline at different levels of threshold.

Metric / Category	Accuracy					
Threshold	*0.30*	*0.35*	*0.40*	*0.50*	*0.60*	*0.70*
Dwelling	0.85	0.87	0.89	0.90	0.90	0.91
Food consumption	0.82	0.84	0.86	0.85	0.78	0.73
Leisure	0.60	0.57	0.56	0.54	0.48	0.37
Mobility	0.81	0.82	0.82	0.82	0.80	0.74
Total	**0.77**	**0.78**	**0.78**	**0.78**	**0.74**	**0.69**

Table 8. Precision and recall values for each energy-consuming activities at varying values of threshold. The values of the precision and recall for the Dwelling category for threshold greater than 0.4 are 0 because no posts were classified in that category.

	Precision						Recall					
Threshold	*0.30*	*0.35*	*0.40*	*0.50*	*0.60*	*0.70*	*0.30*	*0.35*	*0.40*	*0.50*	*0.60*	*0.70*
Dwelling	0.23	0.27	0.20	0.00	0.00	0.00	0.38	0.38	0.13	0.00	0.00	0.00
Food	0.68	0.79	0.95	0.95	0.92	1.00	0.81	0.69	0.59	0.56	0.34	0.16
Leisure	0.80	0.88	0.89	0.87	0.89	1.00	0.61	0.49	0.46	0.45	0.34	0.15
Mobility	0.63	1.00	1.00	1.00	1.00	1.00	0.74	0.33	0.33	0.33	0.26	0.04
Overall	**0.59**	**0.73**	**0.76**	**0.70**	**0.70**	**0.75**	**0.63**	**0.47**	**0.38**	**0.34**	**0.24**	**0.09**

Based on Figure 7 a threshold of either 0.30 or 0.35 appears to result in the best performance. For a threshold of 0.30, a precision of 0.59 is obtained whereas a threshold of 0.35 results in a precision of 0.73. Furthermore, these thresholds (0.30 and 0.35) respectively result in recall scores of 0.63 and 0.47 and in F1-scores of 0.60 and 0.54. Based on the F1-score, a threshold of 0.30 seems to be better performing. Yet, it is dependent on the context whether it is more important to have a higher precision or recall score—i.e., whether it is more important to classify as many social media posts as possible correctly or to discover as many as possible that are referring to energy-consuming activities. In case the quantity of energy (in terms of kWh consumption or CO_2 emission) during an activity is analyzed, a higher precision is considered more beneficial. However, when a qualitative overview of all energy-consuming activities performed by an individual is required, it is more advantageous to have a higher recall score. For our case study, a threshold of 0.35 was selected.

Table 9. The F1-score value for each energy-consuming activity category at varying level of threshold. The values for the Dwelling category for threshold greater than 0.4 are undefined because no posts were classified in that category.

Category \ Metric	F1-Score					
Threshold	*0.30*	*0.35*	*0.40*	*0.50*	*0.60*	*0.70*
Dwelling	0.29	0.32	0.15	-	-	-
Food consumption	0.74	0.73	0.73	0.70	0.50	0.27
Leisure	0.69	0.63	0.61	0.33	0.49	0.26
Mobility	0.68	0.50	0.50	0.50	0.41	0.07
Overall	**0.60**	**0.54**	**0.50**	**0.60**	**0.47**	**0.20**

3.3. Use Case

In this section, we give a deeper look to the posts that were classified in any of the four energy-consuming activities.

We collected the posts regardless of the language. In the analysis, for Amsterdam we consider the terms in English and Dutch, while for Istanbul we consider the terms in English and Turkish. Notice that the terms in different languages are needed only for the textual part of the social media posts, and not for the image labels and place categories.

For the text processing we used three pre-trained embeddings: for the English language we use the model trained on the Google News corpus (https://github.com/mmihaltz/word2vec-GoogleNews-vectors), for Dutch we use a model trained on the combined dataset of Wikipedia (https://dumps.wikimedia.org/nlwiki/20150703), Sonar500 (http://hdl.handle.net/2066/151880) and Roularta corpus (a set of articles form the publishing consortium http://www.roularta.be/en) [47], while for the Turkish language we use a model trained on the Turkish Wikipedia dataset (https://github.com/akoksal/Turkish-Word2Vec).

Table 10 shows the percentage of each category of energy-consuming activities for both cities. In general, we observe that few social media posts are classified to dwelling. Our rule-based classification approach demands evidence for the user being at home before it classifies a post to dwelling. It is very difficult to derive this evidence from the social media post because rarely people check-in at their own home.

Table 10. Percentage of classified social media posts per category of energy-consuming activity.

Category	Amsterdam	Istanbul
Dwelling	3.25% (1326)	4.18% (589)
Food consumption	20.36% (8312)	21.99% (3100)
Leisure	44.75% (18,274)	41.49% (5850)
Mobility	31.64% (12,921)	32.35% (4561)
Total	**100% (40,833)**	**100% (14,100)**

For both Amsterdam and Istanbul, the leisure category has the largest share (approximately 40%) compared to the other categories. The mobility category has the second largest share (approximately 30%). The category of food consumption has a rather small share (approximately 20%). However, nearly all social media posts that are classified to food consumption are also classified to leisure based on the rule-based approach—a food consumption activity that is performed at some other place than home is also considered a leisure activity. This explains why the share of the leisure category is more than twice as large as the share of the food consumption category.

The distribution of social media posts classified to energy-consuming activities cities differs between them. For Amsterdam (Figure 8a), most social media posts are created around the city center—the neighborhood with the highest density (Burgwallen-Nieuwe Zijde) also include the city center. For Istanbul (Figure 8b), multiple neighborhoods share a high amount of energy-consuming activities; Başakşehir and Beşiktaş on the European part of the city and Kadıköy on the Asian part.

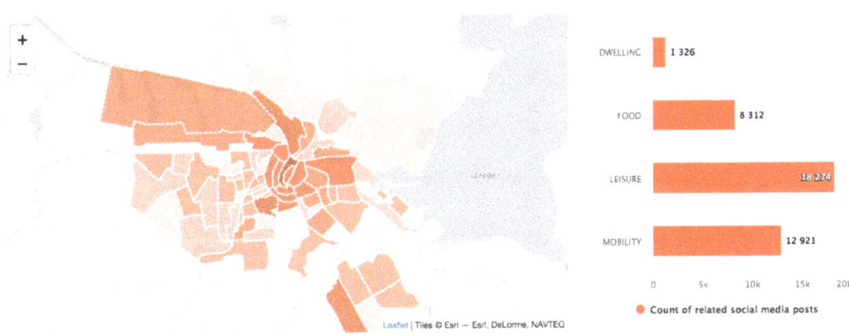

(**a**)

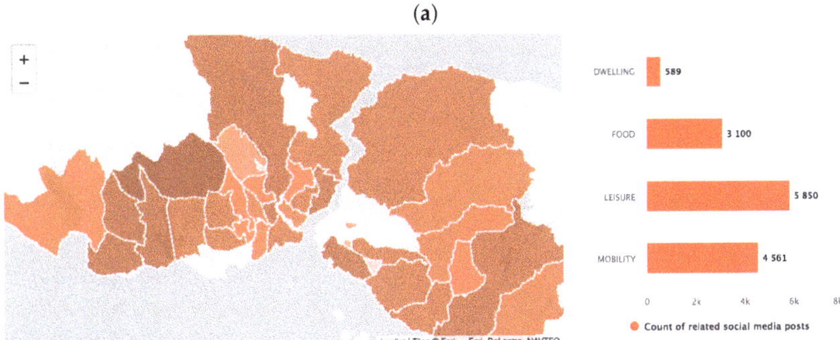

(**b**)

Figure 8. Overall distribution of energy-consuming activities of Amsterdam (**a**) and Istanbul (**b**).

3.3.1. Dwelling

For both cities, few social media posts are classified to dwelling. For Amsterdam (Figure 9a), the posts in this category were mainly created in the city center while in Istanbul (Figure 9b), the posts are more evenly distributed with a higher concentration in the European part of the city (especially in the Başakşehir district).

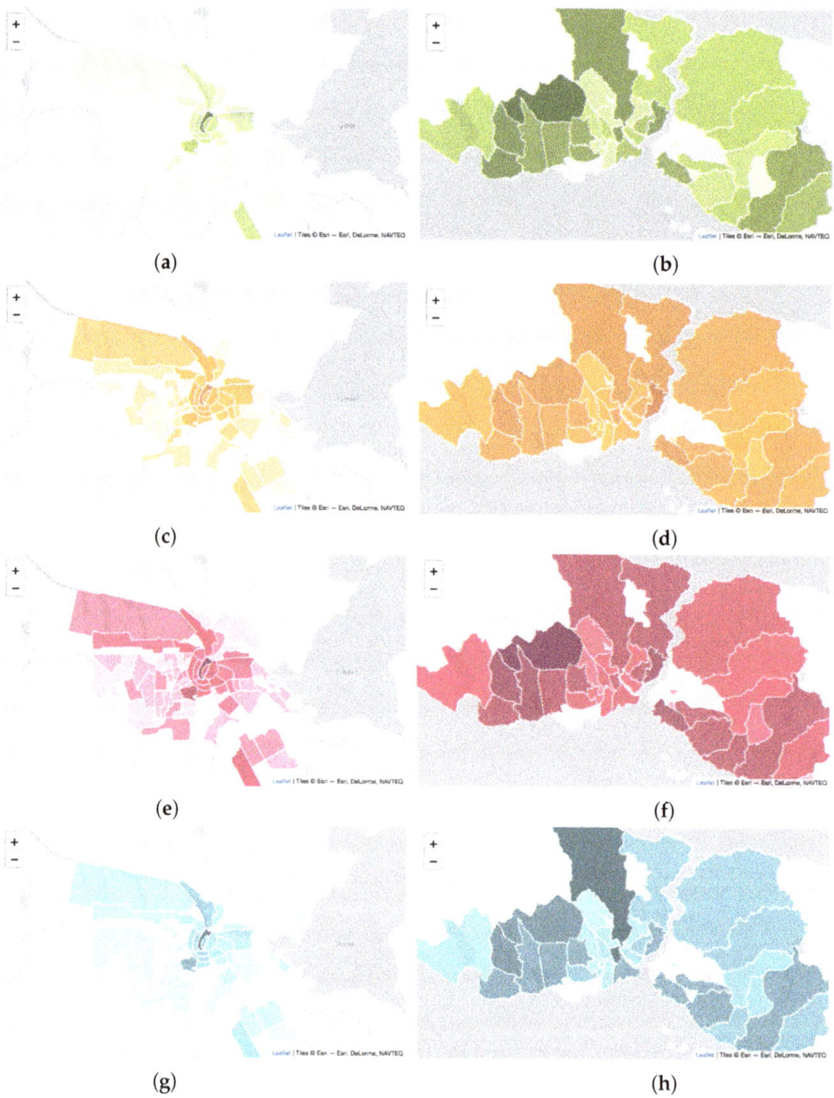

Figure 9. Map visualizing the distribution of social media posts; (**a**,**b**) refer to dwelling, (**c**,**d**) refer to food consumption, (**e**,**f**) refer to leisure and (**g**,**h**) refer to mobility.

As shown in Figure 10, the text terms that are most informative for a dwelling activity in Amsterdam are "House", "TV", and "gaming". In images, "tv", "laptop", and "keyboard" are the most frequently recognized objects that indicate a dwelling activity for both cities. These seem to indicate either recreational or work activities.

There are no place terms related to this type of activity because houses do not have a category in the sources used in the data enrichment phase.

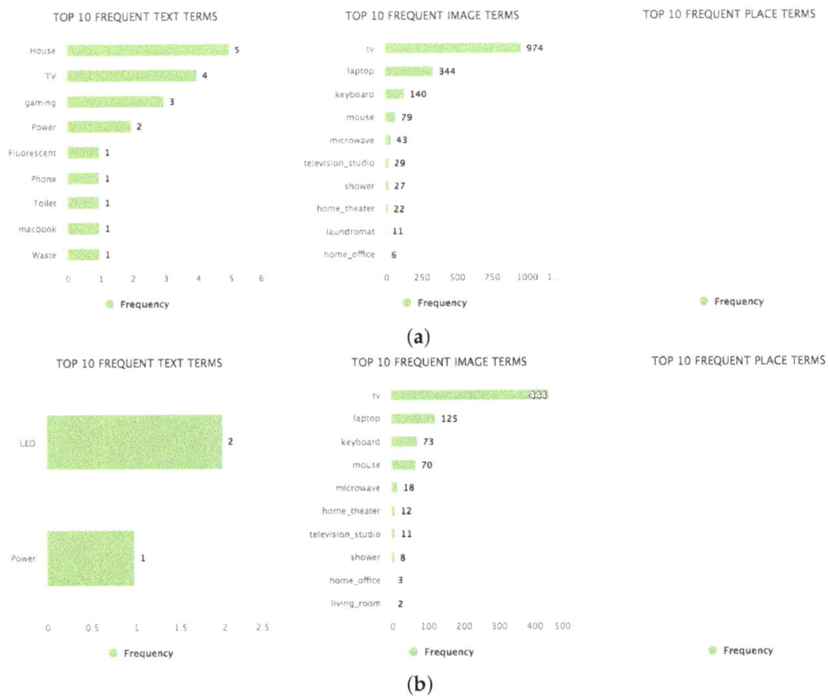

Figure 10. Bar charts visualizing the most occurring terms in social media posts classified to dwelling activities in Amsterdam (**a**) and Istanbul (**b**). For readability purposes, in the figures we show only English terms.

3.3.2. Food Consumption

As shown in Figure 9c, the city of Amsterdam shows the highest concentration of food energy-consuming activities in the city center. On the other hand, Istanbul, as shown in Figure 9d, shows peaks in the Beşiktaş district and in the northern neighborhoods.

Based on the top frequent terms in Figure 11a,b, images seem to be most informative to identify food consumption activities. Furthermore, "food" and "coffee" were the top frequent text terms indicating a food consumption activity in both cities. Besides that, individuals appear to create food consumption-related post most often while checking in at a "Bar" (Amsterdam), "Cafe" (both cities) or "Restaurant" (both cities).

3.3.3. Leisure

In Figure 9e the distribution of social media posts in Amsterdam classified to leisure activities seems to be more distributed over the different neighborhoods. When zooming in on a few neighborhoods (Burgwallen-Nieuwe Zijde, Museumkwartier, and Amstel III/Bullewijk) some interesting observations are made.

In general, the city center (Burgwallen-Nieuwe Zijde) is characterized by many tourists, who are partying, visiting the flower markets, going to museums, or enjoying the canals, among other things. This is reflected in the top frequent text terms: "night", "holiday", "party" (text), "Flower Shop", "Art Museum", and "Hotel" (place) are some terms that comply with these activities.

Museumkwartier is the neighborhood where many of Amsterdam's most famous museums are situated. In fact, we find that the top occurring terms are related to these museums: "museum" (text), "art_gallery" and "museum/indoor" (image), and "Art Museum" (place).

Amstel III/Bullewijk is known for Amsterdam's soccer stadium and the major concert halls. As expected, the top occurring terms are: "concert" and "music" (text), "arena/performance" and "stage/indoor" (image), and "Concert Hall" and "Soccer Stadium" (place).

The distribution of the leisure-related social media posts over Istanbul's neighborhoods (Figure 9f) is rather similar to the food consumption-related one: most dense in the center and west of it (the Başakşehir district, where also the stadium of the homonymous soccer team is present). Interestingly, as shown in Figure 12, it seems that in Istanbul the majority of leisure activities take place in shopping malls.

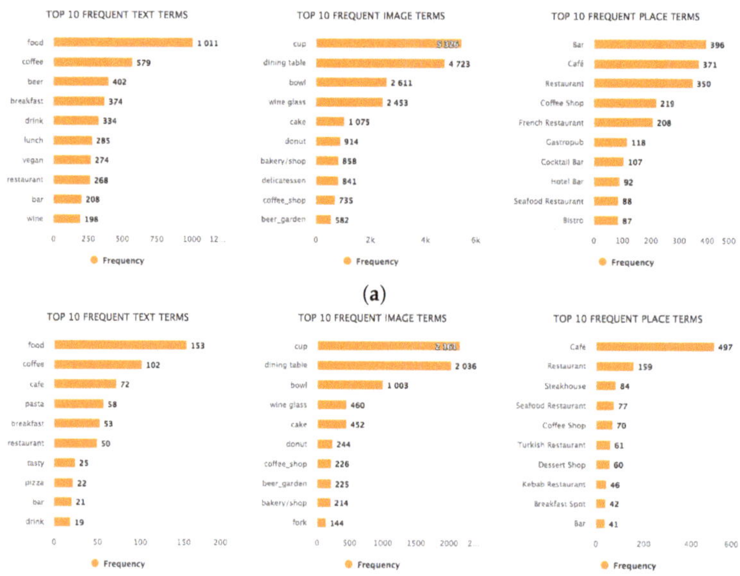

(a)

(b)

Figure 11. Bar charts visualizing the most occurring terms in social media posts classified to food consumption activities in Amsterdam (**a**) and Istanbul (**b**). For readability purposes, in the figures we show only English terms.

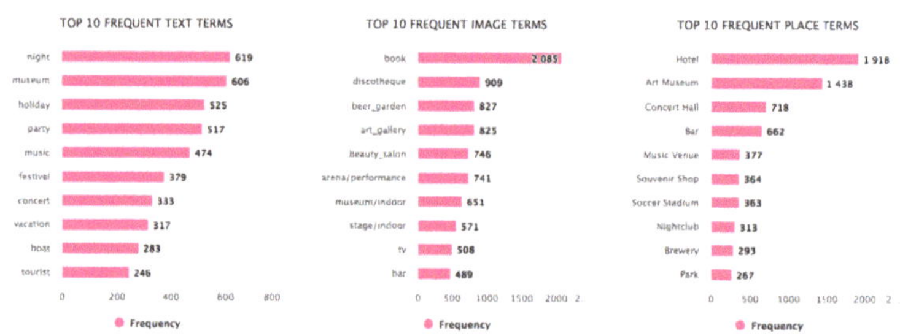

(a)

Figure 12. *Cont.*

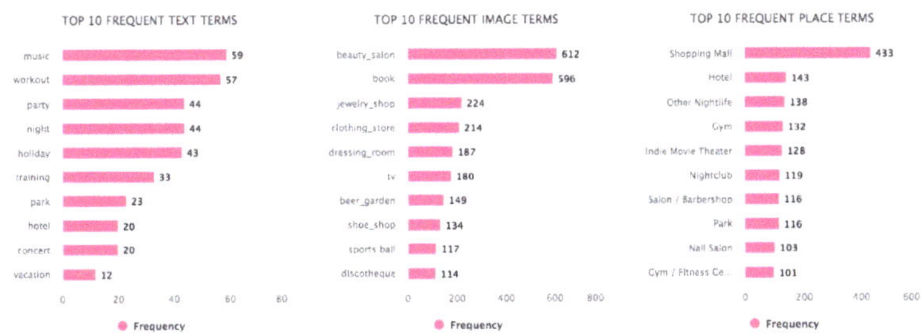

(b)

Figure 12. Bar charts visualizing the most occurring terms in social media posts classified to leisure activities in Amsterdam (**a**) and Istanbul (**b**). For readability purposes, in the figures we show only English terms.

3.3.4. Mobility

Since Amsterdam's train station is situated in the city center, it makes sense that this neighborhood is most dense regarding the count of social media posts classified to mobility (Figure 9g). This is also due to the canal trips in the city center that individuals (mainly tourists) tend to post about.

In Figure 9h two of the western neighborhoods (Başakşehir and Eyüp) are the densest regarding mobility activities. Multiple highways run through these neighborhoods (and particularly Eyüp connects the Black Sea to the Golden Horn) as well as a large highway junction. If we look at the terms (Figure 13), we can notice that in Istanbul are present more term related to transportation by car (e.g., Gas Station, Car Wash, parking_lot, car, etc.).

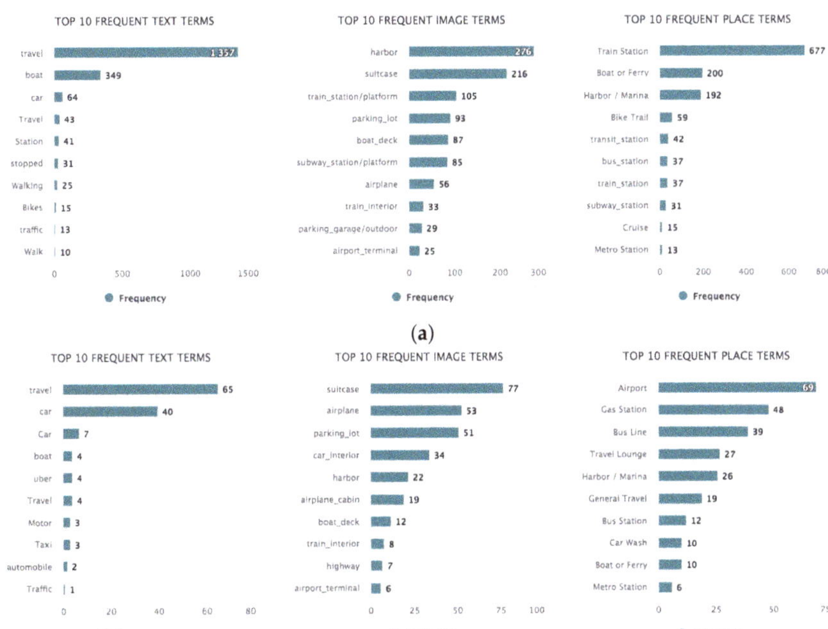

(b)

Figure 13. Bar charts visualizing the most occurring term in social media posts classified to mobility activities in Amsterdam (**a**) and Istanbul (**b**). For readability purposes, in the figures we show only English terms.

If we compare the frequencies of displacements of both cities (Figure 14) we can observe that while in Amsterdam people tend to travel for short distances (between 1 and 5 km), in Istanbul the chart shows a long tail distribution. Since Istanbul is significantly larger in size than Amsterdam, this is in line with our expectations.

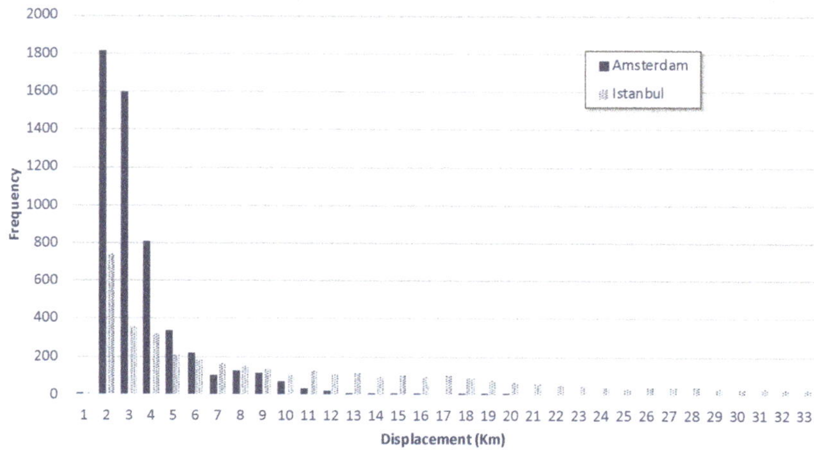

Figure 14. Bar chart visualizing the frequency of displacements (average distance between posts in kilometers) in Amsterdam and Istanbul.

3.3.5. Discussion

In both cities, few social media posts referring to dwelling activities were captured by the framework. This may be because social media users do not consider their regular domestic activities interesting enough to be shared with other social media users.

More posts related to food consumption were captured, but, by looking at the most occurring terms, they seem to occur out of home.

Then, as expected by the typical usage of social media, we detected many posts related to leisure energy-consuming activities. Moreover, they seem to reflect the types of venue present in a particular district, for instance, in the Museumkwartier neighborhood in Amsterdam, we identified many social media posts referring to museums and art.

Finally, people do not create explicit social media content about their mobility activities. When they are traveling, they are more likely to create content about the activities they performed before. However, we can use the distance between posts to detect if a transportation activity occurred.

Even if the two cities present the same ratio of energy-consuming activities, they show a different geographical distribution; while in Amsterdam the activities are localized near the city center and in Amstel III/Bullewijk (where the soccer stadium and the major concert halls are present), in Istanbul the activities are distributed in different neighborhoods, mainly Başakşehir, Beşiktaş, and Kadıköy. Probably, this is due to the different features of the two cities: Amsterdam has a well-defined center, where the main venues are localized; while in Istanbul, also given the different size, have them scattered in various parts of the city.

By looking at the most occurring terms, we notice a small difference between the characterization of the energy-consuming activities in the two cities. In the food category, we can see place categories more related to the Turkish cuisine (e.g., Turkish restaurant and kebab restaurant), and many leisure activities in Istanbul seems to take place in shopping malls. Finally, for the mobility category, in Istanbul, we notice a higher occurrence of terms related to transportation by car.

Summarizing, our pipeline can detect more activities that fall in the broad category of indirect energy-consuming activities, that are, as mentioned in Section 1, activities related to the production, transportation, and disposal of a variety of consumer goods and services [12]. As expected from the typical usage of social media, people post on social media when they are partying, having a fancy dinner out; more rarely they share their domestic activities. Nevertheless, this should not be seen as a flaw of our approach, but it should suggest that indeed social media can be used as a **complementary** source of information regarding energy-consuming activities. In fact, domestic activities are already partially captured by traditional data sources, while the indirect ones are either neglected [11] or the methods used for collecting them have low temporal resolution and are costly (e.g., surveys).

Moreover, our coverage of activity types can be improved by including additional data sources, for instance, the Steam (https://steamcommunity.com/) community for games or the Spotify (https://www.spotify.com/nl/) music stream provider, are more likely to be used for sharing data on dwelling activities, such as gaming or playing music.

3.3.6. Limitations

We acknowledge our approach is not free from limitations. Social media are inherently biased: they are used by only a set of the population (e.g., youths, tourists, etc.) and for purposes different from sharing energy-consuming activities. Moreover, the information shared on social media it is often ambiguous and noisy (e.g., a picture of a tram does not mean that the user is traveling). The issue of ambiguity and noise is partially mitigated by our rule-based approach, which shows promising performance. However, the goal of this work is to investigate to what extent social media can be used as a complementary source of information for energy-consuming activities. A study of demographic representation is left to future work. Language can be an issue when applying our method in areas where English is not the native language. However, this is addressed with multi-language dictionaries and by the use of embeddings trained on the main language spoken in the considered area (e.g., Dutch for Amsterdam). In addition, this issue only concerns the analysis of the text of the social media post, and not the image or the location.

4. Conclusions

In this paper, we proposed a framework to automatically identify and describe energy-consuming activities from social media posts. This framework is composed by an ontology that provides a better understanding of the domain of energy-consuming activities and a data processing pipeline that classify social media posts to the different categories.

Future works will focus on the improvement of the enrichment module of the framework. For instance, entity extraction can be employed to understand whether a word refers to a place (instead of only taking the place check-in into account) to increase the number of geolocated posts processed by the pipeline.

Moreover, our rule-based approach could be used to generate large training sets for a classifier in a distant-supervision fashion.

As mentioned in the previous section, other data sources will be investigated to increase the coverage of types of energy-consuming activity, with a focus on dwelling.

A further validation will be performed by looking at correspondence with more traditional sources (e.g., surveys, smart meter data etc.).

We will also investigate methods to link the information extracted from the social media post to concrete values of energy consumption (in terms of e.g., kWh or CO_2 emissions).

Author Contributions: R.d.K. carried out the design of the framework, the evaluation, and the writing. A.M. helped with the design and contributed to the writing of the article. A.B. supervised all the steps of this work and revised the text.

Funding: This work is partially funded by the JPI Urban Europe Project CODALoop (Project no. 646453).

Acknowledgments: This work is supported the Amsterdam Institute for Advanced Metropolitan Solutions (AMS Institute) and it was carried out on the Dutch national e-infrastructure with the support of SURF Cooperative.

Conflicts of Interest: The authors declare no conflict of interest

Appendix A. Ontology Requirements

Table A1. Competency Questions that form the set of functional requirements for the SSMO ontology.

#	Competency Question (CQ)
1	Does the individual perform an energy-consuming activity?
2	If so, what type (or category) of energy-consuming activity is performed by the individual?
3	At what place is the activity performed by the individual? (i) *To what type (or category) does this place belong?* (ii) *What are the (sets of) coordinates of this place?*
4	At what time is the activity performed by the individual?
5	What is the duration of the activity?
6	Does the individual use an object to perform this activity? (i) *If so, what kind of object?*
7	In case a mobility activity is performed, what kind of mode of transport is used? (i) *What path (composed of different places, among which are the origin and destination) was taken?*
8	In case a leisure activity is performed, what kind of artifact(s) is (are) used? (i) *In case the artifact is an appliance, what is its power?*
9	In case a dwelling activity is performed, what kind of appliance(s) is (are) used? (i) *What is the power of this appliance?*
10	In case of a food consumption activity, what kind of food is consumed? (i) *What ingredients are included in this food?* (ii) *How (= through which process) is this food processed?* (iii) *Does this process require an appliance? If so, what kind of appliance?* (iv) *Where (= at what place) is this food processed?*
11	How many energy-consuming activities are performed at a certain (aggregation of) place(s) during a certain time span?

Appendix B. User Online Survey

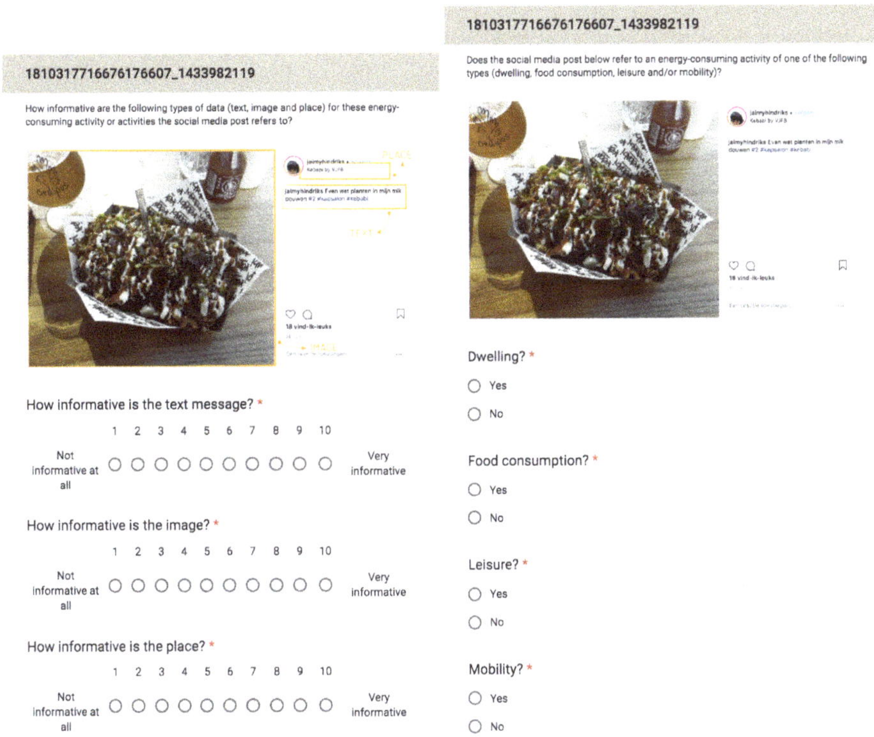

(a) (b)

Figure A1. Example of question for tuning the weights (**a**) and creating the groundtruth (**b**).

References

1. Fraternali, P.; Herrera, S.; Novak, J.; Melenhorst, M.; Tzovaras, D.; Krinidis, S.; Rizzoli, A.E.; Rottondi, C.; Cellina, F. enCOMPASS—An integrative approach to behavioural change for energy saving. In Proceedings of the Global Internet of Things Summit (GIoTS), Geneva, Switzerland, 6–9 June 2017; pp. 1–6.
2. Fischer, C. Feedback on household electricity consumption: A tool for saving energy? *Energy Effic.* **2008**, *1*, 79–104. [CrossRef]
3. Kamilaris, A.; Pitsillides, A.; Fidas, C. Social Electricity: A case study on users perceptions in using green ICT social applications. *Int. J. Environ. Sustain. Dev.* **2016**, *15*, 67–88. [CrossRef]
4. Albertarelli, S.; Fraternali, P.; Herrera, S.; Melenhorst, M.; Novak, J.; Pasini, C.; Andrea-Emilio, A.E.; Rottondi, C. A Survey on the Design of Gamified Systems for Energy and Water Sustainability. *Games* **2018**, *9*, 38. [CrossRef]
5. Froehlich, J.; Larson, E.; Gupta, S.; Cohn, G.; Reynolds, M.; Patel, S. Disaggregated end-use energy sensing for the smart grid. *IEEE Pervasive Comput.* **2011**, *10*, 28–39. [CrossRef]
6. Parsa, A.; Najafabadi, T.A.; Salmasi, F.R. Implementation of smart optimal and automatic control of electrical home appliances (IoT). In Proceedings of the Smart Grid Conference (SGC), Tehran, Iran, 20–21 December 2017; pp. 1–6.
7. Weiss, M.; Helfenstein, A.; Mattern, F.; Staake, T. Leveraging smart meter data to recognize home appliances. In Proceedings of the 2012 IEEE International Conference on Pervasive Computing and Communications (PerCom), Lugano, Switzerland, 19–23 March 2012; pp. 190–197.

8. Bleys, B.; Defloor, B.; Van Ootegem, L.; Verhofstadt, E. The Environmental Impact of Individual Behavior: Self-Assessment Versus the Ecological Footprint. *Environ. Behav.* **2017**, *50*, 187–212. [CrossRef]
9. Torriti, J. Understanding the timing of energy demand through time use data: Time of the day dependence of social practices. *Energy Res. Soc. Sci.* **2017**, *25*, 37–47. [CrossRef]
10. Vassileva, I.; Wallin, F.; Dahlquist, E. Understanding energy consumption behavior for future demand response strategy development. *Energy* **2012**, *46*, 94–100. [CrossRef]
11. Burger, P.; Bezençon, V.; Bornemann, B.; Brosch, T.; Carabias-Hütter, V.; Farsi, M.; Hille, S.L.; Moser, C.; Ramseier, C.; Samuel, R.; et al. Advances in understanding energy consumption behavior and the governance of its change–outline of an integrated framework. *Front. Energy Res.* **2015**, *3*, 29. [CrossRef]
12. Abrahamse, W.; Steg, L.; Vlek, C.; Rothengatter, T. The effect of tailored information, goal setting, and tailored feedback on household energy use, energy-related behaviors, and behavioral antecedents. *J. Environ. Psychol.* **2007**, *27*, 265–276. [CrossRef]
13. Beber, M.A.; Ferrero, C.A.; Fileto, R.; Bogorny, V. Individual and Group Activity Recognition in Moving Object Trajectories. *J. Inf. Data Manag.* **2017**, *8*, 50.
14. Zhu, Z.; Blanke, U.; Tröster, G. Recognizing composite daily activities from crowd-labelled social media data. *Pervasive Mob. Comput.* **2016**, *26*, 103–120. [CrossRef]
15. Bodnar, T.; Dering, M.L.; Tucker, C.; Hopkinson, K.M. Using large-scale social media networks as a scalable sensing system for modeling real-time energy use patterns. *IEEE Trans. Syst. Man Cybern. Syst.* **2017**, *47*, 2627–2640. [CrossRef]
16. Rashidi, T.H.; Abbasi, A.; Maghrebi, M.; Hasan, S.; Waller, T.S. Exploring the capacity of social media data for modelling travel behaviour: Opportunities and challenges. *Transp. Res. Part C Emerg. Technol.* **2017**, *75*, 197–211. [CrossRef]
17. Zhang, Z.; He, Q.; Zhu, S. Potentials of using social media to infer the longitudinal travel behavior: A sequential model-based clustering method. *Transp. Res. Part C Emerg. Technol.* **2017**, *85*, 396–414. [CrossRef]
18. Psyllidis, A.; Bozzon, A.; Bocconi, S.; Bolivar, C.T. A Platform for Urban Analytics and Semantic Data Integration in City Planning. In Proceedings of the 16th International International Conference on Computer-Aided Architectural Design Futures, Sao Paulo, Brazil, 8–10 July 2015; pp. 21–36.
19. Abbar, S.; Mejova, Y.; Weber, I. You tweet what you eat: Studying food consumption through twitter. In Proceedings of the 33rd Annual ACM Conference on Human Factors in Computing Systems, Seoul, Korea, 18–23 April 2015; pp. 3197–3206.
20. Akbari Fard, M.; Hadadi, H.; Tavakoli Targhi, A. Fruits and vegetables calorie counter using convolutional neural networks. In Proceedings of the 6th International Conference on Digital Health Conference, Montréal, QC, Canada, 11–13 April 2016; pp. 121–122.
21. Fried, D.; Surdeanu, M.; Kobourov, S.; Hingle, M.; Bell, D. Analyzing the language of food on social media. In Proceedings of the 2014 IEEE International Conference on Big Data (Big Data), Washington, DC, USA, 27–30 October 2014; pp. 778–783.
22. Backhaus, J.; Breukers, S.; Paukovic, M.; Mourik, R.; Mont, O. *Sustainable Lifestyles. Today's Facts and Tomorrow's Trends. D1. 1 Sustainable Lifestyles Baseline Report*; ECN Policy Studies, Energy research Centre of the Netherlands ECN: Amsterdam, The Netherlands, 2012.
23. Guinée, J.; Heijungs, R.; De Koning, A.; Van, L.; Geerken, T.; Van Holderbeke, M.; Vito, B.J.; Eder, P.; Delgado, L. Environmental Impact of Products (EIPRO) Analysis of the Life Cycle Environmental Impacts Related to the Final Consumption of the EU25. Available online: http://hdl.handle.net/1887/11434 (accessed on 20 December 2018).
24. Mont, O. Concept Paper for the Task Force on Sustainable Lifestyles. In Proceedings of the Expert Meeting on Sustainable Consumption and Production (Technical Report), Stockholm, Sweden, 26–29 June 2007; pp. 1–14.
25. Chandrasekaran, B.; Josephson, J.R.; Benjamins, V.R. What are ontologies, and why do we need them? *IEEE Intell. Syst. Appl.* **1999**, *14*, 20–26. [CrossRef]
26. Fernández-López, M.; Gómez-Pérez, A.; Juristo, N. Methontology: From ontological art towards ontological engineering. In Proceedings of the Ontological Engineering AAAI-97 Spring Symposium Series, Stanford, CA, USA, 24–26 March 1997.

27. Suárez-Figueroa, M.C.; Gómez-Pérez, A.; Villazón-Terrazas, B. How to write and use the ontology requirements specification document. In *OTM Confederated International Conferences "On the Move to Meaningful Internet Systems"*; Springer: Berlin/Heidelberg, Germany, 2009; pp. 966–982.
28. Bonino, D.; Corno, F.; De Russis, L. Poweront: An ontology-based approach for power consumption estimation in smart homes. In *Internet of Things. User-Centric IoT*; Springer: Cham, Switzerland, 2015; pp. 3–8.
29. Breslin, J.G.; Harth, A.; Bojars, U.; Decker, S. Towards semantically interlinked online communities. In *European Semantic Web Conference*; Springer: Berlin/Heidelberg, Germany, 2005; pp. 500–514.
30. Golbeck, J.; Rothstein, M. *Linking Social Networks on the Web with FOAF: A Semantic Web Case Study*; AAAI: Menlo Park, CA, USA, 2008; Volume 8, pp. 1138–1143.
31. Niles, I.; Pease, A. Towards a standard upper ontology. In Proceedings of the International Conference on Formal Ontology in Information Systems-Volume 2001, Ogunquit, ME, USA, 17–19 October 2001; pp. 2–9.
32. Madrazo, L.; Sicilia, A.; Gamboa, G. SEMANCO: Semantic tools for carbon reduction in urban planning. In Proceedings of the 9th European Conference on Product and Process Modelling, Reykjavik, Iceland, 23 July 2012.
33. Burel, G.; Piccolo, L.S.; Alani, H. Energyuse-a collective semantic platform for monitoring and discussing energy consumption. In *International Semantic Web Conference*; Springer: Cham, Switzerland, 2016; pp. 257–272.
34. Bonino, D.; Corno, F. Dogont-ontology modeling for intelligent domotic environments. In *International Semantic Web Conference*; Springer: Berlin/Heidelberg, Germany, 2008; pp. 790–803.
35. BBC Food Ontology. Available online: https://www.bbc.co.uk/ontologies/fo (accessed on 23 October 2018).
36. Travel Ontology. Available online: http://www.cs.man.ac.uk/~stevensr/ontology/c23.owl (accessed on 23 October 2018).
37. Bechhofer, S. OWL: Web ontology language. In *Encyclopedia of Database Systems*; Springer: New York, NY, USA, 2009; pp. 2008–2009
38. Lesk, M. Automatic sense disambiguation using machine readable dictionaries: How to tell a pine cone from an ice cream cone. In Proceedings of the 5th Annual International Conference on Systems Documentation, Toronto, ON, Canada, 8–11 June 1986; pp. 24–26.
39. Banerjee, S.; Pedersen, T. An adapted Lesk algorithm for word sense disambiguation using WordNet. In *International Conference on Intelligent Text Processing and Computational Linguistics*; Springer: Berlin/Heidelberg, Germany, 2002; pp. 136–145.
40. He, K.; Gkioxari, G.; Dollár, P.; Girshick, R. Mask r-cnn. In Proceedings of the 2017 IEEE International Conference on Computer Vision (ICCV), Venice, Italy, 22–29 October 2017; pp. 2980–2988.
41. McKenzie, G.; Janowicz, K.; Adams, B. A weighted multi-attribute method for matching user-generated points of interest. *Cartogr. Geogr. Inf. Sci.* **2014**, *41*, 125–137. [CrossRef]
42. Jiang, S.; Alves, A.; Rodrigues, F.; Ferreira, J., Jr.; Pereira, F.C. Mining point-of-interest data from social networks for urban land use classification and disaggregation. *Comput. Environ. Urban Syst.* **2015**, *53*, 36–46. [CrossRef]
43. Ester, M.; Kriegel, H.P.; Sander, J.; Xu, X. A density-based algorithm for discovering clusters in large spatial databases with noise. *Kdd* **1996**, *96*, 226–231.
44. Mauri, A.; Psyllidis, A.; Bozzon, A. Social Smart Meter: Identifying Energy Consumption Behavior in User-Generated Content. In Proceedings of the Companion of the The Web Conference 2018 on The Web Conference 2018. International World Wide Web Conferences Steering Committee, Lyon, France, 23–27 April 2018; pp. 195–198.
45. Mikolov, T.; Sutskever, I.; Chen, K.; Corrado, G.S.; Dean, J. Distributed representations of words and phrases and their compositionality. In *Advances in Neural Information Processing Systems*; MIT Press Ltd.: Cambridge, MA, USA, 2013; pp. 3111–3119.

46. Mauri, A.; Calbimonte, J.P.; Dell'Aglio, D.; Balduini, M.; Brambilla, M.; Della Valle, E.; Aberer, K. TripleWave: Spreading RDF Streams on the Web. In *The Semantic Web—ISWC 2016*; Groth, P., Simperl, E., Gray, A., Sabou, M., Krötzsch, M., Lecue, F., Flöck, F., Gil, Y., Eds.; Springer International Publishing: Cham, Switzerland, 2016; pp. 140–149.
47. Tulkens, S.; Emmery, C.; Daelemans, W. Evaluating Unsupervised Dutch Word Embeddings as a Linguistic Resource. In *Proceedings of the Tenth International Conference on Language Resources and Evaluation (LREC 2016)*; Chair, N.C.C., Choukri, K., Declerck, T., Grobelnik, M., Maegaard, B., Mariani, J., Moreno, A., Odijk, J., Piperidis, S., Eds.; European Language Resources Association (ELRA): Paris, France, 2016.

energies

MDPI

Article

d2ix: A Model Input-Data Management and Analysis Tool for MESSAGE$_{ix}$

Thomas Zipperle *,† and Clara Luisa Orthofer †

Chair of Energy Economy and Application Technology, Department of Electrical and Computer Engineering, Technical University of Munich, 80333 Munich, Germany; clara.orthofer@tum.de

* Correspondence: thomas.zipperle@tum.de

† These authors contributed equally to this work.

Received: 7 March 2019; Accepted: 15 April 2019; Published: 18 April 2019

Abstract: Bottom-up integrated assessment models, like MESSAGE$_{ix}$, depend on the description of the capabilities and limitations of technological, economical and ecological parameters, and their development over long-time horizons. Even small models of a few nodes, technologies and model years require input-data sets involving several hundred thousand data points. Such data sets quickly become incomprehensible, which makes error detection, collaborative working and the interpretation of results challenging, especially for non-self-created models. In response to the resulting need for manageable, comprehensible, and traceable representation of input-data, we developed a Python-based spreadsheet interface (*d2ix*) that enables presentation and editing of model input-data in a concise form. By increasing accessibility and transparency of the model input-data, *d2ix* reduces barriers to entry for new modellers and simplifies collaborative working. This paper describes the methodology and introduces the open-source Python-package *d2ix*. The package is available under the Apache License, Version 2.0 on GitHub.

Keywords: MESSAGE$_{ix}$; reproducibility; collaborative work; open modelling and data; data-handling; integrated assessment modelling; data pre- and post-processing

1. Introduction

The software package described in the following —*d2ix*— is freely available under the Apache License, Version 2.0 on GitHub under: https://github.com/tum-ewk/d2ix.

1.1. *Input-Data-Handling—The Underrated Modelling Challenge*

Technology-based integrated assessment models, such as MESSAGE$_{ix}$ (formerly known as MESSAGE) have a long history in energy and environmental systems modelling [1,2]. Despite having been developed in times of relatively low computing power, over the last forty years these models have grown in line with multiplying computing capacity, expanding models in dimensions such as coverage and detail [3]. Until the 1990s, the models focused on the energy-system only [4]; however, today's energy-engineering-economic-environment optimising models are designed to describe the full extent of energy-system dynamics, including effects such as polluting greenhouse gas emissions, economic development, land and water use and health implications [5,6]. At the same time, rising computing power allows not only increasing coverage but also magnifies the level of detail represented in models, such as the number of model years, nodes, technologies and technology parameters.

While in line with this structural change, big data not only presents a challenge in terms of energy-systems modelling: here too, the amount of input-data has skyrocketed [7]. Today, one technology in MESSAGE$_{ix}$ is described by forty parameters, of which fourteen are defined not only by the installation year of the technology but also the age of the technology. Thus, in even very simple input-data sets (e.g., describing one node over ten model years), each technology is defined by approximately one thousand input parameters, each again defined by up to twelve sets. Therefore, even a small model of one node over ten years has an input-data set per technology of more than twelve-thousand data points, not including the input-data for describing the ecology or economy (Figure 1).

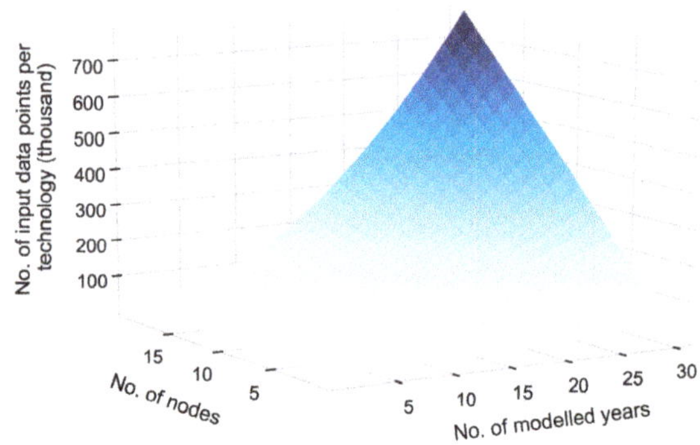

Figure 1. Average number of data points per technology in the input-data set of a MESSAGE$_{ix}$ model in dependence of the number of modelled years and nodes.

1.2. *d2ix—Combining Benefits of Non-Binary and Binary Data Formats*

Currently, most models handle input-data using vast spreadsheets (e.g., MS Excel), csv (comma-separated value) or plain text-files to organise, pre-process and document the model-data. While on the one hand, binary ('higher') formats, such as spreadsheets, provide support with data-handling, (un)intentional changes made to the input files are not trackable and are difficult to retrace. On the other hand, non-binary ('lower') but trackable formats such as csv and text-files lack visual clarity and data-handling support. To ensure transparency in data-handling and reproducibility of model results, the modelling-platform (ixmp), supplies MESSAGE$_{ix}$ users with tools for (i) database communication for version-controlled data management, (ii) a Python/R interface for efficient input-data and results processing and (iii) a web-browser based tool for drag and drop results visualisation [8]. The newly developed 'data to MESSAGE$_{ix}$' (*d2ix*) package adds to this functionality by providing the user with a visually comprehensible overview of the input-data by reducing the dimensions of the input-data set, thereby reducing the number of data points to be handled by the user (Figure 2).

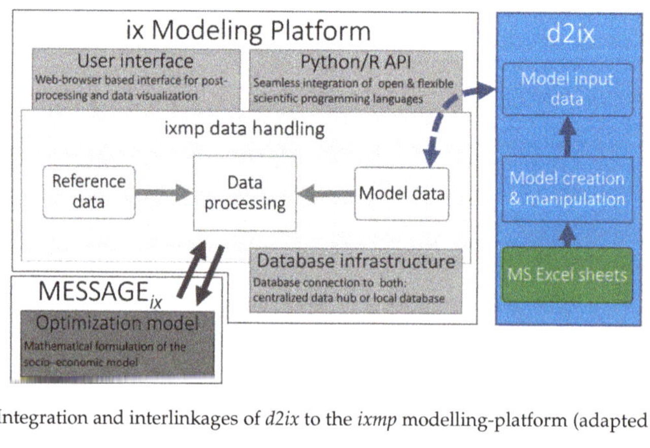

Figure 2. Integration and interlinkages of *d2ix* to the *ixmp* modelling-platform (adapted from [8]).

This model input-data-handling approach, as such, is novel as it is the first to combine reduced form MS Excel spreadsheet data and lucidly change-tackable .yaml files for input-data documentation. By following the FAIR principles of scientific data-handling and analysis, *d2ix* makes data findable, accessible, interoperable and reusable, and thus facilitates collaborative working and can therefore support the energy-modelling community [9]. By enabling new users to quickly become acquainted with existing models, and by simplifying the generation of new scenarios, *d2ix* reduces the barriers to entry into energy- and climate-policy modelling. Furthermore, the synoptic organisation of the input-data set can reduce the risk of errors prone to happen when organising big data sets (Figure 3). Such errors can have detrimental effects such as the data and coding mistakes causing the infamous Reinhart-Roghoff spreadsheet error [10]. Lastly, the interface will be equipped with a unit test that can inspect the model for commodity 'dead ends' and overly restrictive bounds, a feature that can prevent infeasibilities, undesired exceedingly restricted scenarios and the misinterpretation of results. Overall, *d2ix* is a well-suited data-handling tool for large energy-system models such as MESSAGE$_{ix}$. The easy change-trackable framework for transparent model input-data preparation is the first of its kind to be introduced as a standardised model-creation workflow.

	costs					input		primary output			availability				emissions		initial bounds				growth bounds			
technology	inv_cost	fix_cost	var_cost	technical_lifetime	construction_time	commodity_in1	level_in1	efficiency_1	commodity_out1	level_out1	first_year	last_year	capacity_factor	operation_factor	emission_factor_CO2	emission_factor_CH4	initial_activity_up	initial_activity_lo	initial_new_capacity_up	initial_new_capacity_lo	growth_activity_up	growth_activity_lo	growth_new_capacity_up	growth_new_capacity_lo
coal_ppl	500	30	30	20	1			1,0	electricity	secondary	690	720		1	100						0,1			
wind_ppl	1500	10		20	1			1,0	electricity	secondary	690	720		1							0,1			
grid	300		50	20	1	electricity	secondary	1,0	electricity	final	690	720		1										
bulb	5			1	1	electricity	final	1,0	light	useful	690	720		1										

Figure 3. Reduced spreadsheet input for technology specification for the tutorial.

2. Related Work

While, in the light of good scientific practice, model transparency and reproducibility have received wide academic attention, the focus has remained on how to deal with and how to publish raw-data and model code [11]. However, the important link between the two much noted components—the raw-data and the model—the input-data-handling, has so far not been dealt with scientifically [12,13]. In contrast, the major strategies of input-data-handling which established themselves as go-to solutions in energy-system modelling have never been subject to publication but rather research-institution internal, customised, single-user solutions. Thus, most models now provide different data-handling strategies. Four mayor types can be identified among the most commonly used input-data-handling methods. They are:

- **Type I**—Reduced text-file structures: The input-data of such models is handled in several long or one single, even longer, structured text-file. Such text-file-based input-data systems used to find application with most energy-system models. Due to their long history as well as their suitability for synoptic change-tracking, some modellers still rely on Type I input-data-handling strategies (e.g., *Calliope* [14]).
- **Type II**—Full parameter text-files: The input-data handled by this type is organised in a multitude of text-files. Each file contains one parameter in full dimension and shape as required by the database. Despite the low lucidity and the difficulty in tracking any (un-)intentional changes made to the input-data, Type II data-handling schemes are commonly used. Especially community-friendly, open-source models such as *PyPSA* [15] and *oemof* [16] in particular

appreciate the high flexibility of the input-data-handler in combination with the low requirements regarding the programming skills of the modeller.

- **Type III**—Reduced parameter spreadsheets: Here the input-data is organised in MS Excel spreadsheets in reduced dimension. While this dimension reduction increases the lucidity of the input-data, it can at the same time limit flexibility. However, in order to lower barriers to entry for new modellers, several open-source models such as *urbs* [17] and *ficus* [18] rely on Type III input-data structures.
- **Type IV**—Code-based input-data: Code-based input-data can be either hard-coded or predefined and processed in functions. Thus, the input-data is documented and stored together with the code. Such transparent data-handling types allow for the full documentation of code and input-data within one workflow. However, extensive amounts of hard-coded data, can become overwhelming for any new user, just like the text-file-based data. Nevertheless, several renowned models such as MESSAGE$_{ix}$ [8], *Temoa* [19] and *OSEMOSYS* [20] provide interfaces for hard-coded model input-data.

Table 1 summarises and lists the strengths and shortcomings of those four strategies and compares them to the newly developed *d2ix* workflow. It shows that by filling the gaps in documentation, standardisation and transparency, frameworks such as *d2ix* can help improve energy-system modelling by combining the strength of binary and non-binary input-data storage and handling formats.

Table 1. Comparison of the input-data-handling types used so far and the new data-handling framework *d2ix*. The types are described in Section 2.

	Type I	Type II	Type III	Type IV	d2ix
(Git) change-tracking	yes	no	no	yes	yes
Synoptic data presentation	no	no	yes	no	yes
Parameter dimension reduction	no	no	yes	partly	yes
Sub-horizon parameter adaptation	yes	yes	no	yes	partly
Usable without programming knowledge	no	yes	few	no	yes
Dynamic scenario documentation	no	no	no	no	yes
Easy result visualisation	no	no	no	no	yes

3. Methodology

The Python-package we have created, *d2ix*, supports the user in creating new MESSAGE$_{ix}$ models as well as adapting and analysing existing input-data sets and scenarios. The support consists of four main tasks: first, *d2ix* supports the user in organising the input-data for MESSAGE$_{ix}$. For this task, we created an abstracted data model, summarising the reduced model input-data in two spreadsheet files. Secondly, *d2ix* functions as a standardised interface between the spreadsheets and the MESSAGE$_{ix}$ Python API. Third, *d2ix* documents the pre-processed model input-data in yaml text-files. This allows systematic and visual change-tracking of the spreadsheets-based scenario-data using automated change-tracking services such as Git. Lastly, several unit tests implemented in *d2ix* will allow an automated structured inspection of input-data sets to identify commodity 'dead ends' and overly restrictive constraints.

3.1. Class Structure and Definition

The *d2ix* package supports researchers who want to create a MESSAGE$_{ix}$ model, either from scratch or by modifying existing models (Figure 4). This support is supplied by the means of four different classes which handle the data input. In the following, the classes are described in their functionality and structure.

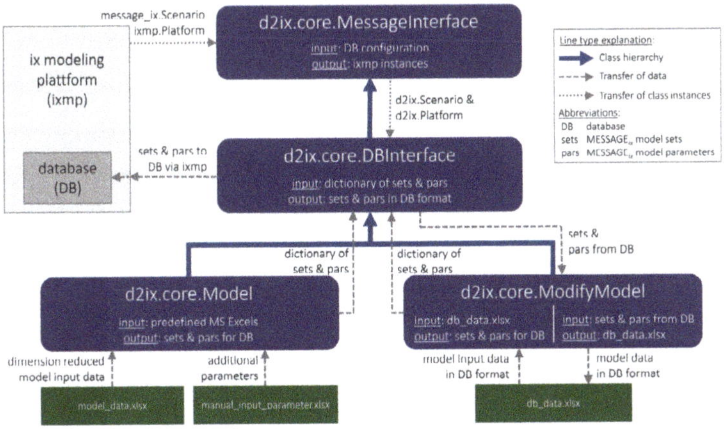

Figure 4. Class hierarchy diagram of the *d2ix* package.

3.1.1. MessageInterface—Communication with the Ix Modelling-Platform

The `MessageInterface` class acts as the interface between *d2ix* and the ix modelling-platform (ixmp). To communicate with the ix modelling-platform, `MessageInterface` applies the MESSAGE$_{ix}$ classes `ixmp.Platform` and `message_ix.Scenario`. While the `Platform` instance contains the connection to the database, the `Scenario` class predefines the format and indexation of the model in- and output-data (parameters, sets and variables) required for running the MESSAGE$_{ix}$ model. The database with which `MessageInterface` establishes a communication with is defined in the run-config file provided in the config folder (..\d2ix\config\run_config.yaml.template). The unique identification of the established `Scenario` instance is defined by the user input (Section 4.2), as is the logger setting of the *d2ix* module.

3.1.2. DBInterface—Data-Handling in *d2ix*

The `DBInterface` class enables data-handling in the *d2ix* package. The `DBInterface` class holds the model input-data in the form of a dictionary containing all model sets and parameters which can be accessed and modified before being transferred to the database via the `MessageInterface` class. The central tasks of this class are (i) to hand over the final input-data created in *d2ix* to the database, (ii) to write the final input-data into text-files for transparency and change-tracking, and (iii) to collect the model results from the database after a model run. Furthermore, the `DBInterface` class will check whether the units used in the input-data are already stored in the database and will add them if they are not.

3.1.3. Model—Data Transformation from Reduced Spreadsheet to Database Format

The `Model` class constitutes the core of the *d2ix* package. Its main task is the pre-processing of the input-data from the reduced *d2ix* spreadsheet format to the expanded final input-data format required by MESSAGE$_{ix}$. Apart from creating all required sets and parameters, the `Model` class automatically adds one slack technology for each demand provided in the input-data set, in order to prevent the model from running into infeasibilities during calibration, and to simplify debugging. After each successful scenario-run in MESSAGE$_{ix}$, the `Model` class reformats the results from database tables into time-series elements optimised for post-processing, applying the `TimeSeries` class from the ixmp package [8].

3.1.4. ModifyModel—From Database to Spreadsheet and Back

The `ModifyModel` class is used to enable the analysis and modification of existing MESSAGE$_{ix}$ models, i.e., models readily available in the database. To do so, the `ModifyModel` class has two main functions: (a) `ModifyModel` allows users to choose a specific MESSAGE$_{ix}$ scenario-run, which is then, first collected from the database, secondly, written to an excel sheet and lastly, made accessible to the user as a Python dictionary. The data can then be analysed and modified either in spreadsheet or through scientific computing (e.g., Python). In the second function (b) the modified data can be returned to the database as a new scenario containing the changes applied by the user.

3.2. Testing and User Experience

In accordance with best-practice collaborative programming [21], we set up a Continuous Integration implementation, with CircleCI and Docker each executing several tasks. Additionally two linters, thus static code analysis segments, are configured for basic code quality checks to ensure long term code maintainability. The coding style is tested with Flake8 and MyPy, the static types in Python. Furthermore the API functionality is tested in a defined environment inside a Docker container using the *d2ix* tutorial and some basic examples.

We tested the functionality of *d2ix* together with various beta users. In a first step, the data transfer from the spreadsheet to the *ixmp* platform and the git-tracked text-files was evaluated, thus proving the data-model functionality of *d2ix*. In a second step, we created three models of different sizes, in order to analyse and improve the runtime performance. The model descriptions and runtime performance are documented in Table 2. Finally, we tested the tool's intuitiveness with users without programming skills. By having such a user without programming experience recreating an existing MESSAGE$_{ix}$ model we succeeded in proving the data-model functionally as well as the coherence of the API. As a test model to recreate, we used the standalone country model of South Africa, which is available under the GNU General Public License, Version 3 on GitHub (https://github.com/tum-ewk/message_ix_south_africa) [22]. Two further MESSAGE$_{ix}$ country models are currently being developed for energy-research purposes.

Table 2. d2ix model-creation performance tests with different models. Calculations were performed on a Intel(R) Core(TM) i7 CPU with 3.2 GHz and 64 GB RAM.

	Nodes	Technologies	Historical Periods	Model Periods	Runtime in sec. *
Test Model 1	2	51	5	8	171
Test Model 2	4	88	65	8	657
Test Model 3	16	698	5	8	2291

* average of 10 runs.

4. Tutorial

4.1. Installation

To start using the open source Python-package *d2ix*, you must to ensure that your environment is equipped with the requirements as described in the README instructions found alongside the `d2ix` repository (https://github.com/tum-ewk/d2ix).

4.2. Running d2ix—Creating a Model from Scratch

The core functionality of the *d2ix* tool is to create a model from scratch. The bases for model creation are two reduced spreadsheets (Figure 3). In this example, we create a new MESSAGE$_{ix}$ scenario—in this case the replica of the 'Westeros' tutorial from the MESSAGE$_{ix}$ repository—using the *d2ix* MS Excel templates. The required parameters, configurations and files with the corresponding path are shown in Listing 1. The code creating the scenario is shown in Listing 2 and is explained below.

Furthermore, an introductory tutorial is provided in the *d2ix* repository under `tutorial.ipynb`.

Listing 1: Defining the *d2ix* model-creation parameters.

```
CONFIG = 'config/run_config.yaml'
BASE_XLS = 'input/modell_data_westeros.xlsx'
MANUAL_PARAMETER_XLS = 'input/manual_input_parameter_westeros.xlsx'
MODEL = 'MESSAGE_Westeros'
```

4.2.1. Creating a Model Instance

The `Model` class provides the functionality to create a model from scratch. The class instance is specified by thirteen parameters which are described in Table 3. Furthermore, the code to create a new instance is provided in Listing 2.

Table 3. Parameters used for creating a model instance.

Parameter	Description
run_config	the path to the run_config.yaml file located in the config folder [1] *The file contains the specifications of the database type.*
base_xls	the path to the *model_data* file, located in the input folder [2] *MS Excel file consists of seven input sheets that contain all necessary information for the model creation. Thus, the demand, the units, the technologies, and the nodes are defined and mapped in model_data. The structure of the file must to remain unchanged, as the input-data expansion done by d2ix depends on the current structure.*
manual_input_parameter	all possible parameters, such as economic parameters and ecological constraints, can be added to the model using the *manual_input_data* file *It provides the option of adding parameters manually by adding a sheet by the name of the parameter which contains the data in the format required by* $MESSAGE_{ix}$. *Thus, parameters in the manual_input_data are not manipulated by d2ix, but simply scanned for new set elements before being passed on to the database.*
$model_{ixmp}$	the name assigned to the model used as an identifier in the ixmp database
$scen_{ixmp}$	the name assigned to the scenario used as an identifier in the ixmp database
historical_data	allows the use of historical data
first_historical_year	defines the first historical year
first_model_year	defines the first model year If not assigned, it is set to the first year of demand.
last_model_year	defines the last model year If not assigned, it is set to the last year of demand.
historical_range_year	defines the the temporal resolution from historical data
model_range_year	defines the the model temporal resolution
verbose	defines the logger level in order to facilitate easy debugging
yaml_export	allows the export of yaml files from the model *The model parameters and sets can be written to structured text-files before being added to the database in order to facilitate change-tracking (e.g., Git) despite the input-data being provided in xlsx format. This can be turned off during model calibration in order to increase speed.*

[1] ..\d2ix\config\run_config.yaml.template; [2] ..\d2ix\input\model_data.xlsx.

In the example shown in Listing 2, we create an instance of the dummy model 'MESSAGE Westeros', which comes as a tutorial in the *d2ix* repository. The run configurations required for scenario creation with *d2ix* as well as the model input-data paths and the model name are defined in Listing 1. The newly created instance is named 'baseline' and spans over a time horizon from the year 690 to the year of 720. The first model year is defined as the year 700. The resulting model-year vector is equal to [690, 700, 710, 720], wherein 690 is a historical year, thus, not considered in the optimisation.

By setting verbose to true, the log-level is set to debug mode which allows for more information to pass from the creation process to the user. Setting the yaml export parameter to true permits the creation of git-trackable yaml files of the input-data. It is recommended to only set it to false during calibration, as this shortens the model creation runtime, though it disables the git-trackability of the input-data set.

Listing 2: Creating a new MESSAGE$_{ix}$ scenario using the *d2ix* spreadsheet templates.

```
1  from d2ix import Model
2
3  # Create a Model instance from the data provided in base_ & manual_parameter_xls
4  d2ix_model = Model(run_config=CONFIG, base_xls=BASE_XLS,
5  manual_parameter_xls=MANUAL_PARAMETER_XLS, model=MODEL, scen='baseline',
6  historical_data=True, first_historical_year=690, first_model_year=700,
7  last_model_year=720, historical_range_year=10, model_range_year=10,
8  verbose=True, yaml_export=True)
9
10 # write data from 'model' dictionary to the database and solve
11 scenario = d2ix_model.model2db()
12 scenario.solve(model='MESSAGE')
13 d2ix_model.close_db()
```

4.2.2. Transferring a Scenario from *d2ix* to the Database—*model2db()*

When the input-data is ready, it can be passed to the database using the *model2db* function, which returns an instance of the `messageix.Scenario` class (Listing 2, line 11).

4.2.3. Solving a Scenario

Using the solve function (from the `messageix.Scenario` class), the database model is dropped to a structured input-gdx file, which is passed on via a solve command to the mathematical model formulation of MESSAGE$_{ix}$. After the successful model run, an output-gdx file is created containing all input and output-data. This file content is automatically passed on to and stored in the database. Further details on the *solve()* function can be found in the MESSAGE$_{ix}$documentation [23]. Sample results of the baseline scenario from the Westeros example are shown in Figure 5.

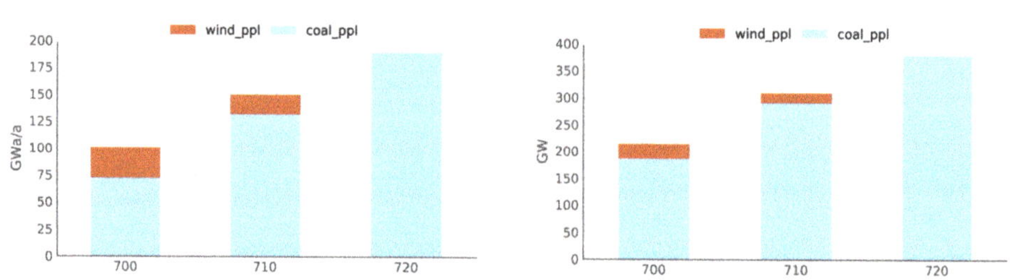

(**a**) Activity (**b**) Installed capacity

Figure 5. Power plant activity and capacity in the 'baseline' scenario (Listing 2).

4.2.4. Modifying the Input-Data—*get_parameter(), set_parameter()*

After creating the class instance, `model` contains a dictionary of all parameters and sets of the expanded input-data, which can now be accessed (Listing 3, line 11), modified (line 12) and returned to the dictionary (line 13). In this scenario we introduce an emission tax using the *d2ix get_parameter*, *set_parameter* procedure. The comparison between Figures 5 and 6 visualises the change in results induced by the introduction of the tax.

Listing 3: Creating a MESSAGE$_{ix}$scenario—with a carbon tax—using the *d2ix get_parameter()* and *set_parameter()* approach.

```
from d2ix import Model

# Create a Model instance from the data provided in base_ & manual_parameter_xls
d2ix_model = Model(run_config=CONFIG, base_xls=BASE_XLS,
manual_parameter_xls=MANUAL_PARAMETER_XLS, model=MODEL, scen='tax-emission',
historical_data=True, first_historical_year=690, first_model_year=700,
last_model_year=720, historical_range_year=10, model_range_year=10,
verbose=True, yaml_export=True)

# Add a emission tax
tax_emission = d2ix_model.get_parameter(par='tax_emission')
tax_emission['value'] = [0.264, 0.429, 0.699]
d2ix_model.set_parameter(par=tax_emission, name='tax_emission')

# write data from 'model' dictionary to the database and solve
scenario = d2ix_model.model2db()
scenario.solve(model='MESSAGE')
d2ix_model.close_db()
```

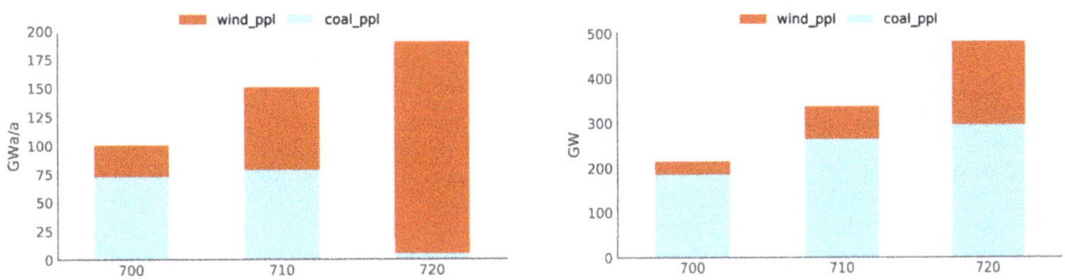

(**a**) Activity (**b**) Installed capacity

Figure 6. Power plant activity and capacity in the 'tax-emission' scenario (Listing 4).

4.3. *Running d2ix— Modifying Existing Models*

The *d2ix* package can also be used to modify existing models. The code required for retrieving, modifying and returning input-data sets to the database is shown in Listing 4, and is explained below.

4.3.1. Creating a ModifyModel Instance

The `ModifyModel` class provides the functionality of collecting models from the database, writing them into a structured spreadsheet file for user modification and returning the modified model to the database. The parameters specifying the `ModifyModel` instance not introduced in Section 4.2.1 (run_config, model, scen and verbose) are described in Table 4.

Table 4. Parameters used for creating a modified model instance.

Parameter	Description
xls_dir	the path to the directory where the MS Excel file will be saved
file_name	the name of the MS Excel that will be created or that the data will be read from

4.3.2. From Database to ModifyModel Instance & Excel Sheet—*scen2xls()*

The *scen2xls()* function (Listing 4, line 8) searches the database for the scenario defined by model and scenario name, and in the `mod_model`. If a scenario with the defined model and scenario name is available, all parameters and sets from the most recent (default) version of the scenario will be written to the spreadsheet. If a version is specified, this version instance of the scenario will be copied.

4.3.3. From the Excel File to ModifyModel Instance—*xls2model()*

The *xls2model()* (Listing 4, line 10) function reads the spreadsheet file specified in the `mod_model` instance and stores the data as a structured dictionary in the instance. The data is then available to modify, analyses and visualise using the Python functionality.

Listing 4: Modifying an existing MESSAGE$_{ix}$ model using spreadsheet inputs.

```
1  from d2ix import ModifyModel
2
3  # Create a ModifyModel instance
4  mod_model = ModifyModel(run_config=CONFIG, model=MODEL, scen=SCEN, xls_dir='xls_folder',
5  file_name='db_data.xlsx', verbose=False)
6
7  # Collecting a scenario from the database and saving it to an MS Excel file
8  mod_model.scen2xls(version=None)
9  # Collection a scenario from a MS Excel file and saving it to the database
10 mod_model.xls2model()
11
12 # write data from 'mod_model' dictionary to the database and solve
13 scenario = mod_model.model2db()
14 scenario.solve(model='MESSAGE')
15 mod_model.close_db()
```

4.4. *Post-Processing a MESSAGEix Scenario*

The *ixmp* package supplies tools for standardised reporting of reference data and results. These tools are documented and described in [8] as well as in the online documentation [24].

5. Conclusions

In *d2ix*, we built a package that supports users in creating, modifying, and analysing MESSAGE$_{ix}$ scenarios. The main benefits of using *d2ix* for scenario creation are threefold. (i) The synoptic input-data supports the transparency and reproducibility of even large models and can thus reduce errors. It further encourages collaborative modelling attempts by making it easier to understand and review model parameters and assumptions implemented by other researchers. (ii) By reducing the dimensions of the input-data, the researchers can easily handle the data using two MS Excel sheets. Hence, *d2ix* reduces barriers to access by reducing input-data complexity and allowing scenario creation without programming knowledge. (iii) *d2ix* permits the combination of the benefits of 'higher' (easy and synoptic data-handling) and 'lower' (change-trackability) data formats. To put it succinctly: by providing a synoptic and easy input-data-handling workflow *d2ix* can support the efforts of the open data movement within the MESSAGE$_{ix}$ modeller community and can serve as an example for data-handling frameworks built for other model types.

However, simplification of input-data does reduce the flexibility of the model, e.g., currently a maximum of two outputs is supplied for each technology. However, this can be bypassed by either adapting the model parameter 'output' using the *get_* and *set_parameter* functionality, or by adapting the input spreadsheet and the underlying code to supply as many outputs as required. An expansion of *d2ix* to increased flexibility could be subject of future work; however, the decision on the specific balance between flexibility and simplicity requires practical experience which still remains to be collected.

Author Contributions: T.Z. and C.L.O. have cooperated in the development of *d2ix*. While T.Z. is the creator of most of the code and software concept, C.L.O. developed the work flow and the overlying package concept, which are described in this paper. Both authors read and approved the final manuscript.

Funding: This work was supported by the German Research Foundation (DFG) and the Technical University of Munich (TUM) in the framework of the Open Access Publishing Program.

Conflicts of Interest: The authors declare no conflict of interest. The funders had no role in the design of the study; in the collection, analyses, or interpretation of data; in the writing of the manuscript, or in the decision to publish the results.

References

1. Schrattenholzer, L. *The Energy Supply Model Message*; Number 81-31 in Research Report; OCLC: 254145200; International Institute for Applied Systems Analysis (IIASA): Laxenburg, Austria, 1981.
2. Messner, S.; Schrattenholzer, L. MESSAGE–MACRO: Linking an energy supply model with a macroeconomic module and solving it iteratively. *Energy* **2000**, *25*, 267–282. [CrossRef]
3. Koomey, J.; Berard, S.; Sanchez, M.; Wong, H. Implications of Historical Trends in the Electrical Efficiency of Computing. *IEEE Ann. Hist. Comput.* **2011**, *33*, 46–54. [CrossRef]
4. Messner, S.; Strubegger, M. The energy model MESSAGE III. In *Advances in Systems Analysis: Modelling Energy-Related Emissions on a National and Global Scale*; Hake, J.F., Kleemann, M., Kuckshinrichs, W., Martinsen, D., Walbeck, M., Eds.; Konferenzen des Forschungszentrums Juelich: Juelich; Germany; 1994.
5. Huppmann, D.; Rogelj, J.; Kriegler, E.; Krey, V.; Riahi, K. A new scenario resource for integrated 1.5 °C research. *Nat. Clim. Chang.* **2018**, *8*, 1027–1030. [CrossRef]
6. Fricko, O.; Havlik, P.; Rogelj, J.; Klimont, Z.; Gusti, M.; Johnson, N.; Kolp, P.; Strubegger, M.; Valin, H.; Amann, M.; et al. The marker quantification of the Shared Socioeconomic Pathway 2: A middle-of-the-road scenario for the 21st century. *Glob. Environ. Chang.* **2017**, *42*, 251–267. [CrossRef]
7. Baker, T.; Asim, M.; Tawfik, H.; Aldawsari, B.; Buyya, R. An energy-aware service composition algorithm for multiple cloud-based IoT applications. *J. Netw. Comput. Appl.* **2017**, *89*, 96–108. [CrossRef]
8. Huppmann, D.; Gidden, M.; Fricko, O.; Kolp, P.; Orthofer, C.; Pimmer, M.; Kushin, N.; Vinca, A.; Mastrucci, A.; Riahi, K.; et al. The MESSAGE Integrated Assessment Model and the ix modeling platform (ixmp): An open framework for integrated and cross-cutting analysis of energy, climate, the environment, and sustainable development. *Environ. Model. Softw.* **2019**, *112*, 143–156. [CrossRef]
9. Wilkinson, M.D.; Dumontier, M.; Aalbersberg, I.J.; Appleton, G.; Axton, M.; Baak, A.; Blomberg, N.; Boiten, J.W.; da Silva Santos, L.B.; Bourne, P.E.; et al. The FAIR Guiding Principles for scientific data management and stewardship. *Sci. Data* **2016**, *3*, 160018. [CrossRef] [PubMed]
10. Reinhart, C.M.; Rogoff, K.S. Growth in a Time of Deb—Errata. *Am. Econ. Rev.* **2010**, *100*, 573–578. [CrossRef]
11. Pfenninger, S. Energy scientists must show their workings. *Nature* **2017**, *542*, 393–393. [CrossRef] [PubMed]
12. Pfenninger, S.; Hirth, L.; Schlecht, I.; Schmid, E.; Wiese, F.; Brown, T.; Davis, C.; Gidden, M.; Heinrichs, H.; Heuberger, C.; et al. Opening the black box of energy modelling: Strategies and lessons learned. *Energy Strategy Rev.* **2018**, *19*, 63–71. [CrossRef]
13. Cao, K.K.; Cebulla, F.; Vilchez, J.J.G.; Mousavi, B.; Prehofer, S. Raising awareness in model-based energy scenario studies—Transparency checklist. *Energy Sustain. Soc.* **2016**, *6*. [CrossRef]
14. Pfenninger, S.; Pickering, B. Calliope: A multi-scale energy systems modelling framework. *J. Open Source Softw.* **2018**, *3*, 825. [CrossRef]
15. Brown, T.; Hörsch, J.; Schlachtberger, D. PyPSA: Python for Power System Analysis. *J. Open Res. Softw.* **2018**, *6*. [CrossRef]
16. Hilpert, S.; Kaldemeyer, C.; Krien, U.; Günther, S.; Wingenbach, C.; Plessmann, G. The Open Energy Modelling Framework (oemof) - A new approach to facilitate open science in energy system modelling. *Energy Strategy Rev.* **2018**, *22*, 16–25. [CrossRef]
17. Dorfner, J. Open Source Modelling and Optimisation of Energy Infrastructure at Urban Scale. Ph.D. Thesis, Technical University of Munich, Munich, Germany, 2016.
18. Atabay, D. An open-source model for optimal design and operation of industrial energy systems. *Energy* **2017**, *121*, 803–821. [CrossRef]
19. Decarolis, K.H.S.S.J.F. Modeling for insight using Tools for Energy Model Optimization and Analysis (Temoa). *Energy Econ.* **2013**, 339–349. [CrossRef]
20. Howells, M.; Rogner, H.; Strachan, N.; Heaps, C.; Huntington, H.; Kypreos, S.; Hughes, A.; Silveira, S.; DeCarolis, J.; Bazillian, M.; et al. OSeMOSYS: The Open Source Energy Modeling System. *Energy Policy* **2011**, *39*, 5850–5870. [CrossRef]

21. Latte, B.; Henning, S.; Wojcieszak, M. Clean code: On the use of practices and tools to produce maintainable code for long-living. In Proceedings of the Workshops of the Software Engineering Conference 2019, Stuttgart, Germany, 18 February 2019.
22. Orthofer, C.L.; Huppmann, D.; Krey, V. South Africa After Paris—Fracking Its Way to the NDCs? *Front. Energy Res.* **2019**, *7*, 20. [CrossRef]
23. International Institute for Applied Systems Analysis (IIASA). The MESSAGEix Framework Documentation. 2018. Available online: http://messageix.iiasa.ac.at (accessed on 8 April 2019).
24. International Institute for Applied Systems Analysis (IIASA). The MESSAGEix Framework. 2018. Available online: https://github.com/iiasa/message_ix (accessed on 8 April 2019).

energies

MDPI

Article

Towards an Automated, Fast and Interpretable Estimation Model of Heating Energy Demand: A Data-Driven Approach Exploiting Building Energy Certificates

Antonio Attanasio [1,†], Marco Savino Piscitelli [2,†], Silvia Chiusano [3,*,†], Alfonso Capozzoli [2,†] and Tania Cerquitelli [1,†]

1 Department of Control and Computer Engineering, Politecnico di Torino, 10129 Turin, Italy; antonio.attanasio@polito.it (A.A.); tania.cerquitelli@polito.it (T.C.)
2 Department of Energy, Politecnico di Torino, 10129 Turin, Italy; marco.piscitelli@polito.it (M.S.P.); alfonso.capozzoli@polito.it (A.C.)
3 Interuniversity Department of Regional and Urban Studies and Planning, Politecnico di Torino, 10129 Turin, Italy
* Correspondence: silvia.chiusano@polito.it; Tel.: +39-011-090-7176
† These authors contributed equally to this work.

Received: 22 February 2019; Accepted: 27 March 2019; Published: 2 April 2019

Abstract: Energy performance certification is an important tool for the assessment and improvement of energy efficiency in buildings. In this context, estimating building energy demand also in a quick and reliable way, for different combinations of building features, is a key issue for architects and engineers who wish, for example, to benchmark the performance of a stock of buildings or optimise a refurbishment strategy. This paper proposes a methodology for (i) the *automatic estimation* of the building *Primary Energy Demand* for space heating (PED_h) and (ii) the *characterization of the relationship* between the PED_h value and the main building features reported by Energy Performance Certificates (EPCs). The proposed methodology relies on a two-layer approach and was developed on a database of almost 90,000 EPCs of flats in the Piedmont region of Italy. First, the *classification layer* estimates the segment of energy demand for a flat. Then, the *regression layer* estimates the PED_h value for the same flat. A different regression model is built for each segment of energy demand. Four different machine learning algorithms (Decision Tree, Support Vector Machine, Random Forest, Artificial Neural Network) are used and compared in both layers. Compared to the current state-of-the-art, this paper brings a contribution in the use of data mining techniques for the asset rating of building performance, introducing a novel approach based on the use of independent data-driven models. Such configuration makes the methodology flexible and adaptable to different EPCs datasets. Experimental results demonstrate that the proposed methodology can estimate the energy demand with reasonable errors, using a small set of building features. Moreover, the use of Decision Tree algorithm enables a concise interpretation of the quantitative rules used for the estimation of the energy demand. The methodology can be useful during both designing and refurbishment of buildings, to quickly estimate the expected building energy demand and set credible targets for improving performance.

Keywords: energy performance certificate; heating energy demand; buildings; data mining; classification; regression; decision tree; support vector machine; random forest; artificial neural network

---

1. Introduction

Energy efficiency is a growing policy priority for many countries around the world, for both economic and environmental reasons. In the 28 countries that are part of the International Energy Agency (IEA), buildings are responsible for about the 21% of total final energy consumption (26% in Italy) [1]. The amount of this energy used for heating and cooling systems is about 55% in the residential sector (74% in Italy) [1]. Regulatory bodies in several countries took actions to reduce wasteful energy consumption and greenhouse gas emissions and to encourage the use of renewable sources and the design of energy efficient buildings [2].

In most cases, the building energy performance rating has been indicated as a cornerstone to pursue the aforementioned aims. For instance, the *Energy Performance of Buildings Directive* (EPBD), issued by the European Commission, makes the evaluation of energy performance compulsory for new and existing buildings [2].

The EPBD provides member states with guidelines for the building *energy performance certification* process, which includes *energy performance rating* and *energy labeling*. The former is based on a scale of values referred to one or more significant parameters like Energy Use Intensity (EUI) and Primary Energy Demand (PED), while the latter consists in the assignment of an energy performance class (or label) to the building, based on the energy performance rating value. The EPBD lets member states to define the actual implementation of its directives. In Italy the EPBD is currently implemented by various national legislative decrees and technical standards, but there are different rating schemes developed in local areas (regions and autonomous provinces) [3].

Among the existing rating systems worldwide, the *Building Research Establishment's Environmental Assessment Method* (BREEAM) developed in the United Kingdom in 1990, is the first and leading assessment method. *Leadership in Energy and Environmental Design* (LEED) developed in the United States in 1998, is nearly the dominant building assessment system (implemented in more than 40 countries). Other well-known methods include *Comprehensive Assessment System for Building Environmental Efficiency* (CASBEE) of Japan, *National Australian Built Environment Rating System* (NABERS), *Building Environmental Assessment Method of Hong Kong* (HK-BEAM), *Green Mark* of Singapore, *EcoProfile* of Norway, *Deutche Gesellschaft fur Nachhaltiges Bauen* (DGNB) of Germany, *Green Building Label* (GBL) of China [4–6].

The interest in buildings energy performance assessment is increased in the last years, especially to estimate how different features affect the building efficiency. Indeed, from a design perspective, it is very important to determine the effect of the building features on its future energy performance in the early designing phase [7]. Similarly, for existing buildings, it could be useful to evaluate the suitability of a refurbishment plan [8,9]. Whatever the used approach, estimating building energy performance in a quick and reliable way, for different combinations of building features, is a key issue for different actors including public authorities [10]. In this context, Energy Performance Certificate (EPC) provides theoretical measure of how efficient a building could be if operated in standard conditions. However, the performance gap, i.e., the difference between estimated and actual energy performance could be significant. For instance, in [11] is stated that for the Swedish EPCs dataset the assessed performance gap is about the 20% for energy consumption assessments. An EPC is therefore not fully representative of the actual performance during operation but makes it possible to perform comparisons and benchmarking analysis between buildings.

In this paper we propose the *Heating Energy Demand Estimation for Building Asset Rating* (HEDEBAR) methodology providing the following features. (i) HEDEBAR allows the *automatic estimation* of the *Primary Energy Demand for space heating* (PED_h) reported by Energy Performance Certificates (EPCs) (calculated in "standard rating" conditions, according to EN ISO 13790 [12], UNI TS 11300-1 [13], and UNI TS 11300-2 [14]). (ii) Moreover, HEDEBAR allows to unfold the criteria adopted during the asset rating of real buildings, through the extraction of the *principal building features* that contribute *to estimate the building energy demand*. The purpose is twofold: (i) *predictive*, as we define models for the robust energy rating of residential buildings, through the estimation of their PED_h;

(ii) *descriptive*, as we provide an interpretation of the method used to issue EPCs, by highlighting the main features that determine the energy demand of buildings.

The HEDEBAR methodology uses data from EPCs to learn the criteria used by the rating system to issue them. It is based on the hypothesis that building features affect the energy demand in different ways for different classes of building energy efficiency. Therefore, a *two-layer approach* is defined to differentiate the analysis of buildings that belong to distinct *segments of energy demand* (i.e., distinct ranges of PED_h value) and to eventually increase the precision in predicting the PED_h value. In the first layer a *classification* problem is considered to estimate the segment of energy demand of the building to be analyzed. Then, in the second layer a *regression* problem is considered to estimate the PED_h value for the same building. We build a different regression model for each segment of energy demand. The proposed two-layer approach allows us to increase the prediction accuracy with respect to a single layer model, which disregards the possible segment of energy demand of the building.

As a case study, the HEDEBAR methodology has been validated on a dataset of real EPCs of almost 90,000 flats in the Piedmont region of Italy [15–17] released as open data by the Piedmont region. These data are available on a Web platform developed by *CSI Piemonte* (the Information System Consortium) and are regulated by the *Piedmont Region* authority (Sustainable Energy Development Sector).Experimental results obtained on such open data demonstrated that HEDEBAR allows estimating PED_h with a reasonable error by only analyzing a small set of 10 building features. Extracted knowledge, human-readable, can be easily exploited by different stakeholders during the decision making process, e.g., public authorities and regulatory bodies should plan future energy policies that leverage on specific building features [18].

The proposed methodology can be useful for designers and building stakeholders to estimate PED_h and to set reference threshold values for physical input variables. Due to the large dimension of the adopted dataset, the information provided can be considered representative of residential dwelling stock in Piedmont. Moreover, the proposed models are based on statistical variables easy to be adaptable to different datasets. Moreover the developed models can be profitably used by local authorities for a preliminary and quick estimation of PED_h as a function of different values of few influencing attributes in order to perform benchmarking analysis or energy savings scenario analyses.

The paper is organized as follows: Section 2 analyses relevant works in the analysis of data from energy performance assessment; Section 3 describes the HEDEBAR methodology adopted to find a model for the characterization of heating energy demand; Section 4 shows the experimental results, which are then discussed in Section 5.

2. Related Work

Three main types of buildings *energy performance assessment* are commonly acknowledged [19]: *Energy benchmarking*, i.e., the comparison of Energy Performance Indicators (EPIs) of a building with a sample representative of similar buildings; *Energy rating*, i.e., the evaluation and classification of the building energy performance according to predefined criteria; and *Energy labeling*, i.e., the assignment of an energy performance class (or label) to the building, according to a scale of values defined for some relevant parameter (e.g., EUI, PED).

Energy rating can be implemented in the following ways: (i) *measured* (or *operational*) *rating*, based on real metering on-site [20] and (ii) *calculated rating*, based on ideal energy use. Measured rating is mostly used in the operation and maintenance phases of existing buildings [21]. Calculated rating is more suitable in the design phase of new buildings, in particular with the aid of Building Energy Simulation (BES) software like in the case of LEED and BREEAM rating systems [6,22]. Calculated rating is further divided into *asset rating* and *tailored rating*. While asset rating methods consider standard usage patterns and climatic conditions and can be shaped either to building designs or to existing buildings, tailored ratings consider actual conditions and usage patterns for the buildings under analysis.

Within the scientific context, several research activities have been carried out on buildings energy performance assessment, for: (i) predicting energy demand [7,10,23] and energy class [24], (ii) rating and benchmarking [25–28], (iii) individuating representative buildings for different classes of energy performance [29–31], (iv) characterizing the relationship between energy demand and relevant building features [32–34], and (v) improving existing methods, also using new model based on data mining algorithms like regression models, decision trees, neural networks, and clustering [24,32,35–38].

Several works have proposed a benchmarking of different types of buildings. Dall'O' et al. [25] analyse a real data set of energy certificates to assess the energy performance, to detect anomalies in the registered certificates and to quantify the energy retrofit potential in existing buildings. Chung et al. [26] developed a benchmarking process for energy efficiency of commercial buildings by means of Multiple Regression Analysis (MRA). Gao and Malkawi [29] use clustering to classify buildings according to multiple features, like physical properties, environmental conditions, occupancy. Lara et al. [30] adopt the cluster analysis to find out a few samples representative of about 60 buildings, in order to optimize the energy retrofit measures. Hong et al. [27] use an approach based on case-based reasoning, MRA, ANN and GA, to produce a methodology for operational rating with higher explanatory power and higher prediction accuracy at the same time. A parallel research effort by Acquaviva et al. [39] has been devoted to efficiently compute inter- and intra-building performance indicators on fine-grained thermal energy consumption data for a large set of buildings located in a major Italian city. Tso and Yau [37] compared the accuracy of linear regression, ANN, and decision tree in predicting average weekly electricity consumption during both summer and winter in Hong Kong. Koo et al. [7] use the finite element method to estimate the heating and cooling energy demand of buildings, using data about building envelope design. In [10] a decision tree is used to model the real consumption of residential buildings in order to predict the energy use of newly designed buildings. Melo et al. [24] use ANN to improve the accuracy of surrogate models for labeling purposes, based on simulations results. Khayatian et al. [35] tackle the problem of uniformity of criteria among different certificates, therefore they use ANNs to predict the heating energy demand and to validate a dataset of energy certificates.

The analysis of real data from EPC databases has been performed in various countries [11]. The authors in Fabbri et al. [40] discuss about the effects of EPBD Directive and Italian EPC system on the real estate market prospective. The study presented in Hjortling et al. [41] provides an energy consumption baseline for buildings in Sweden, using data from 186k energy performance certificates issued for commercial buildings and based on energy bills rather than on theoretical calculations. The paper shows that real energy consumption is often higher than the one stipulated by the building code. The methodology presented in Xiao et al. [42] exploits a cluster analysis of the energy consumption (EUI excluding District Heating) of office buildings in China, to study its statistical distribution characteristics. It was found that the distribution of energy consumption has quite different characteristics than in Japan and the US. Other analyses of EPCs aimed at defining the current energy consumption baseline of existing buildings in Greece and Spain are presented respectively in Dascalaki et al. [43] and Gangolells et al. [44].

Compared to the current state-of-the-art, this paper brings a contribution in the use of data mining techniques for the asset rating of buildings, both in methodological and analytical terms. From the *methodological* perspective, the paper proposes a novel approach to characterize the heating energy demand of buildings using multiple independent models for different building segments. From the *analytical* perspective, the proposed approach estimates the heating energy demand with reasonable errors, using a small set of building features and generating interpretable models that provide useful information about the most relevant features affecting energy demand.

3. Data Analysis Methodology

The HEDEBAR (*Heating Energy Demand Estimation for Building Asset Rating*) methodology estimates the *heating energy demand* of residential flats as a model of a few influencing features available within Energy Performance Certificates (EPCs).

HEDEBAR considers different building features that affect the energy demand. It is based on the hypothesis that the impact of each feature over the energy demand varies for different *segments* of values of the same energy demand. Hence, a *two-layer approach* has been defined to model this aspect. The logical components of HEDEBAR are represented in Figure 1 and they are briefly described below.

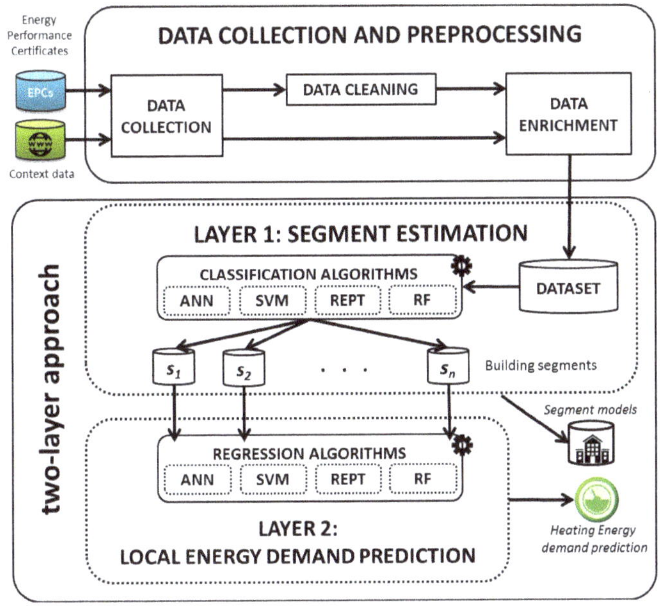

Figure 1. The proposed HEDEBAR methodology for automatic asset rating.

Data collection and preprocessing includes all the preliminary tasks necessary to provide the proper data set to the algorithms that operate in the later phases. Specifically, the *Data collection* component takes data from the energy certificates and other contextual information. *Data preprocessing* includes removing records with errors and missing values; discarding features that are useless to energy demand modeling; and enriching the resulting data set with contextual information not included in EPCs. These steps are better described in Section 3.2.

The *Segment estimation* is the first phase of the two-layer approach. Different classification algorithms have been trained during this step, to learn a classification model that properly assigns flats to different predefined segments of energy demand, considering only the selected features.

The *Local energy demand prediction* is the second phase of the two-layer approach. It uses regression algorithms to learn a regression model for estimating the *heating energy demand* considering only the selected features. An independent regression model for each segment of the first layer has been trained and tested.

During the two phases of the two-layer approach, the performance of each algorithm has been assessed in order to select the best one. When two or more algorithms have similar prediction performances, the one generating the most interpretable, i.e., human-readable model is preferred.

The two-layer approach provides a twofold output: the *classification and regression models* for the analyzed flats, useful to understand the features with the highest explanatory power with respect to the energy demand and to highlight the differences among the segments; the *heating energy demand prediction* for new flats.

3.1. Flat Characterization

The EPC includes the different features of a building affecting its energy performance as well as the variables used to quantify its energy demand. The feature selection process has been driven by previous experiences on EPCs datasets analysed by the authors [15,16] with the aim of using few input variables that are also easy to be collected. The following four main categories of input variables were identified for the purpose of the analysis: (i) *geometry*, (ii) *envelope*, (iii) *time*, and (iv) *system*. The categories are briefly described below, while Table 1 reports the relevant features for each of them.

Geometry. The variables in this category describe the different geometric features of the flat, which have an impact on its energy performance. The category includes variables such as average ceiling height, heat transfer surface and heated gross volume of the flat.

Envelope. The features in this category are related to the physical properties of the building (i.e., the thermal transmittance values of the opaque and transparent building envelope). In this category are also considered the dynamic characteristics of the building envelope through the variable q_{env}. This variable is expressed as an ordinal attribute that ranges from 1 to 5. The five quality classes are related to specific numerical ranges of time lag and decrement factor that can be extracted from a table provided in DM 26/6/2009 [45].

Time. This category includes time variables such as the building construction year.

System. This category includes features related to the heating system (i.e., the average system global efficiency for space heating). The average global efficiency of the heating system is calculated on the basis of the standard values of efficiency for each sub-system (generation, distribution, control, emission) according to UNI TS 11300-2 [14].

Among all the variables considered in this study, the *Primary Energy Demand for space heating* PED_h has been selected as the *target variable* of the analysis. PED_h (expressed in kWh/m^{2}y) is an energy related variable defined for benchmarking purposes. It is an estimation of the amount of real energy consumption of a flat in standard use conditions and it contributes to assign an energy class label to the flat. The PED_h value is estimated starting from the remaining *explanatory variables* included in Table 1 and can be used to compare different flats. In particular, similar pools of input variables proved to be robust enough for modeling in an effective way the building energy demand [15,16]. The PED_h value refers to the period of a heating season and it is normalized by the flat floor area. PED_h contributes to the evaluation of the overall Primary Energy Demand of flats (PED) together with the Primary Energy Demand for domestic hot water (PED_w). The heating energy demand is evaluated considering a building energy balance. The modelling of the building geometry considers real shapes and self or over shading of other buildings. The quasi steady-state calculation method is based on the monthly balance of heat losses (transmission and ventilation) and heat gains (solar and internal) evaluated in monthly average conditions. Transmission heat losses are estimated taking into account opaque and transparent surfaces and as well as the thermal bridging effect. In "Standard Rating", parametric values depending on floor area or heated net volume are taken into account when evaluating the ventilation rate and internal heat gains. The dynamic effects on the net heating energy demand are taken into account by introducing the dynamic parameters, utilization factor and an adjustment of the set-point temperature for intermittent heating/cooling or set-back. These parameters depend on the thermal inertia of the building, on the ratio of heat gains to heat losses and on the occupancy/system management schedules. The annual PED for space heating is calculated from the net energy demand through different system efficiencies (emission, control, distribution, generation) considering the thermal losses in the various sub-systems. For the heating season, the average system efficiency is defined as the ratio between the annual net energy and the annual PED for heating. The PED includes also the electrical energy demand of auxiliary systems.

Table 1. List of features selected to characterize and estimate the heating energy demand with the HEDEBAR asset rating methodology.

Category	Name	Symbol	Unit	Range
	Explanatory variables			
Geometry	Floor area	A	m^2	$\mathbb{R}^+$
	Heat transfer surface	S	m^2	$\mathbb{R}^+$
	Average ceiling height	H	m	$\mathbb{R}^+$
	Gross Heated Volume	V	m^3	$\mathbb{R}^+$
	Aspect ratio	R	m^{-1}	$\mathbb{R}^+$
Envelope	Average U-value of vertical opaque envelope	U_o	W/m^2K	$\mathbb{R}^+$
	Average U-value of the windows	U_w	W/m^2K	$\mathbb{R}^+$
	Quality of building envelope	q_{env}	-	$\{1,2,3,4,5\} \subset \mathbb{N}$
Time	Construction year	y_c	y	$\mathbb{N}$
System	Average global efficiency for space heating	η_h	-	$[0,1] \subset \mathbb{R}$
	Target variable			
Energy	Normalized primary energy demand for space heating	PED_h	kWh/m^2y	$\mathbb{R}^+$

3.2. Data Preprocessing

The whole raw data set gathered from EPCs usually includes many building features, represented through variables of different data types such as numeric (integer or real), nominal, textual, and boolean. However, some features could be not relevant for the subsequent data analysis and their inclusion in the features set would increase the complexity of the generated models. Most of the not selected variables are poorly related with the PED_h (e.g., textual descriptions, address of the flat) or include attributes with a high explanatory potential that are not so easy to be assessed without running a simulation in advance (e.g., heat losses for transmission, ventilation and infiltration). Moreover, data sets derived from energy certificates filled by auditors could contain imputation errors which can badly affect the quality of the extracted knowledge.

To address the above issues and to improve both accuracy and usefulness of the data analytics phase, HEDEBAR includes a preprocessing step. This step aims to (i) *clean* the original data collection to remove outliers and errors in data and (ii) *enrich* data with additional *contextual information* to cope with external environmental conditions that could differently affect the estimation of the PED_h value for each flat. These steps are better described below.

Data cleaning. The whole data set is firstly inspected based on the advice of domain experts to remove the less relevant features. In addition, on the selected input variables a data cleaning analysis was performed. The data cleaning phase is crucial in order to ensure the robustness of the analysis. In fact, EPCs datasets can be characterized low quality (in terms of attribute inconsistencies) [11]. However, the domain expertise in the energy and buildings field can prevent or at least limit inconsistency issues. According to [11] the consistency checks considered in this study are:

(i) Constraint rules for columns (e.g., area or volume cannot be negative); (ii) Domain expert analysis of values of the attributes (e.g., physical thresholds of system efficiency or thermal transmittance); (iii) Statistical checks (e.g., outlier detection though box plots).

Data enrichment. Data collected from the energy certificates are enriched with additional contextual information acquired from external data sources. To cope with external environmental conditions that could differently affect the estimation of the PED_h value for each flat, PED_h has been recalculated according to a reference standard climatic condition. In particular, all the EPCs issued in Piedmont region are evaluated for both the standard climatic conditions of the actual city (in which the building is located), and the one of Turin. The PED_h considered as target variable in this study is then expressed for all flats as if they were located in Turin considering the same standard monthly outdoor temperature and solar radiation. Therefore, comparisons among flats can be done regardless of their actual location. However, if it is necessary to assess the performance of a flat in a city different from Turin, a data scaling based on standard Degree Days (DD) can be considered a valuable procedure. Specifically, to scale the estimated PED_h it is possible to multiply it for the ratio between the standard DD value of the city where the flat is located and the ones of Turin.

3.3. Two-Layer Approach for the Estimation of Heating Energy Demand

The HEDEBAR methodology makes use of the features from energy certificates as explanatory variables to predict the PED_h value of a flat.

The impact of each feature on the PED_h value can vary over different classes of energy efficiency. To cope for this aspect, distinct ranges of PED_h value, called *segments of energy demand* or simply *segments*, can be defined to partition the data set into groups of flats with more uniform energy efficiency. This segmentation allows HEDEBAR to analyze independently the different classes of flat energy demand (e.g., low, medium, and high).

The estimation of the PED_h value is structured in HEDEBAR as a *two-layer approach*, including two phases named *Segment estimation* and *Local energy demand prediction*. The two phases are applied in sequence to accurately predict the PED_h value of a flat:

- Firstly, the *Segment estimation* phase identifies the expected (discrete) *segment of energy demand* of the flat. The approach considers a set of reference segments of energy demand of a flat. This task has been modeled as a classification problem. A classifier is used to assign each flat to the corresponding (discrete) segment of energy demand based on its features.
- Then, the *Local energy demand prediction* phase predicts the (continuous) numeric value of PED_h for the flat, based on its features. This second task is formulated as a regression problem. A different regression model is trained in advance for each segment of energy demand.

Thus, in HEDEBAR a new flat (with unknown energy demand PED_h) is first classified into a segment of energy demand through the *Segment estimation* phase. Then, the PED_h value of the flat is estimated through the *Local energy demand prediction* phase, using the regression model assigned to that segment.

To generate the classification and regression models used in the two phases, the HEDEBAR system can easily integrate most classification and regression algorithms currently available in literature. To select the most appropriate algorithms, two complementary aspects were considered: (i) the ability of the algorithm to *accurately predict* the *segment of energy demand* and the PED_h value for a flat, and (ii) the *interpretability of the model* it generates. Based on these criteria, we selected four reference algorithms to be evaluated for integration in the two phases of HEDEBAR: *Artificial Neural Network* (ANN), *Support Vector Machine* (SVM) [46], *Reduced Error Pruning Tree* (REPT), and *Random Forest* (RF). ANN and SVM methods provided good performances for both classification and regression tasks in several applications. However, these methods generate non-interpretable models and are usually characterized by high computational cost for building the model. REPT and RF methods have good performances as well, but with overall lower computational costs. Moreover, REPT algorithm generates an interpretable model, which makes possible a better understanding of the relationship between the features and the energy demand. Finally, all the four algorithms have a good degree of robustness to outliers and missing values in the data set, even if in HEDEBAR these issues are handled in advance in the data preprocessing phases. The open source Rapid Miner v5.3.0 toolkit [47] and the statistical software R [48] have been used for the development of the classification and regression algorithms. The following paragraphs provide an overview of the main characteristics of four algorithms.

Artificial Neural Network (ANN). Inspired by the structure and behavior of biological neural networks, *Artificial Neural Networks* (ANNs) are often used to model complex relationships between input and output variables or to find patterns in data. An ANN consists of an interconnected group of nodes (neurons), organized in different layers, which receive inputs from other nodes and return as output a value computed as a function of suitably weighted inputs. A very popular type of ANN is the *feed-forward* neural network, where information moves through neurons only in forward direction, from the input to the output nodes.

The training of ANN is usually performed through *back-propagation* algorithm: the final outputs are compared with the correct values of training samples to compute the value of a predefined error-function. The error is then fed back through the network to adjust the weights of each connection in order to reduce the value of the error function. After repeating this process for a sufficiently large number of training cycles, the network usually converges to some state where the error of the calculations is small [49].

Support Vector Machine (SVM). Based on the work of Vladimir Vapnik in statistical learning theory [50], *Support Vector Machines* (SVMs) are a set of supervised learning methods, which can be used for classification or regression. A SVM model represents data samples as points in space, separated by a set of hyperplanes, so that the samples of the different categories are divided by a clear gap that is as wide as possible. Intuitively, a good separation is achieved by the hyperplane that has the largest distance to the nearest training data points of any class (*functional margin*), since, in general, the larger is the margin the lower is the generalization error of the classifier. When the samples are not linearly separable, *soft-margin* SVMs allow for classification errors during the training, to produce a more generic model for new data [51].

SVMs map samples into a higher-dimensional space, where presumably the separation is easier. However, the computational and storage requirements of SVMs increase rapidly with the number of training vectors and with the space dimension. To keep the computational load reasonable, SVMs use a kernel function K(x,y) that simplifies the computation of dot products in terms of the variables in the original space. The kernel function can be of different type such as linear, polynomial, sigmoid [49].

Reduced Error Pruning Tree (REPT). *Reduced Error Pruning Tree* (REPT) [52] is a fast decision tree learning algorithm that builds classification or regression trees using information gain or variance reduction as splitting criterion. More specifically, it generates multiple trees and it picks the best one, that will be considered as the representative. REPT uses *reduced error pruning* with *back fitting* method to prune the tree. At each iteration, a validation subset is used to estimate the Mean Square Error (MSE) on the predictions made by the tree. Starting at the leaves, each node is replaced with its most popular class and if the prediction accuracy is not affected then the change is kept.

Optimized for speed, REPT only sorts values of numeric attributes once at the beginning of the model preparation. Reduced error pruning has the advantage of simplicity and speed, moreover the representation of the data in form of a tree has the advantage, compared with other approaches, of being meaningful and easy to interpret.

Random Forest (RF). *Random Forest* is a combination of tree predictors such that each tree depends on the values of a random vector sampled independently and with the same distribution for all trees in the forest [53]. The generalization error for forests converges almost surely to a limit as the number of trees in the forest becomes large. RF is based on *bagging*, a technique for reducing the variance of an estimated prediction function. Indeed, RF fits a number of decision tree classifiers on various sub-samples of the data set (and also on various subsets of features) and uses averaging to improve the predictive accuracy and to control over-fitting. The resulting model is a voting model of all the random trees in the forest.

4. Case Study

In this section we validate the effectiveness and the usability of the proposed HEDEBAR methodology focusing on the following aspects: (i) the ability to correctly estimate the segment of energy demand for each flat, and (ii) the ability to accurately predict the PED_h value for each flat. The experimental analysis also addresses (iii) the selection of the classification and regression algorithms integrated in the two layers of the system, (iv) the comparison with a single layer approach in terms of prediction error and overall execution time, (v) the impact of the system configuration parameters, and (vi) the explanation of the main variables that determine the membership of flats into segments and their PED_h values.

We experimentally evaluated HEDEBAR on a real data collection of EPCs issued in 2013 for buildings located in the Piedmont region, North West of Italy. The data set includes approximately 90,000 energy certificates, of flats located across the 8 provinces of the Piedmont region.

4.1. Characterization of Flat Segments

As explained in the methodology section, the data set has been partitioned into different segments according to the values of variable PED_h with the aim of grouping together flats with similar energy efficiency.

Specifically, three reference segments have been considered representing respectively *low energy demand flats* (segment s_1), *high energy demand flats* (s_2), and *very high energy demand flats* (s_3). Data set splitting into segments has been done considering also the reference value range of PED_h specified in [15,16]. Segment s_1 includes flats with PED_h values between 0 and 100 kWh/m^{2}y, while flats in segment s_2 have 100 kWh/m^{2}y $\leq PED_h \leq$ 300 kWh/m^{2}y, and in segment s_3 $PED_h \geq$ 300 kWh/m^{2}y.

The three segments result into sets with the following cardinalities. The larger segment is s_2 including 39,003 flats, followed by s_1 with 25,930 flats, and the s_3 with 21,176 flats.

The dataset has been split into three segments to identify representative groups of energy performance certificates representing flats with similar performances. Specifically, a group represents flats with low energy demand, the second includes flats with medium-high energy demand, while the last one includes flats with very high energy demand. The three segments also allow guaranteeing a significant number of flats in each group together with a variable distribution for each feature under analysis. A number of segment higher than three should lead to very small groups of energy performance certificates with a limited data variability for each variable. In this case an estimation model for a segment should not be general (i.e., data overfitting). A small number of segments should lead to the definition of complex estimation models of heating energy demand. In this case derived models could not be easily understood and quickly exploited by a domain expert.

Box plots in Figure 2 show the distribution for some interesting variables (i.e., average U-value of vertical transparent envelope, average global efficiency for space heating, construction year, and aspect ratio) separately for each segment under analysis. In general, all segments present a good variability range for each variable under analysis. Specifically, segment s_1 includes a set of residential flats characterized by a low energy demand. In fact, flats in this group are characterized by the lowest values of U_w (median 2.11, IQR $[1.75, 2.76]$), U_o (median 0.45, IQR $[0.33, 0.67]$) and R (median 0.6, IQR $[0.4, 0.7]$); and the highest values of η_h (median 0.81, IQR $[0.73, 0.87]$) and y_c (median 2004, IQR $[1970, 2009]$). On the other hand, segment s_3 includes flats characterized by a very high energy demand, represented by the highest values of U_w (median 3.66, IQR $[2.80, 4.62]$), U_o (median 0.98, IQR $[0.83, 1.04]$) and R (median 0.9, IQR $[0.7, 1.0]$); and the lowest values of η_h (median 0.68 range $[0.60, 0.73]$) and of y_c (median 1962, IQR $[1940, 1973]$. Finally, segment s_2 is characterized by median values and IQRs of the five variables that lie between those of the two previous segments.

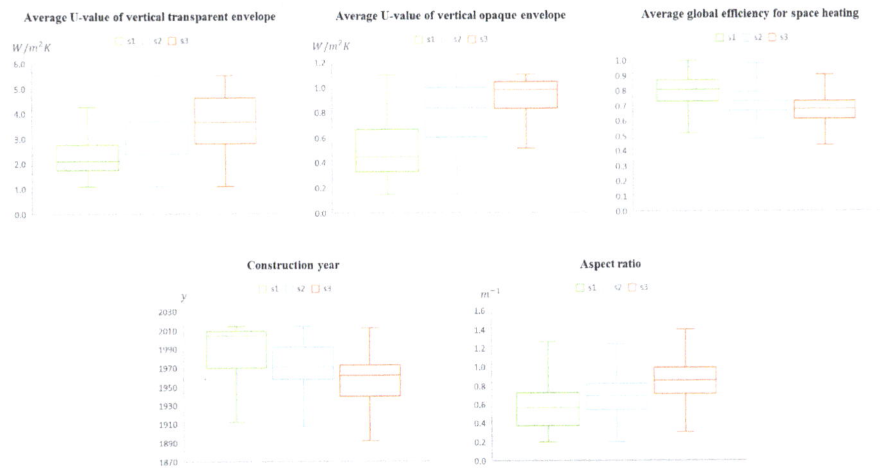

Figure 2. Box plots of the values of 5 input variables evaluated for each of the three different segments of energy demand.

Figure 3 shows the distribution of the certificates across the 8 Piedmont provinces, separately for each segment. The three charts are quite similar to each other, demonstrating that the geographical distribution is very similar across the three segments.

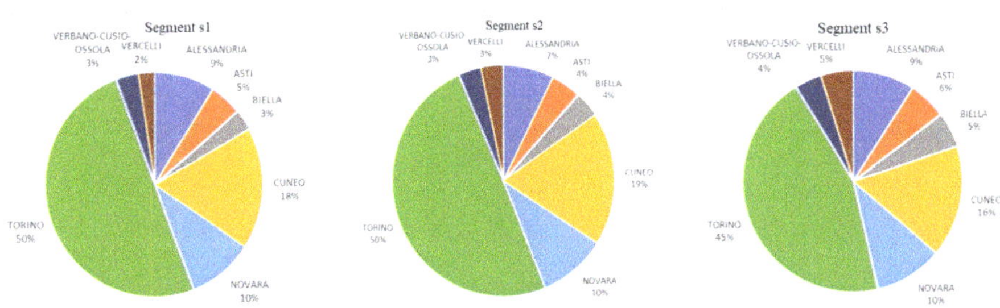

Figure 3. Distribution of the buildings across the 8 provinces of the Piedmont region for each of the three different segments of energy demand.

4.2. Segment Estimation

The classification task aims at assigning each new flat into the correct segment of energy demand. The classes of the classification task are the three segments presented in Section 4.1, identified by the nominal labels s_1, s_2, and s_3. All the four classification algorithms integrated in HEDEBAR (i.e., ANN, REPT, RF and SVM) have been experimentally evaluated for the classification of flats. The algorithm providing the classification model with the best classification performance has been selected as reference for this phase.

To validate the results of the classification process four established performance measures [54] have been considered. The overall quality of the classification model is evaluated in terms of *accuracy*. This measure counts the total number of flats correctly assigned to their corresponding segment. However, the unbalanced distribution of flats in the three segments could lead to a biased value of accuracy, as it could be mostly influenced by bigger segments. Therefore, other measures have been also used for a more accurate evaluation of the classification model. Per-class classifier predictions were evaluated according to *precision*, *recall*, and *F1-measure*. *Precision*(s_i) indicates the percentage of flats that are correctly revealed as in segment s_i. $Recall(s_i)$ indicates the number of flats assigned to segment s_i with respect to the total number of flats actually in s_i. The *F1-measure*(s_i), which is computed as the harmonic average of $Precision(s_i)$ and $Recall(s_i)$, quantitatively estimates the balancing between $Recall(s_i)$ and $Precision(s_i)$. In the experiment evaluation, we computed the precision, recall, and F1-measure values for each class label corresponding to each of the three segments.

A good trade-off between recall and precision is needed to properly predict the PED_h values for a new flat. On the one side, high precision values on most (all) segments are crucial to foster an accurate prediction of the PED_h values in the subsequent regression task. Indeed, the correct classification of a flat into the corresponding segment facilitates the subsequent prediction of the PED_h value for the flat. In fact, this prediction is performed through a model trained using data of flats with similar energy performance. A low $Precision(s_i)$ value indicates that many flats were mistakenly classified into segment s_i. This would result in erroneous predictions of PED_h values in the second step. On the other hand, achieving high recall values on most segments is desirable as well. A low $Recall(s_i)$ indicates that few flats of segment s_i are correctly classified into s_i, and they have been wrongly assigned to a segment other than s_1. This wrong assignment would result into an erroneous predictions of PED_h values due to the selection of a less appropriate prediction model in the second step.

Table 2 reports the results achieved by the four classification algorithms integrated into HEDEBAR. It shows the accuracy on the overall data set as well as precision, recall, and F1-measure for the three segments.

Table 2. Overall classification accuracy and precision, recall and F1-measure for each segment of ANN, REPT, RF and SVM algorithms.

	ANN	REPT	RF	SVM
	Overall			
Accuracy (%)	67.51	82.03	85.67	67.24
	Segment s_1			
Precision (%)	77.71	87.70	90.52	82.49
Recall (%)	70.03	83.84	87.27	61.97
F1-measure (%)	73.67	85.73	88.87	70.77
	Segment s_2			
Precision (%)	62.11	80.40	82.65	60.68
Recall (%)	75.54	80.56	85.49	81.74
F1-measure (%)	68.17	80.48	84.05	69.56
	Segment s_3			
Precision (%)	68.65	78.60	83.58	70.62
Recall (%)	49.62	82.53	81.96	46.98
F1-measure (%)	57.60	74.93	82.76	56.42

The RF classifier provides the highest accuracy value (85.67%) followed by REPT (82.03%), ANN (67.51%) and SVM (67.24%). Moreover, RF achieves also the best F1-measure on all segments (88.87%, 84.05%, and 82.76% in segments s_1, s_2 and s_3 respectively). More in detail, RF obtains the highest precision value for all segments (90%, 82.65%, and 83.58% for segments s_1, s_2, and s_3 respectively). RF also provides the highest recall values for two segments (87.27% and 85.49% for segments s_1 and s_2 respectively), while the recall obtained on segment s_3 (81.96%) is very close to the value provided by algorithm REPT (82.53%), which is the highest recall value over the four algorithms. Since the RF classifier achieves the highest values for almost all performance parameters, we chose it as reference algorithm for creating the model which classifies a new flat into the corresponding segment.

REPT is the second best algorithm for almost all performance parameters, providing accuracy, precision and recall values lower than those of RF, but still more than acceptable. An additional key point of REPT is the fact that this algorithm builds an interpretable classification model. This model is a decision tree from which human-readable classification rules can be extracted. Thus, domain experts can use the model not only to automatically classify a flat into the corresponding segment but also to analyze the most relevant properties that characterize each segment as well as to understand why a flat has been classified into a segment (see Section 4.5.1).

The SVM and ANN algorithms provide the worst values for all performance parameters, which are significantly lower than those obtained with RF and REPT algorithms.

Therefore, according with the experimental evaluation we decided to include two different classification models into the *Segment estimation* layer of the HEDEBAR framework. The RF classifier is used to automatically label a new flat with the corresponding segment. Based on the assigned segment, the proper regression model is selected in the subsequent layer (*Local energy demand prediction*) to predict the PED_h value for the flat. Instead, the REPT model is used to provide domain experts with a qualitative analysis of the impact of variables characterizing flats on the primary heating energy demand. This aspect is further discussed in Section 4.5.1.

4.3. Local Energy Demand Prediction (PED_h)

The regression task aims at estimating the value of PED_h for a flat. In HEDEBAR a different regression model for PED_h prediction is created for each of three segments s_1, s_2, and s_3. The ANN, REPT, RF and SVM algorithms have been experimentally evaluated for the creation of the regression model for each segment.

Table 3 displays the mean prediction errors of the four algorithms in predicting PED_h for each segment as well as the mean errors averaged over the three segments. The *prediction error* is the difference between the real value and the predicted value of PED_h. Three different measures of prediction error, among those commonly used in literature, have been calculated: (i) *Mean Absolute Error* (MAE) is the mean of all the absolute values of the errors obtained with the test samples; (ii) *Mean Absolute Percentage Error* (MAPE) expresses the mean absolute error in percentage terms; (iii) *Root Mean Square Error* (RMSE) is the square root of the mean of the square of all the errors obtained with the test samples. While MAE refers only to the mean value of the distribution of absolute errors, RMSE is affected also by the standard deviation of such distribution. Compared to MAE, RMSE amplifies and severely punishes large errors.

Table 3. Errors in predicting PED_h for ANN, REPT, RF, and SVM algorithms and for each flat segment.

	ANN	REPT	RF	SVM
	Overall			
RMSE (kWh/m^2)	39.85	33.12	33.83	38.40
MAE (kWh/m^2)	29.67	22.21	22.35	27.41
MAPE (%)	27.02	16.64	16.89	21.52
	Segment s_1			
RMSE (kWh/m^2)	30.99	21.99	22.16	28.95
MAE (kWh/m^2)	23.04	13.45	13.88	18.83
MAPE (%)	40.76	20.25	20.47	27.32
	Segment s_2			
RMSE (kWh/m^2)	37.80	29.72	30.87	37.03
MAE (kWh/m^2)	28.23	20.57	21.52	28.02
MAPE (%)	22.33	14.75	15.62	20.37
	Segment s_3			
RMSE (kWh/m^2)	49.76	47.69	49.84	50.31
MAE (kWh/m^2)	38.78	36.26	37.53	37.76
MAPE (%)	20.87	15.90	17.18	17.19

The REPT algorithm produces the overall lowest error values for the three measures (MAPE = 16.64%, RMSE = 33.12 kWh/m^{2}y, MAE = 22.21 kWh/m^{2}y) and it has also the best performance in each segment. In relative terms, REPT performs better in segments s_2 and s_3, where MAPE is 14.75%, and 15.90% respectively, while it has a substantially lower performance in segment s_1, where MAPE = 20.25%. The second best algorithm is RF, with an overall MAPE of 16.89%, while SVM and ANN provide higher error values (MAPE = 21.52% and MAPE = 27.02% respectively). Therefore, the REPT algorithm has been selected for local energy demand prediction, in order to better characterize groups of flats with similar features.

Figure 4 analyses more in depth the distribution of prediction errors, by reporting the box plots for *absolute error* and *percentage error* of the four algorithms over the three segments. The difference between REPT and the other algorithms is clear especially in segments s_1 and s_2.

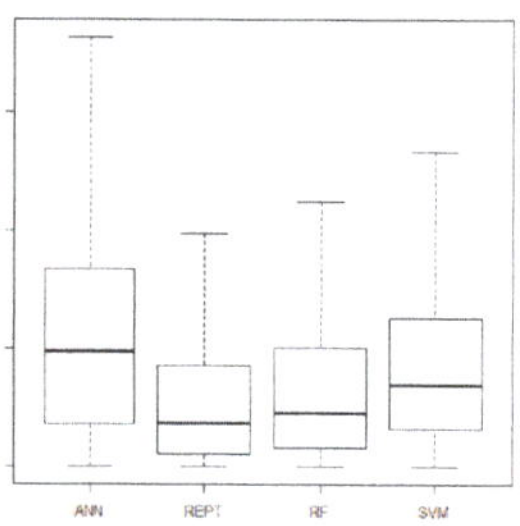

(**a**) Absolute error for segment s_1

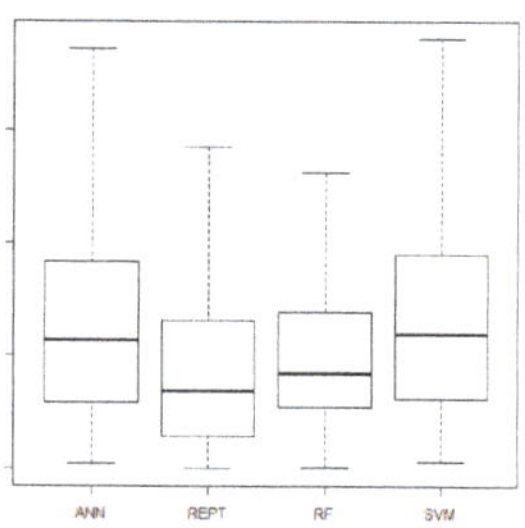

(**b**) Absolute error for segment s_2

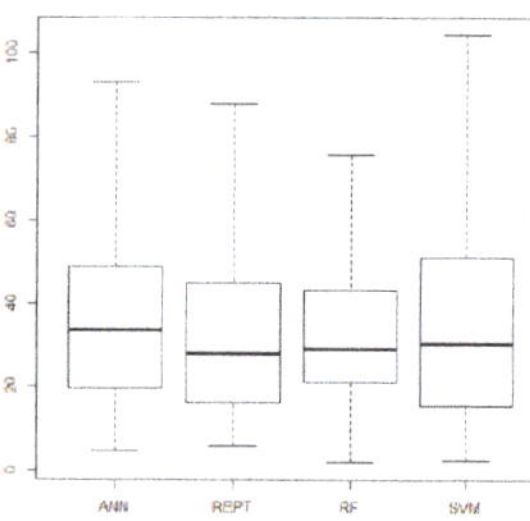

(**c**) Absolute error for segment s_3

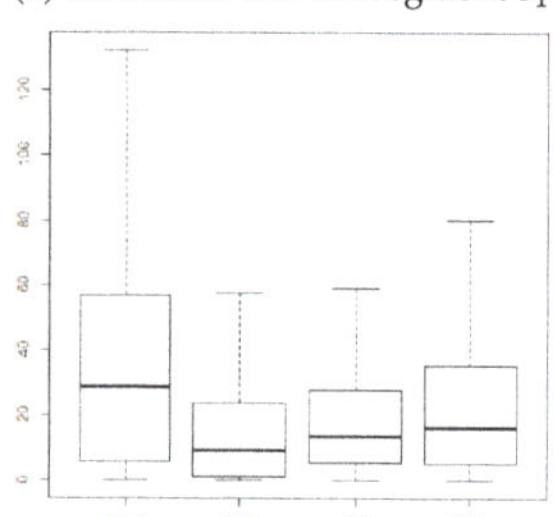

(**d**) Percentage error for segment s_1

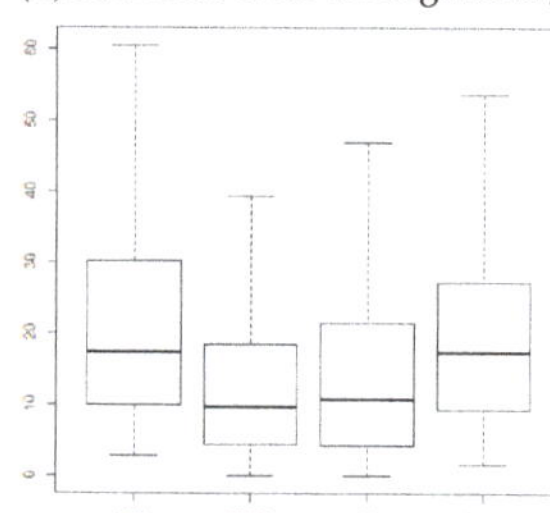

(**e**) Percentage error for segment s_2

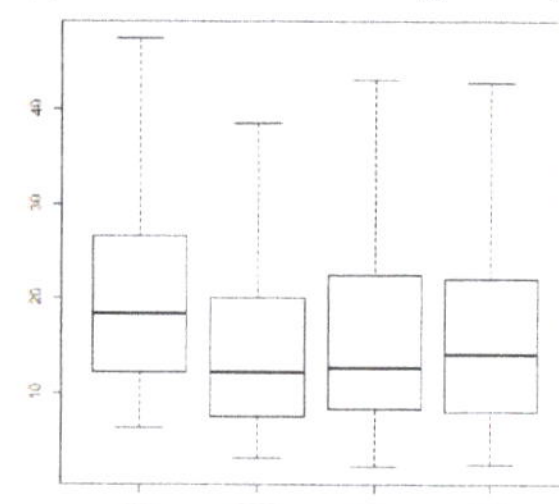

(**f**) Percentage error for segment s_3

Figure 4. Box plots of absolute error and percentage error of estimation of energy demand for each algorithm and for the three different flat segments.

4.4. Performance Comparison with a Single Layer Approach for PED_h Prediction

In this section we compare the performance in the prediction of the PED_h value between the *two-layer approach* used in HEDEBAR and a *single layer approach*. This latter approach exploits a unique regression model for all three segments, instead of building different models tailored to each segment. The ANN, REPT, RF and SVM algorithms have been evaluated to build the regression model for PED_h prediction with the single layer approach. The configuration setting for the single layer approach is discussed in Section 4.6.

Results for the two-layer and single layer approaches are reported in Tables 3 and 4, respectively. The experimental evaluation showed that, as for the two-layer approach, also for the single layer approach the best performance for PED_h prediction is obtained using the REPT algorithm. However, the REPT algorithm applied to the overall data set provides a model with MAPE value equal to 21.26% (see Table 4). Instead, using the two-layer approach the REPT models tailored to each segment result into a significantly lower overall MAPE value, equal to 16.64% (see Table 3). Also the RMSE and MAE values are significantly higher with the single layer approach (respectively, 37.37 kWh/m^2 and 26.10 kWh/m^2) than with the two-layer approach (respectively, 33.12 kWh/m^2 and 22.21 kWh/m^2). These results demonstrate the suitability of the two-layer approach used in HEDEBAR. In fact, the segmentation of the entire data set into groups of flats with similar energy demand allows to build differentiated models, which can more precisely predict the PED_h value for a flat in the segment.

Table 4. Errors in predicting PED_h for ANN, REPT, RF and SVM algorithms using a single step regression.

	ANN	REPT	RF	SVM
RMSE (kWh/m^2)	45.33	37.37	38.03	42.65
MAE (kWh/m^2)	30.01	26.10	26.36	28.34
MAPE (%)	27.46	21.26	21.53	23.67

4.5. Interpretation of the Energy Demand Estimation Models

This section provides a qualitative analysis of the impact of explanatory variables (building features) on the dependent variable, (heating energy demand). The analysis makes use of the REPT model, which has the advantage of providing interpretable decision trees. To better understand how the REPT algorithm models the relationship between input variables and the heating energy demand, we illustrate the first levels of the obtained decision trees.

4.5.1. Segment Estimation Model

The descriptive power of the REPT model comes from its capacity of putting in evidence the features that mostly affect the energy demand, according to the analyzed certification system.

The REPT model is represented by a tree graph, made of nodes and leaves connected by edges. In the REPT model built in HEDEBAR for segment estimation, each path of the tree includes a subset of building features. The leaf node of a path represents the predicted class label, corresponding to the energy demand segment s_1, s_2 or s_3 in this study. Therefore, each tree path includes a subset of features describing the buildings in one of the three segments.

A common way to build such trees is based on a recursive partitioning method. It consists in a forward step-wise approach where at each node the best split (according to input split variable, and the split value) is automatically evaluated by the algorithm for maximizing homogeneity in its child nodes. In this way the selection of split variables and split values consists in a data-driven process that does not require a manual selection by the analyst. As an example, the node including the *construction year* feature (y_c) can include the value 2007 as splitting value. The two outgoing edges for the node are associated to two distinct sets of values for y_c such as for example $y_c < 2007$ and $y_c \geq 2007$. Thus, each path includes a subset of variables, together with their corresponding ranges of values, describing the buildings associated with the segment label appearing in the leaf node of the path. For the classification of a new flat, the tree path composed of all the edges with splitting rules satisfying the features of the flat is selected. The segment label appearing in the leaf node of the path is used to estimate the segment of energy demand for the flat.

The first four levels of the REPT model are illustrated in Figure 5 (please refer to Table 1 for the interpretation of input variable symbols). It is possible to observe that the *average U-value of vertical opaque envelope* parameter (U_o) is the one mostly affecting the energy demand. Also the *aspect ratio* (R) and the *construction year* (y_c) appear at the first three levels of the tree. *Average U-value of the windows* (U_w) and *average global efficiency for space heating* (η_h) appear only at the fourth level. In general, the splits closest to the root node are the most important ones. This is the reason why only the upper portion of the classification tree is shown in Figure 5.

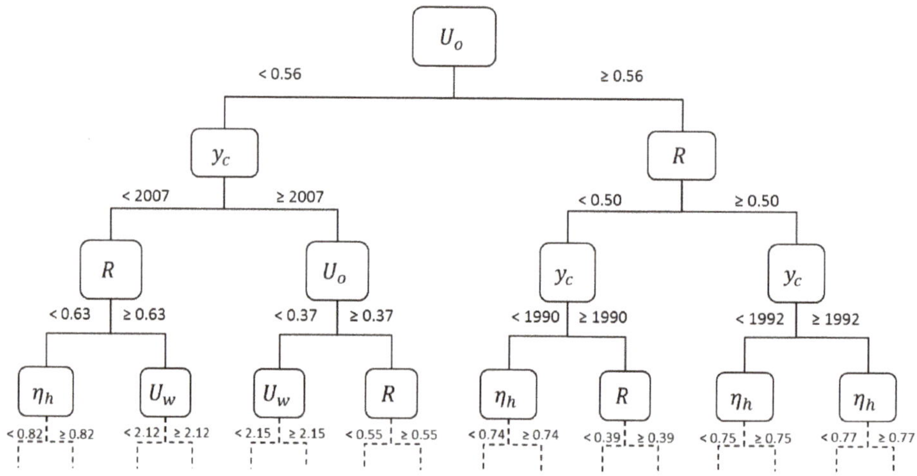

Figure 5. REPT model of the classification phase. The first four levels of the tree are illustrated and, for each path, the histogram illustrates the number of leaves assigned to each segment.

To further facilitate the interpretation of the tree model and to highlight the characteristics of each segment, the classification rules that summarize the main paths of the tree were extracted. The model developed for segment estimation has an overall size of 342 nodes with a maximum depth of 20 levels. Identify the most significant paths of the tree means to extract from the set of decision rules the ones that involve a significant number of records and reach high values of accuracy. These rules bring out the most representative building properties of each segment together with their ranges of values. Rules are extracted by traversing tree paths and they are structured in two parts: (i) the *rule antecedent* includes the buildings features and the corresponding ranges of values; (ii) the *rule consequent* includes the energy demand segment associated to flats that satisfy the conditions of the rule antecedent. Table 5 resumes the subset of rules selected as reference example from the REPT model. Specifically, for each segment we selected the rules with the highest classification accuracy among those that classify at least 500 flats. For the selected paths, the classification accuracy, i.e., the percentage of flats classified into the correct segment, ranges from 74.7% to 93.7%.

Table 5. Main rules of the REPT model for classification. For each row, intervals are specified only for the variables used by the corresponding rule. The last column contains the segment assigned by the rule.

		Rule	Antecedent			Rule Consequent
U_o	y_c	R	U_w	η_h	q_{env}	Segment
$[0, 0.37[$	$[2007, +\infty[$		$[0, 2.15[$			$\Rightarrow s_1$
$[0.56, +\infty[$	$[1992, +\infty[$	$[0.5, 0.68[$		$[0, 0.77[$		$\Rightarrow s_2$
$[0.78, +\infty[$	$]-\infty, 1991]$	$[0.63, 0.98[$	$[3.41, +\infty[$	$[0, 0.75[$	$[2, 5]$	$\Rightarrow s_3$

Rules like those in Table 5 are an important source of information about the classification model. Therefore, by examining these decision rules, the significant factors influencing PED_h can be identified also by a non-expert user and it is possible to roughly estimate the segment of a new flat.

For instance, the rule for segment s_1 is based on the average U-values of vertical opaque envelope (U_o) and of the windows (U_w) and on the construction year (y_c). More specifically, the rule states that, if $U_o < 0.37$ W/m^{2}K and $U_w < 2.15$ W/m^{2}K, the building envelope guarantees a very high level of thermal insulation and low heat dissipation. Moreover, flats that satisfy this rule were built with construction standards adopted from 2007 onwards, thus guaranteeing an overall energy efficiency that is classified into segment s_1.

The rule for segment s_2 includes also the aspect ratio (R) and the average global efficiency for space heating (η_h). This rule shows that, for high energy demand flats, R has intermediate values, while the η_h is always lower than 0.77. The average U-value of vertical opaque envelope (U_o) has a minimum value of 0.56 W/m^{2}K, which is higher than the maximum value used in the previous rule of s_1 (0.37 W/m^{2}K), thus implying always a higher thermal transmittance. Moreover, the rule includes high energy demand flats constructed since 1992, i.e., the minimum construction year for this rule is 15 years lower than the one for the previous rule (2007).

The rule selected for segment s_3 has very high values of aspect ratio (R), starting from a minimum of 0.63 m^{-1} which is almost equal to the maximum value for s_2 (0.68 m^{-1}). Additional negative factors are represented by the high lower bounds for U-values (U_o, U_w) and the construction year (y_c) always before 1991.

4.5.2. Local Energy Demand Prediction Models

Figure 6 depicts the first three levels of the REPT regression models of *Local energy demand estimation* for the three flat segments. Variables of splitting rules associated to the tree nodes are almost the same of the classification model represented in Figure 5, however their importance vary according to the segment. The tree for segment s_1 has a single variable for each level, i.e., *U-value of vertical opaque envelope* (U_o) at the first, *aspect ratio* (R) at the second, and *U-value of the windows* (U_w) at the third, thus providing a simple and easily interpretable model. In segment s_2 the *average global efficiency for space heating* (η_h) has a higher importance than in s_1, as it appears at the third level of the tree. The same variable appears in most of the rules of the same level in segment s_3. Here *average U-value of the windows* (U_w) is considered only for the most efficient flats (with $U_o < 0.76$ W/m^{2}K and $R < 0.89$ m^{-1}), while for those with higher energy demand, the *average global efficiency for space heating* (η_h) becomes more significant.

The splitting value of *average U-value of vertical opaque envelope* (U_o) increases from segment s_1 to segment s_3, meaning that flats belonging to the first segment are characterized by higher thermal insulated walls.

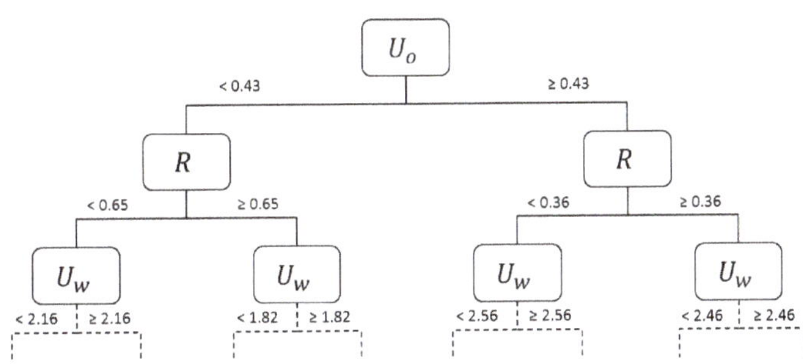

(**a**) Segment s_1

Figure 6. *Cont.*

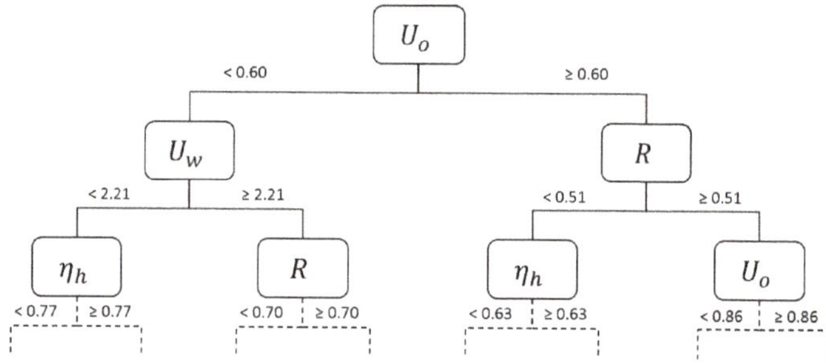

(**b**) Segment s_2

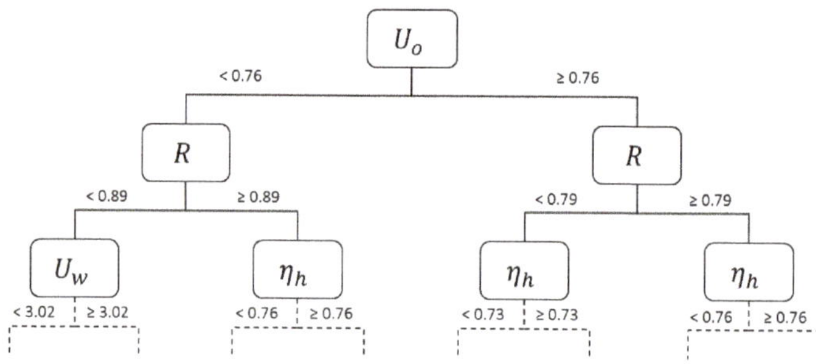

(**c**) Segment s_3

Figure 6. REPT models for each of the 3 flat segments.

4.6. Parameter Tuning of Algorithms

This section describes how the main parameters of the four algorithms considered in this study were tuned in order to reach the lowest values of prediction error both in the *Segment estimation* and *Local energy demand prediction* phases in the HEDEBAR framework. The same tuning procedure has been used also for the configuration of the single layer approach considered for performance comparison and described in Section 4.4.

For both phases, the prediction error was assessed using the k-fold cross-validation method, with $k = 10$. Therefore, the input dataset for the target phase has been split into k subsets of the same size. In turn, 1 subset is used for testing and the remaining $k - 1$ are used for training. Hence, k independent training and test iterations are performed. For each iteration, the training set is used by the four algorithms to generate the classification or regression models, according to target phase in the HEDEBAR framework. Then, the test set is used to evaluate the capacity of each classification and regression model to predict respectively the segment of energy demand and the PED_h value of new flats. The overall error value after the k iterations is computed as the mean of the errors of the k tests.

The procedure for tuning the optimal configuration for each of the four algorithms used in HEDEBAR produced similar values of parameter settings for the creation of the classification and regression models. These parameter settings turned out to be the optimal configuration even for the single layer approach. As an example, this section describes the results of parameters tuning for the creation of the regression model used in the *Local energy demand prediction* phase. The parameter tuning procedure is aimed at minimizing the values of the prediction errors MAPE, MAE, and RMSE (Figure 7).

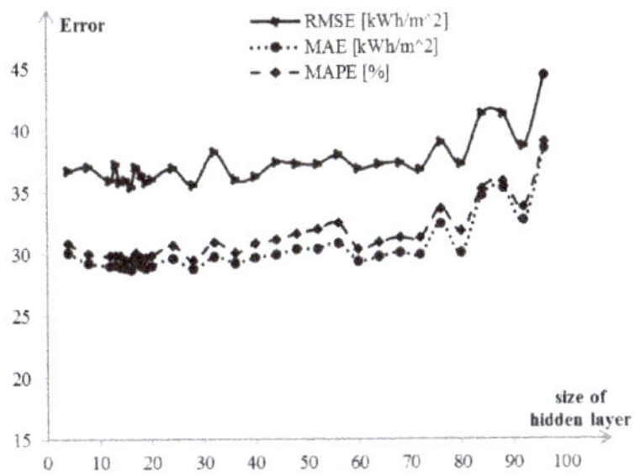

(**a**) ANN algorithm with respect to the size of the hidden layer.

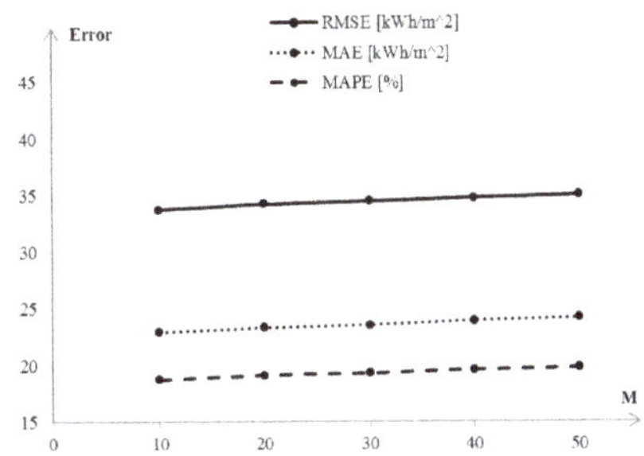

(**b**) REPT algorithm with respect to the minimum number of instances per leaf *M*.

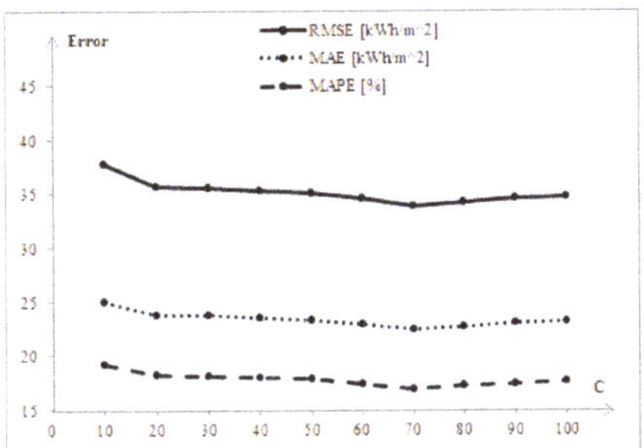

(**c**) RF algorithm with respect to the number of trees.

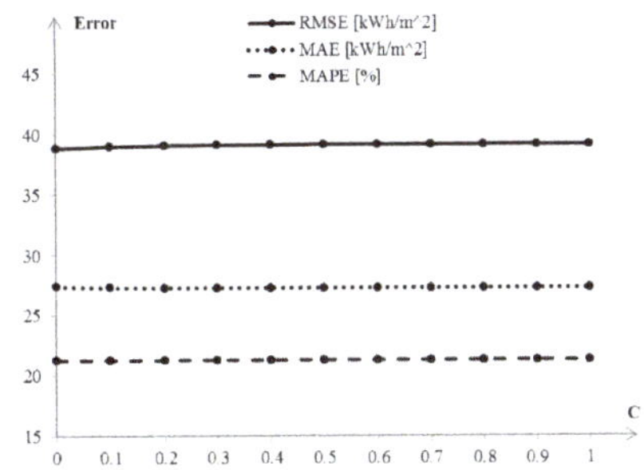

(**d**) SVM algorithm with respect to the complexity constant *C*.

Figure 7. Overall Local energy demand prediction errors of the algorithms for different values of their parameters.

For the ANN algorithm, a single hidden layer of variable size was considered, since using more than one layer did not provide any significant improvement of accuracy. Some common rules of thumb for the size of the hidden layer in the ANN are suggested by different works like [55], where the number of neurons are related to the number of input and output variables. Overall, the size of the hidden layer should be high enough to let the ANN model the problem correctly, but also low enough to ensure generalization. An increasing number of neurons was used during the tests, ranging in the interval $[4, 100]$ until the prediction error starts to grow due to over-fitting. The other parameters of the ANN are: $learning\_rate = 0.3$, $training\_cycles = 10^3$, $\epsilon = 1 \times 10^{-5}$. The values of RMSE, MAE and MAPE for different sizes of the hidden layer are reported in Figure 7a. 16 neurons for the hidden layer provide the lowest values of the three errors.

In the REPT algorithm, the dimension of the pruning subset was set to one third of the training set, hence with three folds in the algorithm ($N = 3$). No maximum tree depth has been set instead. The *information gain* was used as splitting criterion. The REPT algorithm was tuned by varying the minimum number of instances per leaf ($M \in \{10, 20, 30, 40, 50\}$). The values of RMSE, MAE and MAPE are reported in Figure 7b. The three error measures slightly, yet constantly, increase together with M. Therefore M was set equal to 10.

In the RF algorithm, the previous settings of REPT was used for all the decision trees. The variation of prediction error was assessed with respect to the number of trees I in the range [10, 100]. The values of RMSE, MAE and MAPE are reported in Figure 7c. $I = 70$ provides the lowest error values.

For SVM regression, a linear kernel function was considered and the variation of prediction errors, with respect to the complexity constant C, was assessed. This variable is used to set a degree of tolerance for misclassification of training samples. A too large value of complexity constant can lead to over-fitting, while too small values may result in over-generalization. Values for C have been selected in the range $[0, 10]$. The other parameter settings of the SVM are: $max\_iterations = 10^4$, convergence $\epsilon = 1 \times 10^{-3}$. The values of RMSE, MAE and MAPE are reported in Figure 7d. The trends of the three error measures are nearly constant with a slightly lower value of RMSE for $C = 0$.

5. Discussion and Conclusions

In this paper, the HEDEBAR methodology for the automatic asset rating of flats energy efficiency has been described. We recall that the analysis has been possible thanks to the availability of open data of Energy Performance Certificates. HEDEBAR proposes a two-layer approach to compute the ideal *Primary Energy Demand for space heating* (PED_h) of flats according to the certification scheme used to issue their EPCs. In this section we discuss the results obtained through HEDEBAR, addressing the results achieved using the proposed two-layer approach, and the interpretation and the possible exploitation of the extracted knowledge.

Accurate estimation of the flat energy demand with a reduced features set. Experimental results demonstrated the ability of the HEDEBAR methodology to estimate the PED_h value for a flat. PED_h is not the actual energy consumption of a flat, but its primary energy demand calculated in standard conditions. It is a significant parameter for the comparison of flats based on their features. The estimated values of PED_h are precise enough to provide a dependable assessment of flat energy efficiency for different values of the features characterizing flats.

From a methodological perspective, the experimental evaluation demonstrated that the two-layer approach used in HEDEBAR performs significantly better than a single layer algorithm in estimating the PED_h (MAPE values are respectively 16.64% and 29.82%). Therefore the segmentation of the initial data collection into different groups of flats with similar energy demand allows to produce differentiated models, which fit better the specific features of the respective segments.

The predictive performance of the HEDEBAR methodology is similar to the one of Khayatian et al. [35], where ANNs are used to predict the PED_h value, using EPCs related to the Lombardy region. Indeed, even if the experimental evaluation has been conducted on different datasets, HEDEBAR and the approach in [35] provide comparable results (MAPE equal to 16.64% HEDEBAR and to 14.44% in [35]). However, differently from [35], HEDEBAR estimates the value of PED_h in two steps using the REPT algorithm, which provides an interpretable model.

Modular approach able to integrate various algorithms and applicable to EPCs from other certification schemes. The HEDEBAR approach can make use of various classification and regression algorithms and can be used also to analyze data of EPCs issued according to other certification schemes.

The performed experimentation puts in evidence the algorithms with the best performances among those which were tested. In the *Segment estimation* phase, RF algorithm has the highest classification accuracy, while, in the *Local energy demand prediction* phase, REPT algorithm has the lowest error values in predicting PED_h. REPT also has a good classification accuracy. Therefore, RF in the first and REPT in the second phase turned out to be the most suitable combination of algorithms for the estimation of PED_h from the variables included in the EPC data set.

Interpretation of the energy demand estimation models. A key advantage of HEDEBAR is the use of REPT algorithm, whose decision tree models make results understandable and exploitable also for non-domain experts. Useful information can be obtained from this model as it helps to discover in a straightforward way energy patterns among large dataset. The algorithm automatically selects the different attributes for generating split rules and the ones closest to the root node can be assumed as the most influencing attributes. Therefore, the performance improvement brought by the two-layer approach, especially to the REPT algorithm, provides the HEDEBAR methodology with both a good estimation precision and a set of interpretable models of energy demand. Resulting models pointed out the most relevant features according to the considered rating system.

In the *Segment estimation* layer, 5 features out of 10 (*average U-values of opaque envelope and of the windows, aspect ratio, construction year*, and *average global efficiency for space heating*) appear in the first four levels of the decision tree and can be considered as the most relevant ones of the model. Indeed, they were preferred to other variables for splitting the initial flat set since they generate more homogeneous subsets in terms of PED_h value, thus allowing the overall model to reach a more accurate segmentation of the flat set. The characteristics of the three segments of energy demand are also summarized by means of short *decision rules*, which bring out the most representative building properties and their ranges of values for each segment. With a view to improving the efficiency of a flat, the model makes possible to individuate the features that mostly cause its membership to a specific energy demand segment. A proper change of their values, when possible (e.g., by means of targeted refurbishment actions), can substantially increase the energy efficiency of the flat. For some flats, bringing the values of few features within the appropriate ranges causes their reassignment to a lower segment.

In the *Local energy demand prediction* layer, 4 features out of 10 appear in the first three levels of the three decision tree models (the same as in Segment estimation except *construction year*). The differentiated analysis highlighted the main features impacting on PED_h for different segments of energy demand. In this case, the *U-value of vertical opaque envelope* (U_o) has demonstrated to be one of the most important variables for all segments. Indeed U_o is at the first level of all the three REPT models, with increasing splitting values from s_1 to s_3. The aspect ratio (R) is also a significant variable, as it appears in the second level of all the three REPT models. The *average U-value of windows* (U_w) is more important for low levels of energy demand (segment s_1), where the contribution of heat loss through windows can make the difference. On the other hand, the relevance of the *overall efficiency of the heating system* (η_h) is evident only for *high* and *very high* energy demand flats (segments s_2 and s_3).

Possible exploitation of HEDEBAR findings. Energy demand estimation is crucial to assess the energy performance in buildings and represents the first step to make any decision for enhancing their efficiency. The proposed approach has the advantage of learning a model from data about previous certificates that is then applied to new flats. The methodology can concretely help domain experts to evaluate the possible improvements of energy efficiency of flats. To this purpose, data driven models are useful for quickly estimating the expected building energy demand and in setting credible targets for improving performance [56]. In general, designers and authority planners should exploit such tools capable to suggest them where put their effort, among large stocks of buildings, and which could be the most convenient retrofitting strategies. In this way it is possible to plan future financial investment policies that leverage on specific building features and help devising more targeted actions to improve energy efficiency for different segments of buildings. Moreover the proposed methodological process allows to extract, by means of interpretable models (i.e., decision trees), useful and understandable knowledge regarding the expected energy performance of buildings according to few physical driving variables . Such benchmarks should be the reference for the building owners to improve the energy performance when it is poor and for technicians to identify the optimal cost-effective energy saving opportunities.

Author Contributions: The research presented in this paper was a collaborative effort made by all the authors. All the authors contributed to the literature review, methodology, implementation and experimental analyses, as well as to the writing and reviewing of the paper.

Funding: This research received no external funding.

Acknowledgments: The authors express their gratitude to Giovanni Nuvoli (Settore Sviluppo Energetico Sostenibile Regione Piemonte) and to CSI Piemonte.

Conflicts of Interest: The authors declare no conflict of interest.

Abbreviations

The following abbreviations are used in this manuscript:

ANN	Artificial Neural Network
BREEAM	Building Research Establishment Environmental Assessment Method
CASBEE	Comprehensive Assessment System for Building Environmental Efficiency
DD	Degree Days
DGNB	Deutche Gesellschaft fur Nachhaltiges Bauen of Germany
EPBD	Energy Performance of Buildings Directive
EPC	Energy Performance Certificate
EPI	Energy Performance Indicator
EUI	Energy Use Intensity
GA	Genetic Algorithm
GBL	Green Building Label of China
HK-BEAM	Hong Kong-Building Environmental Assessment Method
IEA	International Energy Agency
IQR	Interquartile Range
LEED	Leadership in Energy and Environmental Design
MAE	Mean Absolute Error
MAPE	Mean Absolute Percentage Error
MRA	Multiple Regression Analysis
MSE	Mean Square Error
NABERS	National Australian Built Environment Rating System
PED	Primary Energy Demand
PED_h	Primary Energy Demand for space heating
PED_w	Primary Energy Demand for hot water
REPT	Reduced Error Pruning Tree
RF	Random Forest
RMSE	Root Mean Square Error
SVM	Support Vector Machines

References

1. IEA. *International Energy Agency, Energy Efficiency Indicators Highlights*; OECD/IEA: Paris, France, 2016.
2. European Parliament CotEU. Directive 2010/31/EU of 19 May 2010 on the Energy Performance of Buildings (Recast). *Off. J. Eur. Union* **2010**, *53*, L153/13.
3. Andaloro, A.P.; Salomone, R.; Ioppolo, G.; Andaloro, L. Energy certification of buildings: A comparative analysis of progress towards implementation in European countries. *Energy Policy* **2010**, *38*, 5840–5866. [CrossRef]
4. Li, Y.; Chen, X.; Wang, X.; Xu, Y.; Chen, P.H. A review of studies on green building assessment methods by comparative analysis. *Energy Build.* **2017**, *146*, 152–159. [CrossRef]
5. Darko, A.; Chan, A.P. Critical analysis of green building research trend in construction journals. *Habitat Int.* **2016**, *57*, 53–63. [CrossRef]
6. Wang, S.; Yan, C.; Xiao, F. Quantitative energy performance assessment methods for existing buildings. *Energy Build.* **2012**, *55*, 873–888. [CrossRef]
7. Koo, C.; Park, S.; Hong, T.; Park, H.S. An estimation model for the heating and cooling demand of a residential building with a different envelope design using the finite element method. *Appl. Energy* **2014**, *115*, 205–215. [CrossRef]
8. Fan, Y.; Xia, X. An optimization model for building envelope retrofit considering energy performance certificate. In Proceedings of the 2017 36th Chinese Control Conference (CCC), Dalian, China, 26–28 July 2017; pp. 2750–2755.
9. Prieler, M.; Leeb, M.; Reiter, T. Characteristics of a database for energy performance certificates. *Energy Procedia* **2017**, *132*, 1000–1005. [CrossRef]

10. Yu, Z.; Haghighat, F.; Fung, B.C.; Yoshino, H. A decision tree method for building energy demand modeling. *Energy Build.* **2010**, *42*, 1637–1646. [CrossRef]
11. Pasichnyi, O.; Wallin, J.; Levihn, F.; Shahrokni, H.; Kordas, O. Energy performance certificates—New opportunities for data-enabled urban energy policy instruments? *Energy Policy* **2019**, *127*, 486–499. [CrossRef]
12. ISO 13790. *Thermal Performance of Buildings, Calculation of Energy Use for Space Heating*; International Organization for Standardization: Geneva, Switzerland, 2008.
13. UNI TS 11300-1. *Prestazioni energetiche degli edifici—Parte 1: Determinazione del fabbisogno di energia termica dell'edificio per la climatizzazione estiva ed invernale*; Standard, UNI—Ente Nazionale Italiano di Unificazione: Italy, 2014.
14. UNI TS 11300-2. *Prestazioni energetiche degli edifici—Parte 2: Determinazione del fabbisogno di energia primaria e dei rendimenti per la climatizzazione invernale, per la produzione di acqua calda sanitaria, per la ventilazione e per l'illuminazione in edifici non residenziali*; Standard, UNI—Ente Nazionale Italiano di Unificazione: Italy, 2014.
15. Di Corso, E.; Cerquitelli, T.; Piscitelli, M.S.; Capozzoli, A. Exploring energy certificates of buildings through unsupervised data mining techniques. In Proceedings of the 2017 IEEE International Conference on Internet of Things (iThings) and IEEE Green Computing and Communications (GreenCom) and IEEE Cyber, Physical and Social Computing (CPSCom) and IEEE Smart Data (SmartData), Exeter, UK, 21–23 June 2017; pp. 991–998.
16. Capozzoli, A.; Serale, G.; Piscitelli, M.S.; Grassi, D. Data mining for energy analysis of a large data set of flats. *Proc. Inst. Civ. Eng. Eng. Sustain.* **2017**, *170*, 3–18. [CrossRef]
17. Cerquitelli, T.; Corso, E.D.; Proto, S.; Capozzoli, A.; Bellotti, F.; Cassese, M.G.; Baralis, E.; Mellia, M.; Casagrande, S.; Tamburini, M. Exploring energy performance certificates through visualization. In Proceedings of the Workshops of the EDBT/ICDT 2019 Joint Conference, EDBT/ICDT 2019, Lisbon, Portugal, 26–29 March 2019.
18. Fabbri, K. Planning a Regional Energy System in Association with the Creation of Energy Performance Certificates (EPCs), Statistical Analysis and Energy Efficiency Measures: An Italian Case Study. *Buildings* **2013**, *3*, 545–569. [CrossRef]
19. Pérez-Lombard, L.; Ortiz, J.; Gonzàlez, R.; Maestre, I.R. A review of benchmarking, rating and labelling concepts within the framework of building energy certification schemes. *Energy Build.* **2009**, *41*, 272–278. [CrossRef]
20. Nikolaou, T.; Kolokotsa, D.; Stavrakakis, G.; Apostolou, A.; Munteanu, C. Review and State of the Art on Methodologies of Buildings' Energy-Efficiency Classification. In *Managing Indoor Environments and Energy in Buildings with Integrated Intelligent Systems*; Springer International Publishing: Cham, Switzerland, 2015; pp. 13–31.
21. Lu, X.; Lu, T.; Kibert, C.J.; Viljanen, M. A novel dynamic modeling approach for predicting building energy performance. *Appl. Energy* **2014**, *114*, 91–103. [CrossRef]
22. Tronchin, L.; Fabbri, K. Energy performance building evaluation in Mediterranean countries: Comparison between software simulations and operating rating simulation. *Energy Build.* **2008**, *40*, 1176–1187. [CrossRef]
23. Patiño-Cambeiro, F.; Bastos, G.; Armesto, J.; Patiño-Barbeito, F. Multidisciplinary Energy Assessment of Tertiary Buildings: Automated Geomatic Inspection, Building Information Modeling Reconstruction and Building Performance Simulation. *Energies* **2017**, *10*, 1032. [CrossRef]
24. Melo, A.; Cóstola, D.; Lamberts, R.; Hensen, J. Development of surrogate models using artificial neural network for building shell energy labelling. *Energy Policy* **2014**, *69*, 457–466. [CrossRef]
25. Dall'O', G.; Sarto, L.; Sanna, N.; Tonetti, V.; Ventura, M. On the use of an energy certification database to create indicators for energy planning purposes: Application in northern Italy. *Energy Policy* **2015**, *85*, 207–217. [CrossRef]
26. Chung, W.; Hui, Y.; Lam, Y.M. Benchmarking the energy efficiency of commercial buildings. *Appl. Energy* **2006**, *83*, 1–14. [CrossRef]
27. Hong, T.; Koo, C.; Kim, D.; Lee, M.; Kim, J. An estimation methodology for the dynamic operational rating of a new residential building using the advanced case-based reasoning and stochastic approaches. *Appl. Energy* **2015**, *150*, 308–322. [CrossRef]
28. De Ruggiero, M.; Forestiero, G.; Manganelli, B.; Salvo, F. Buildings Energy Performance in a Market Comparison Approach. *Buildings* **2017**, *7*, 16. [CrossRef]

29. Gao, X.; Malkawi, A. A new methodology for building energy performance benchmarking: An approach based on intelligent clustering algorithm. *Energy Build.* **2014**, *84*, 607–616. [CrossRef]
30. Lara, R.A.; Pernigotto, G.; Cappelletti, F.; Romagnoni, P.; Gasparella, A. Energy audit of schools by means of cluster analysis. *Energy Build.* **2015**, *95*, 160–171. [CrossRef]
31. Collins, M.; Curtis, J. Bunching of residential building energy performance certificates at threshold values. *Appl. Energy* **2018**, *211*, 662–676. [CrossRef]
32. Lin, M.; Afshari, A.; Azar, E. A data-driven analysis of building energy use with emphasis on operation and maintenance: A case study from the UAE. *J. Clean. Prod.* **2018**, *192*, 169–178. [CrossRef]
33. van den Brom, P.; Meijer, A.; Visscher, H. Performance gaps in energy consumption: household groups and building characteristics. *Build. Res. Inf.* **2018**, *46*, 54–70. [CrossRef]
34. Droutsa, K.G.; Kontoyiannidis, S.; Dascalaki, E.G.; Balaras, C.A. Mapping the energy performance of hellenic residential buildings from EPC (energy performance certificate) data. *Energy* **2016**, *98*, 284–295. [CrossRef]
35. Khayatian, F.; Sarto, L.; Dall'O', G. Application of neural networks for evaluating energy performance certificates of residential buildings. *Energy Build.* **2016**, *125*, 45–54. [CrossRef]
36. Park, H.S.; Lee, M.; Kang, H.; Hong, T.; Jeong, J. Development of a new energy benchmark for improving the operational rating system of office buildings using various data-mining techniques. *Appl. Energy* **2016**, *173*, 225–237. [CrossRef]
37. Tso, G.K.; Yau, K.K. Predicting electricity energy consumption: A comparison of regression analysis, decision tree and neural networks. *Energy* **2007**, *32*, 1761–1768. [CrossRef]
38. Magalhães, S.M.; Leal, V.M.; Horta, I.M. Modelling the relationship between heating energy use and indoor temperatures in residential buildings through Artificial Neural Networks considering occupant behavior. *Energy Build.* **2017**, *151*, 332–343. [CrossRef]
39. Acquaviva, A.; Apiletti, D.; Attanasio, A.; Baralis, E.; Bottaccioli, L.; Castagnetti, F.B.; Cerquitelli, T.; Chiusano, S.; Macii, E.; Martellacci, D.; et al. Energy Signature Analysis: Knowledge at Your Fingertips. In Proceedings of the 2015 IEEE International Congress on Big Data, New York, NY, USA, 27 June–2 July 2015; pp. 543–550.
40. Fabbri, K.; Tronchin, L.; Tarabusi, V. Real Estate market, energy rating and cost. Reflections about an Italian case study. *Procedia Eng.* **2011**, *21*, 303–310. [CrossRef]
41. Hjortling, C.; Björk, F.; Berg, M.; af Klintberg, T. Energy mapping of existing building stock in Sweden—Analysis of data from Energy Performance Certificates. *Energy Build.* **2017**, *153*, 341–355. [CrossRef]
42. Xiao, H.; Wei, Q.; Jiang, Y. The reality and statistical distribution of energy consumption in office buildings in China. *Energy Build.* **2012**, *50*, 259–265. [CrossRef]
43. Dascalaki, E.G.; Kontoyiannidis, S.; Balaras, C.A.; Droutsa, K.G. Energy certification of Hellenic buildings: First findings. *Energy Build.* **2013**, *65*, 429–437. [CrossRef]
44. Gangolells, M.; Casals, M.; Forcada, N.; Macarulla, M.; Cuerva, E. Energy mapping of existing building stock in Spain. *J. Clean. Prod.* **2016**, *112*, 3895–3904. [CrossRef]
45. MISE. *Decreto Ministeriale 26/6/2009—Ministero dello Sviluppo Economico. Linee guida nazionali per la certificazione energetica degli edifici*; MISE-Ministero dello Sviluppo Economico: Roma, Italy, 2009.
46. Hastie, T.; Tibshirani, R.; Friedman, J. Boosting and Additive Trees. In *The Elements of Statistical Learning: Data Mining, Inference, and Prediction*; Springer: New York, NY, USA, 2009; pp. 337–387.
47. Hofmann, M.; Klinkenberg, R. *RapidMiner: Data Mining Use Cases and Business Analytics Applications*; Chapman & Hall/CRC: Boca Raton, FL, USA, 2013.
48. R Core Team. *R: A Language and Environment for Statistical Computing*; R Foundation for Statistical Computing: Vienna, Austria, 2017.
49. RapidMiner. RapidMiner Operator Reference Manual. Available online: https://docs.rapidminer.com/latest/studio/operators/ (accessed on 1 April 2019).
50. Vapnik, V.N.; Vapnik, V. *Statistical Learning Theory*; Wiley: New York, NY, USA, 1998; Volume 1.
51. Ben-Hur, A.; Weston, J. A User's Guide to Support Vector Machines. In *Data Mining Techniques for the Life Sciences*; Carugo, O., Eisenhaber, F., Eds.; Humana Press: Totowa, NJ, USA, 2010; pp. 223–239.
52. Thaseen, S.; Kumar, C.A. An analysis of supervised tree based classifiers for intrusion detection system. In Proceedings of the 2013 International Conference on Pattern Recognition, Informatics and Mobile Engineering, Salem, India, 21–22 February 2013; pp. 294–299.
53. Breiman, L. Random Forests. *Mach. Learn.* **2001**, *45*, 5–32. [CrossRef]

54. Pang-Ning T.; Steinbach M.; Kumar V. *Introduction to Data Mining*; Addison-Wesley: Boston, MA, USA, 2006.
55. Rafiq, M.; Bugmann, G.; Easterbrook, D. Neural network design for engineering applications. *Comput. Struct.* **2001**, *79*, 1541–1552. [CrossRef]
56. Capozzoli, A.; Grassi, D.; Causone, F. Estimation models of heating energy consumption in schools for local authorities planning. *Energy Build.* **2015**, *105*, 302–313. [CrossRef]

energies

MDPI

Article

Parametric Performance Analysis and Energy Model Calibration Workflow Integration—A Scalable Approach for Buildings

Massimiliano Manfren [1] and Benedetto Nastasi [2,3,*]

[1] Faculty of Engineering and Physical Sciences, University of Southampton, Boldrewood Innovation Campus, Burgess Rd, Southampton SO16 7QF, UK; M.Manfren@soton.ac.uk

[2] Department of Planning, Design and Technology of Architecture, Sapienza University of Rome, Via Flaminia 72, 00196 Rome, Italy

[3] Department of Architectural Engineering & Technology, TU Delft University of Technology, Julianalaan 134, 2628BL Delft, The Netherlands

* Correspondence: benedetto.nastasi@outlook.com

Received: 25 November 2019; Accepted: 27 January 2020; Published: 1 February 2020

Abstract: High efficiency paradigms and rigorous normative standards for new and existing buildings are fundamental components of sustainability and energy transitions strategies today. However, optimistic assumptions and simplifications are often considered in the design phase and, even when detailed simulation tools are used, the validation of simulation results remains an issue. Further, empirical evidences indicate that the gap between predicted and measured performance can be quite large owing to different types of errors made in the building life cycle phases. Consequently, the discrepancy between a priori performance assessment and a posteriori measured performance can hinder the development and diffusion of energy efficiency practices, especially considering the investment risk. The approach proposed in the research is rooted on the integration of parametric simulation techniques, adopted in the design phase, and inverse modelling techniques applied in Measurement and Verification (M&V) practice, i.e., model calibration, in the operation phase. The research focuses on the analysis of these technical aspects for a Passive House case study, showing an efficient and transparent way to link design and operation performance analysis, reducing effort in modelling and monitoring. The approach can be used to detect and highlight the impact of critical assumptions in the design phase as well as to guarantee the robustness of energy performance management in the operational phase, providing parametric performance boundaries to ease monitoring process and identification of insights in a simple, robust and scalable way.

Keywords: building performance simulation; parametric modelling; energy management; model calibration; energy efficiency; Passive House

---

1. Introduction

The increasing effort towards resource efficiency and sustainability in the building sector [1] is progressively changing the way buildings are designed and managed. The decarbonisation of the built environment is a key objective for energy and environmental policies in the EU [2,3] and worldwide [4]. New efficiency paradigms (i.e., NZEBs) regarding existing and new buildings [5] have been introduced in recent years in the EU and other countries, at the global level. Passive design strategies making use of solar energy and internal gains are well established [6]. However, optimistic assumptions are often made in the design phase and semi-stationary calculation methodologies are still commonly employed [7]. Further, the gap between simulated and measured performance is a general issue [8] and the benefits of "green" design practices should be critically evaluated [9,10], by assessing transparently the impact of human and technical factors [11]. With respect to human factors in particular, the effects of occupants' behaviour [12] and of their comfort preferences [13] on building performance are generally overlooked in the design phase. This paper aims to present a way to integrate modelling methodologies used across building life cycle phases, from design to operation, in a simple and scalable way. A residential building has been chosen as a case study. The building is a detached single family certified Passive House built in Italy, in the Province of Forlì-Cesena, in the Emilia Romagna region. It has been monitored for three years, learning incrementally insights by comparing the original design phase simulation data with actual measured data.

2. Background and Motivation

The research work answers to the necessity of linking parametric performance analysis and model calibration from a conceptual and practical point of view. Building performance parametric and probabilistic analysis is an essential tool today to ensure robustness of performance and the importance of the Design of Experiments (DOE) is becoming clear [14–17], both for new and retrofitted buildings [18,19]. For example, accounting for the robustness of performance estimates with respect to economic indicators (e.g., in cost-optimal analysis [20–22]) is important because uncertainty can affect the credibility and, consequently, hinder the success of policies oriented to investments on efficiency in the built environment. In this research, baseline design simulation, i.e., original design simulation for the building project, was used as baseline and multiple Design of Experiments (DOE) simulations were run to compute the impact of the variability of multiple inputs (envelope components performance, operational settings, occupant's behaviour and comfort preferences, etc.), as specified in detail in Section 3.1. The parametric approach aims at detecting critical assumptions in the preliminary design stage, to guarantee a more robust evaluation of performance [15,23]. In simpler terms, the objective of the parametric simulation is to include from the very beginning more realistic, and possibly less optimistic, assumptions, and use the simulation outcomes as boundaries for comparative performance analysis during the operation phase. In order to reduce the computational effort, meta-modelling techniques can be used [24] (i.e., surrogate, reduced-order). The choice of meta-modelling techniques depends on several factors [25]: they are very flexible and they can be employed for different uses such as the optimization of design [26], model calibration [24] and control [27]. Additionally, different meta-models can give similar performance on the same problem [24,28]. In this research regression models were tested for performance prediction, using energy signatures [29,30] regressed against weather data [24,31–33]. Therefore, multiple piecewise linear multivariate regression models are trained first on simulation data, as described in Section 3.2. These models are, then, updated and calibrated on measured data during three years of operation. Visualization and numerical techniques are combined to allow an intuitive results interpretation as well as to facilitate human interaction in the calibration process, encompassing model training and testing phases. While being less sophisticated than other machine learning techniques available today, multivariate regression models have been chosen because of a set of important features. First of all, standardization [29,30], temporal [34,35] and spatial scalability [36,37], weather normalization using Variable Base Degree-Days (VBDD) [38,39]. After that, the applicability to multiple types of building end-uses [33] and the flexibility with respect to diverse operational strategies and conditions [12,40,41], e.g., accounting for different levels of thermal inertia [42]. Further, the possibility to easily extend their applicability using techniques such as Monte Carlo simulation [41], Bayesian analysis [43,44], eventually exploiting the approximated physical interpretation of coefficients [33,45]. Finally, this technique is suitable for performance tracking with periodic recalibration in changing climate conditions [46,47] and can complement the analysis of performance of technologies such as heat pumps and cooling machines [48,49], considering also exergy balance [50,51]. In the next Section the research methodology is explained, starting from parametric simulation and, then, moving to regression analysis on energy signatures.

3. Research Methodology

In the original design of the building, Passive House Planning package (PHPP) [52] was used for simulation. Instead, in this study we used a validated grey-box dynamic model [53,54] to perform multiple simulation runs in a reduced time frame. Indeed, grey-box models are very flexible and can be used in the inverse mode to estimate lumped properties of the actual building, eventually extending their applicability with Bayesian analysis [55,56] or Dempster-Shafer theory of the evidence [57]. In this case, the original building design configuration was considered as a baseline. Then parametric simulations were run using the Design of Experiments (DOE) methodology [58], similarly to other research studies on the variability in building performance simulation [15,16,59]. Variations and multiple runs are meant to reproduce the actual variability of the performance of envelope components, air-change rates and of occupants' behaviour and comfort preferences. As described before, these variations in the operation phase (generally) entail a significant gap between simulation and actual measured performance.

3.1. Parametric Performance Analysis of the Case Study

The case study chosen is a single family detached Passive House built in Italy, in the Province of Forlì-Cesena, in the Emilia Romagna region. The case study was chosen because it represents an example of a high efficiency building design and we wanted to analyse its actual performance in operation (as well as its evolution in time) together with the applicability of the approach proposed, based on an extension of well-established M&V techniques. The approach proposed substantially anticipates the use of inverse modelling at the design stage and the goal of parametric simulation is that of creating an envelopment of data to be considered as possible scenarios of actual building performance in operation. The building has a high level of insulation of envelope components and it is equipped with mechanical ventilation with heat recovery (air/air heat exchanger), a solar thermal to integrate DHW production, a ground-source heat pump system (GSHP) and a PV plant for local electricity production. Simulation input data are summarized in Table 1, reporting baseline configuration with respect to the two level Design of Experiments (DOE) configurations. The U values in Table 1 were averaged with respect to the external surface of components (summarized then in the heat loss surface area) and considered the impact of thermal bridges. Technical systems data are synthesized hereafter in Table 2.

Table 1. Simulation data from baseline and two-level Design of Experiments (DOE).

Group	Type	Unit	Baseline	Design of Experiment Levels −1	Design of Experiment Levels +1
Climate	UNI 10349:2016	-			
Geometry	Gross volume	m^3	1557		
	Net volume	m^3	1231		
	Heat loss surface area	m^2	847		
	Net floor area	m^2	444		
	Surface/volume ratio	1/m	0,54		
Envelope	U value external walls	W/(m^{2}K)	0.18	0.23	0.27
	U value roof	W/(m^{2}K)	0.17	0.21	0.26
	U value transparent components	W/(m^{2}K)	0.83	1.04	1.25
Activities	Internal gains (lighting, appliances and occupancy, daily average)	W/m^2	1	1	1.5
	Occupants	-	5	5	5
Control and operation	Heating set-point temperature	°C	20	20	22
	Cooling set-point temperature	°C	26	26	28
	Air-change rate (infiltration and mechanical ventilation with heat recovery in heating mode)	vol/h	0.2	0.2	0.4
	Shading factor (solar control summer mode)	-	0.5	0.5	0.7
	Domestic hot water demand	l/person/day	50	50	70
	Schedules—DOE constant operation	-	0.00–23.00	0.00–23.00	0.00–23.00
	Schedules—DOE behaviour 1	-	7.00–22.00	7.00–22.00	7.00–22.00
	Schedules—DOE behaviour 2	-	7.00–9.00, 17.00–22.00	7.00–9.00, 17.00–22.00	7.00–9.00, 17.00–22.00

Table 2. Technical systems data.

Technical System	Technology	Type	Unit	Value
Heating/Cooling system	Ground Source Heat Pump	Brine/Water Heat Pump	kW	8.4
	Ground heat exchanger	Borehole Heat Exchanger (2 double U boreholes)	m	100
On-site energy generation	Building Integrated Photo-Voltaic (BIPV)	Polycrystalline Silicon	kWp	9.2
	Solar Thermal	Glazed flat plate collector Domestic Hot Water storage	m^2 m^3	4.32 0.74

In order to simulate realistically multiple operating conditions, different schedules for internal gains (lighting, appliances and people), heating, cooling and air-exchange rates (ventilation/infiltration) have been created. Three DOE simulation runs were performed, one for each operational schedule, (simulating diverse occupants' behaviour) namely continuous operation (constant operation profile), operation mainly from 7.00 to 22.00 (behaviour 1), operation mainly from 7.00 to 9.00 and from 17.00 to 22.00 (behaviour 2).

The typical Key Performance Indicators (KPIs) considered in building energy analysis are final energy use (e.g., thermal demand for heating, cooling and domestic hot water), energy demand (e.g., energy carriers such as electricity, natural gas, etc.), cost of energy services, primary energy use and CO_2 emissions. In this study, we concentrate on aggregated electricity demand for heating, cooling, domestic hot water (DHW), lighting and appliances, because all these services are supplied by electricity.

3.2. Parametric Performance Analysis and Model Calibration Integrated Workflow

The choice in this research is to adopt a piecewise linear multivariate regression approach, using energy signature technique [29] to analyse both data generated by means of parametric simulation in the design phase and monitored data during the calibration phase. As a matter of fact, for the calibration purpose, many types of meta-models are available. A regression approach is proposed in this study following the arguments presented in Section 2. Table 3 shows the piecewise linear multivariate regression models [30] implemented. Three linear sub-models compose the overall predictive model, each one defined between specific boundaries for heating and cooling and baseline demand, respectively. Dummy variables are added to enable a piecewise linear model formulation. Dummy variables are binary (0,1) and are multiplied by the original independent variable to obtain interaction variables, in such a way that the total model is the sum of heating, cooling and base load components (piecewise linear components). Regression models consider only external temperature dependence, in the case of model type 1 while, external temperature together with solar radiation dependence, in the case of model type 2.

Table 3. Regression models for heating, cooling and baseline demand analysis.

Demand	Model Type 1	Model Type 2
Heating	$q_{h,1} = a_0 + a_1\theta_e + \varepsilon$	$q_{h,2} = b_0 + b_1\theta_e + b_2 I_{sol} + \varepsilon$
Cooling	$q_{c,1} = c_0 + c_1\theta_e + \varepsilon$	$q_{c,2} = d_0 + d_1\theta_e + d_2 I_{sol} + \varepsilon$
Base load	$q_{b,1} = e_0 + e_1\theta_e + \varepsilon$	$q_{b,2} = f_0 + f_1\theta_e + f_2 I_{sol} + \varepsilon$

To assess and compare the simulation data in the design phase and measured data in the operation phase, basic statistical indicators are used together with statistical indicators specific for state-of-the-art model calibration procedures [30,60,61]. The basic statistical indicators chosen were R^2 and Mean Absolute Percentage Error (*MAPE*). The determination coefficient R^2 expresses the goodness of a regression model fit, varying from 0 to 1 (or 0% to 100%), where the maximum values indicate that the model fits perfectly the data. The R^2 was calculated as 1 minus the ratio between the sum of the squares of residuals and total sum of the squares using Equation (1). Mean Absolute Percentage Error (*MAPE*) represents the average absolute value of the difference between measured and predicted data, normalized to measured data. Equation (2) reports the *MAPE* calculation (we can substitute M_i with S_i when simulated data are used instead of measured ones).

$$R^2 = 1 - \frac{SS_{res}}{SS_{tot}} = 1 - \frac{\sum_i (y_i - \hat{y}_i)^2}{\sum_i \left(y_i - \overline{y}_i\right)^2} \quad (1)$$

$$MAPE = \frac{1}{n}\sum_i \left|\frac{M_i - P_i}{M_i}\right| \cdot 100 \quad (2)$$

Going to the specific indicators for calibration, Normalized Mean Bias Error (*NMBE*) and *Cv(RMSE)* Coefficient of Variation of Root Mean Squared Error (*RMSE*) were used. *NMBE* is the total sum of the differences between measured (or simulated in the case of design phase, replacing M_i with S_i) and predicted energy consumption at the calculated time intervals, in this case monthly, divided by the sum of the measured (or simulated) energy consumption. *NMBE* is reported in Equation (3). An overestimation of energy consumption determines a positive value of *NMBE* while an underestimation determines a negative one.

$$NMBE = -\frac{\sum_i (M_i - P_i)}{\sum_i M_i} \cdot 100 \quad (3)$$

Cv(RMSE) is the normalized measure of the differences between measured M_i (or simulated S_i in the case of design phase) and predicted data P_i. It is based on *RMSE*, a measure of the sample deviation of the differences among values measured and predicted by the model divided by *A*, which represents measured (or simulated in the case of design phase, replacing M_i with S_i) average energy consumption. The lower the *Cv(RMSE)* value the better calibrated the model is. *Cv(RMSE)* calculation is illustrated in Equations (4), (5) and (6).

$$Cv(RMSE) = \frac{RMSE}{A} \cdot 100 \quad (4)$$

$$RMSE = \sqrt{\frac{\sum_i (M_i - P_i)^2}{n}} \quad (5)$$

$$A = \frac{\sum_i M_i}{n} \quad (6)$$

The threshold metrics considered in different protocols for M&V and calibration at the state-of-the-art [23,44,45], are discussed in the literature [62] and reported in Table 4 for calibration with monthly data.

Table 4. Threshold limits of metrics for model calibration with monthly data.

Metric	ASHRAE Guidelines 14	IPMVP	FEMP
NMBE (%)	±5	±20	±5
Cv(RMSE) (%)	15	-	15

Finally, the analysis of deviations (differences) between measurements and predictions can be useful to discover hidden patterns in data. Equation (7) was used for this purpose. The energy consumption is underestimated when a positive deviation occurs at a certain point (i.e., measured consumption M_i is higher than predicted P_i) while an overestimation takes place when a negative deviation derives from calculation (i.e., measured consumption M_i is lower than predicted P_i).

$$D_i = M_i - P_i \quad (7)$$

4. Results and Discussion

This study aimed to illustrate an integrated workflow from the parametric performance analysis to model calibration through its essential steps, using a Passive House case study as example. First, the results obtained from the baseline and DOE simulations, performed according to the input data reported in Section 3.1 in Table 1, were used to calculate Key Performance Indicators (KPIs) on a yearly base. These indicators serve as a basis for the comparison of parametric simulation output data. In Figure 1 we report a summary of the weather data used for simulation (design weather data file) and during model calibration (the monitoring period). More specifically, weather data reported are monthly average external air temperatures and daily average global solar radiation on the horizontal surface. These data are representative of typical average days for every month.

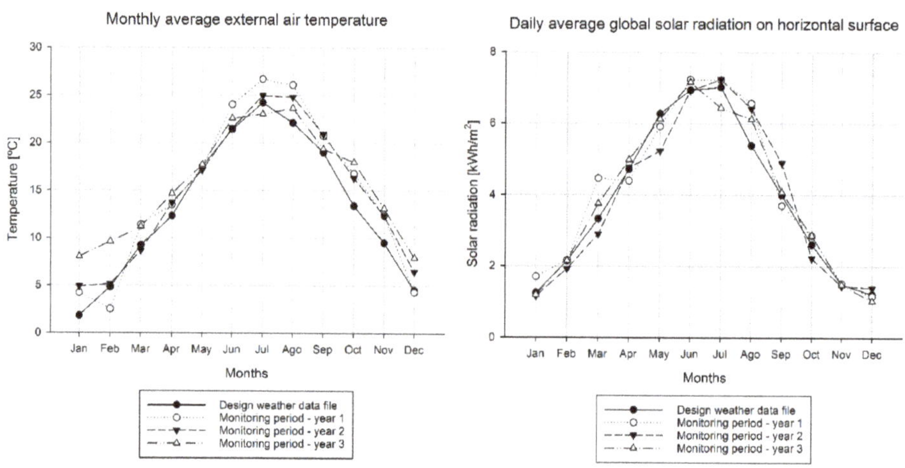

Figure 1. Weather data used in the study

While the integrated workflow presented could be applied in a more general way, following the arguments reported in Section 2, the focus of this study was put on analysing the aggregated electricity demand data. Electricity demand was divided by the square meters of the net floor area and reported hereafter in Table 5 for the baseline, lower bound (LB) and upper bound (UB), which corresponded to the envelopment of outputs from the DOE simulation.

Table 5. Comparison of the baseline and two-level DOE simulation data—lower bound and upper bound of Key Performance Indicator (KPI) yearly values with respect to the baseline.

KPI	Unit	Baseline	Design of Experiment	
			LB	UB
Electricity consumption	kWh/m²	20.8	16.9	31.7

In Figure 2 the detailed composition of electricity demand for baseline simulation configuration (input configuration is provided in Table 1) is shown. The electricity demand for domestic hot water service was negligible in the summer months as it was supplied by the solar thermal system (Table 2).

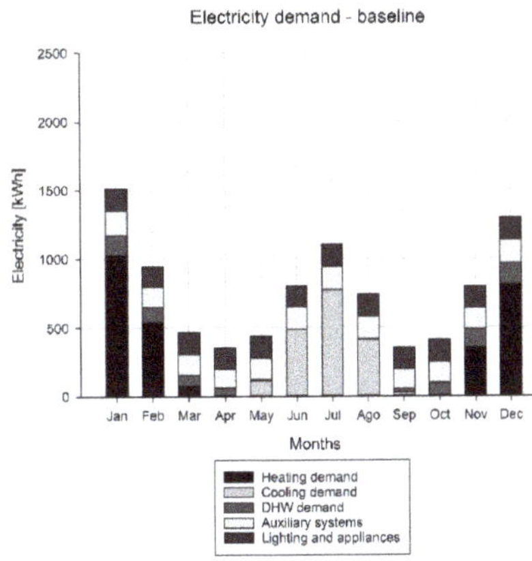

Figure 2. Electricity demand—baseline model results

Simulation data were then used to train regression models type 1 and type 2, as explained in Section 3.2. In this phase models were still uncalibrated, i.e., they were not calibrated on measured data but simply trained on simulation data, in order to verify their applicability and goodness of fit (i.e., the ability to approximate the results of dynamic simulations). The statistical indicators obtained, introduced in Section 3.2, are reported in Table 6, showing that both model types can fit simulation data reasonably well, even though the performance of model type 2 was comparatively higher.

Table 6. Model training on the design phase data (uncalibrated models, a priori data)

Model Type	Calibration Process Stage	Training Dataset	Testing Dataset	Statistical Indicators			
				R^2	*MAPE*	*NMBE*	*Cv(RMSE)*
				%	%	%	%
Type 1	Uncalibrated	DOE - Overall LB	-	93.65	9.34	0.06	13.58
Type 1	Uncalibrated	DOE - Overall UB	-	96.64	7.33	0.02	9.01
Type 2	Uncalibrated	DOE - Overall LB	-	99.90	1.42	−0.02	1.65
Type 2	Uncalibrated	DOE - Overall UB	-	99.78	1.93	−0.01	2.36

Subsequently, the first step of the parametric analysis corresponds to the comparison of monthly electric energy demand data for the baseline and DOE lower bound and upper bound configurations, as in Table 1 (parametric simulation input). The comparison is reported in Figure 3, showing on the left side the monthly energy values obtained by simulation and on the right side the corresponding parametric energy signatures (expressed as average power). Energy signatures enable the comparison between simulated and measured data during the subsequent monitoring process and represent the a priori knowledge we have about the building performance, which we could use to identify anomalies visually and numerically. Indeed, the regression models developed were independent of the specific weather data used, as weather data were the independent variables (air temperature and solar radiation in this case), while the average power was the dependent variable. As shown in Figure 1, 4 years of weather data were considered in this study, 1 design weather data file and 3 years of monitoring data.

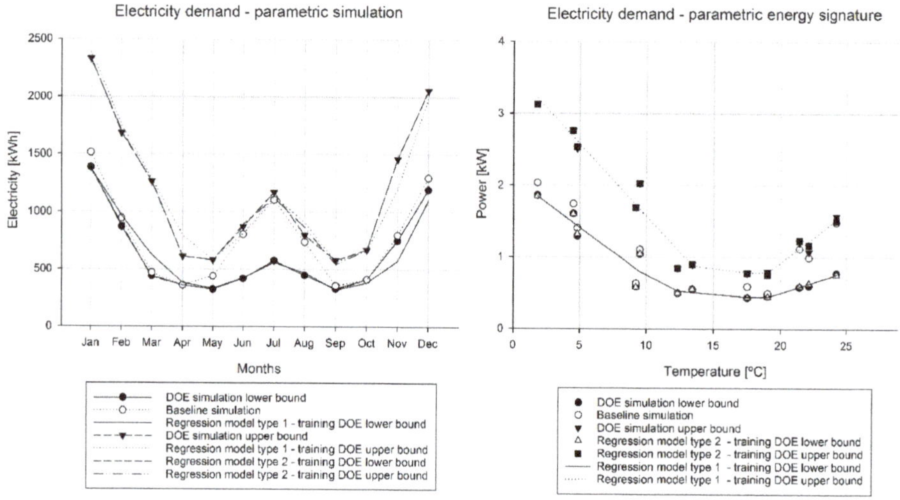

Figure 3. Electricity demand—DOE parametric model simulation and training (a priori knowledge) for monitoring purpose.

After that, the results of the incremental model calibration process during the three year monitoring period are reported for both model types in Table 7. The measured data were more scattered compared to the simulated ones, leading to higher R^2, *MAPE* and *Cv(RMSE)*. In this phase, the type 1 model did not reach the calibration threshold with 2 years of monthly data because *Cv(RMSE)* was 19.75%, higher than 15% threshold reported in Table 4. So, it could be defined as partially calibrated. Instead, model type 2 was calibrated, as confirmed by statistical indicators in training and testing phases.

Table 7. Model training and testing on the operation phase data (calibrated models, a posteriori knowledge).

Model Type	Calibration Process Stage	Training Dataset	Testing Dataset	Statistical Indicators			
				R^2	*MAPE*	*NMBE*	*Cv(RMSE)*
				%	%	%	%
Type 1	Partial Calibrated	Measured data—Year 1 and 2		82.64	11.44	0.04	13.44
			Measured data—Year 3	69.74	18.40	−6.95	19.75
Type 2	Calibrated	Measured data—Year 1 and 2	-	86.07	9.97	0.05	12.02
		-	Measured data—Year 3	87.54	11.97	−2.21	12.50

In any case, a reasonable amount of data and a corresponding time span are needed. In this case study, two years of monthly data to reach calibration or partial calibration of regression models were necessary. As described before, uncalibrated design models, reported in Table 6 and depicted in Figure 3, could provide a useful support in the monitoring process, as they represent estimated bounds of performance (lower and upper bounds of a data envelopment) determined by means of parametric simulation. The assumptions that characterize parametric building performance simulation themselves can be updated based on experience gained in model calibration processes in real buildings, e.g., by reducing or increasing the level of variability of a certain input quantity (Table 1) when more detailed information is available. For this purpose, a priori knowledge represented by simulated data, i.e., uncalibrated models can be compared with a posteriori knowledge, represented by measured data, as shown on the left side of Figure 4. In the same figure, on the right side, a posteriori knowledge, i.e., calibrated models with measured data (at the end of the monitoring period) are reported for comparison.

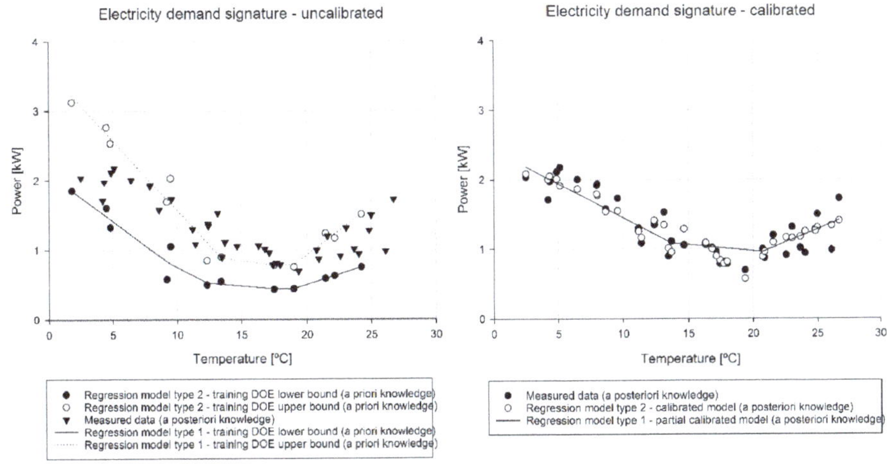

Figure 4. Monitoring the electricity demand—overall analysis of a priori and a posteriori knowledge from uncalibrated to calibrated models.

The analysis of the changes of models' regression coefficients during the calibration process and, in particular, changes of slopes and break points for piecewise linear energy signature models, constitute starting points for a more in depth analysis, based on approximate physical interpretation as explained in Sections 2 and 3. Hereafter, we illustrate how the monitoring process evolved in time. By plotting the data with respect to time, i.e., months of monitoring, we obtained Figures 5 and 6 for uncalibrated (a priori knowledge) and calibrated (a posteriori knowledge) models, respectively.

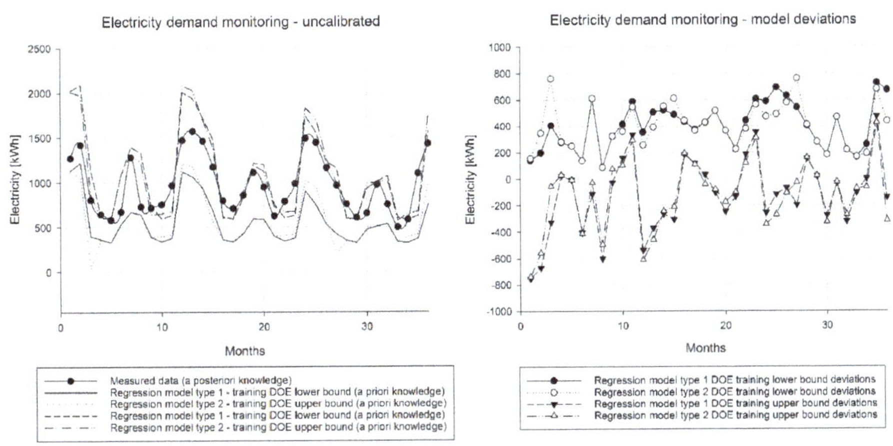

Figure 5. Electricity demand monitoring—comparison between measured data, regression model type 1 and type 2 lower bound and upper bound and deviations between measured and predicted data.

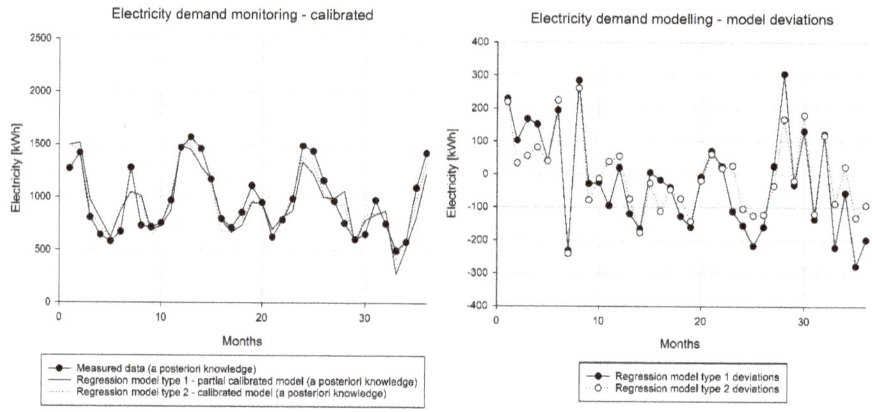

Figure 6. Electricity demand monitoring—comparison of measured data, partial calibrated regression model type 1, calibrated regression model type 2 and deviations between measured and predicted data.

As reported in Table 7, the calibrated models were trained on the first two years of data and then tested on the third year of data (36 months of the total monitoring period). In Figure 5 we could observe the evolution of building performance in time with respect to our (pre)established performance boundaries, while in Figure 6 we could verify how measured data and calibrated (or partially calibrated in the case of type 1) models data reasonably overlaps on a monthly base. On the right side of both Figures 5 and 6, the deviations between measured and predicted data are plotted. Deviations in Figure 5 indicate that, at many points in time, the building had an energy consumption near to the upper bound of simulated electricity consumption while, in just a few points in time, it has an energy consumption near to the lower bound of simulation.

Finally, in Figure 6 the deviations between measured and predicted data exhibited a pattern in time (similar for both types of models). Variations can depend on multiple factors and, among them, on behavioural change of occupants that may have determined different values of internal gains and/or differences in operation schedules and settings of technical systems. Understanding this requires a more in depth analysis that will be part of future research, together with the application of the same methodology for a multi-level (regression-based) model calibration with physical interpretation of regression coefficients, as reported before.

5. Conclusions

Rigorous normative standards for new and existing buildings are an essential part of energy and sustainability policies today. The effort put in modelling in the design phase is not, by itself, a guarantee of optimal measured performance. Optimistic assumptions and simplifications are often considered in the design phase and the validation of simulation results represents an issue, as well as model calibration on measured data and long-term monitoring. In this research a simple and scalable way to validate and monitor building performance using monthly data was proposed. It uses an envelopment of data generated in the design phase by means of the Design Of Experiment (DOE) technique together with multivariate regression models, periodically retrained during building operation. In this way, a continuous improvement in design and operation practices becomes possible by linking parametric performance analysis to model calibration, i.e., using inverse modelling already in the design phase, considering multiple configurations. In fact, the assumptions that characterize building performance analysis can be updated based on the experience gained in model calibration, e.g., by reducing or increasing the level of variability of a certain input quantity when more detailed information is available. Further research should be devoted, on the one hand, to the creation of a transparent connection between this approach and ongoing technical standardization, using verification and validation standards for forward models. On the other hand, the use of inverse modelling techniques, i.e., surrogate models, meta-models, in Measurement and Verification (M&V) during the operation phase should become increasingly common, making use of the current state-of-the art of technical standardization. All these elements are scientifically and empirically consolidated but their integration and synthesis are still open issues. Therefore, we believe that future research efforts should be oriented in this direction, in particular with respect to the robustness of performance estimates, i.e., identification of realistic boundaries for performance at multiple levels such as building zones, technical systems and meters under realistic operating conditions. It must be also considered the possibility to scale models from single buildings to building clusters and stock for large scale performance benchmarking. In fact, scalability of analysis techniques can greatly contribute to the definition of effective policies in energy and sustainability transition in the future, supported by large scale data analytics.

Author Contributions: Conceptualization, M.M.; methodology, M.M.; investigation, M.M.; writing—original draft preparation, M.M. and B.N.; writing—review and editing, B.N. All authors have read and agreed to the published version of the manuscript.

Funding: This research received no external funding.

Conflicts of Interest: The authors declare no conflict of interest. The funders had no role in the design of the study; in the collection, analyses, or interpretation of data; in the writing of the manuscript, or in the decision to publish the results.

Nomenclature

Variables and Parameters	
A	average value
a,b,c,d,e,f	regression coefficients
$Cv(RMSE)$	coefficient of variation of RMSE
D	deviation, difference between measured and simulated data
I	radiation
M	measured data
$MAPE$	mean absolute percentage error
$NMBE$	normalized mean bias error
q	specific energy transfer rate (energy signature)
P	predicted data
R^2	determination coefficient
RD	relative deviation
$RMSE$	root mean square error
S	simulated
SS	sum of the squares
y	numeric value
θ	temperature
Subscripts and Superscripts	
-	average
^	predicted value
b	baseline
c	cooling
h	heating
i	index
res	residual
sol	solar

References

1. Dodd, N.; Donatello, S.; Garbarino, E.; Gama-Caldas, M. *Identifying Macro-Objectives for the Life Cycle Environmental Performance and Resource Efficiency of EU Buildings*; JRC EU Commission: Seville, Spain, 2015.
2. BPIE. *Europe's Buildings Under the Microscope*; Buildings Performance Institute Europe (BPIE): Brussels, Belgium, 2011.
3. Saheb, Y. *Energy Transition of the EU Building Stock—Unleashing the 4th Industrial Revolution in Europe*; OpenExp: Paris, France, 2016; p. 352.
4. Berardi, U. A cross-country comparison of the building energy consumptions and their trends. *Resour. Conserv. Recycl.* **2017**, *123*, 230–241. [CrossRef]
5. D'Agostino, D.; Zangheri, P.; Cuniberti, B.; Paci, D.; Bertoldi, P. *Synthesis Report on the National Plans for Nearly Zero Energy Buildings (NZEBs)*; JRC EU Commission: Ispra, Italy, 2016.
6. Chwieduk, D. *Solar Energy in Buildings: Thermal Balance for Efficient Heating and Cooling*; Academic Press: Cambridge, MA, USA, 2014.
7. ISO. *Energy Performance of Buildings—Energy Needs for Heating and Cooling, Internal Temperatures and Sensible and Latent Head Loads—Part 1: Calculation Procedures*; Technical Report No. ISO 52016-1:2017; ISO: Geneva, Switzerland, 2016.
8. Imam, S.; Coley, D.A.; Walker, I. The building performance gap: Are modellers literate? *Build. Serv. Eng. Res. Technol.* **2017**, *38*, 351–375. [CrossRef]
9. Scofield, J.H.; Cornell, J. A critical look at "Energy savings, emissions reductions, and health co-benefits of the green building movement". *J. Expo. Sci. Environ. Epidemiol.* **2019**, *29*, 584–593. [CrossRef] [PubMed]
10. MacNaughton, P.; Cao, X.; Buonocore, J.; Cedeno-Laurant, J.; Sprengle, J.; Bernstein, A.; Allen, J. Energy savings, emission reductions, and health co-benefits of the green building movement. *J Expo. Sci. Environ. Epidemiol.* **2018**, *28*, 307–318. [CrossRef]
11. Yoshino, H.; Hong, T.; Nord, N. IEA EBC annex 53: Total energy use in buildings—Analysis and evaluation methods. *Energy Build.* **2017**, *152*, 124–136. [CrossRef]
12. Tagliabue, L.C.; Manfren, M.; Ciribini, A.L.C.; De Angelis, E. Probabilistic behavioural modeling in building performance simulation—The Brescia eLUX lab. *Energy Build.* **2016**, *128*, 119–131. [CrossRef]
13. Fabbri, K.; Tronchin, L. Indoor environmental quality in low energy buildings. *Energy Procedia* **2015**, *78*, 2778–2783. [CrossRef]
14. Jaffal, I.; Inard, C.; Ghiaus, C. Fast method to predict building heating demand based on the design of experiments. *Energy Build.* **2009**, *41*, 669–677. [CrossRef]
15. Kotireddy, R.; Hoes, P.-J.; Hensen, J.L.M. A methodology for performance robustness assessment of low-energy buildings using scenario analysis. *Appl. Energy* **2018**, *212*, 428–442. [CrossRef]
16. Schlueter, A.; Geyer, P. Linking BIM and design of experiments to balance architectural and technical design factors for energy performance. *Autom. Constr.* **2018**, *86*, 33–43. [CrossRef]
17. Shiel, P.; Tarantino, S.; Fischer, M. Parametric analysis of design stage building energy performance simulation models. *Energy Build.* **2018**, *172*, 78–93. [CrossRef]
18. EEFIG. *Energy Efficiency—the First Fuel for the EU Economy, how to Drive New Finance for Energy Efficiency Investments*; Energy Efficiency Financial Institutions Group: Brussels, Belgium, 2015.
19. Saheb, Y.; Bodis, K.; Szabo, S.; Ossenbrink, H.; Panev, S. *Energy Renovation: The Trump Card for the New Start for Europe*; JRC EU Commission: Ispra, Italy, 2015.
20. Aste, N.; Adhikari, R.S.; Manfren, M. Cost optimal analysis of heat pump technology adoption in residential reference buildings. *Renew. Energy* **2013**, *60*, 615–624. [CrossRef]
21. Tronchin, L.; Tommasino, M.C.; Fabbri, K. On the "cost-optimal levels" of energy performance requirements and its economic evaluation in Italy. *Int. J. Sustain. Energy Plan. Manag.* **2014**, *3*, 49–62.
22. Fabbri, K.; Tronchin, L.; Tarabusi, V. Energy retrofit and economic evaluation priorities applied at an Italian case study. *Energy Procedia* **2014**, *45*, 379–384. [CrossRef]
23. Ligier, S.; Robillart, M.; Schalbart, P.; Peuportier, B. Energy performance contracting methodology based upon simulation and measurement. In Proceedings of the IBPSA Building Simulation Conference 2017, San Francisco, CA, USA, 7–9 August 2017.
24. Manfren, M.; Aste, N.; Moshksar, R. Calibration and uncertainty analysis for computer models—A meta-model based approach for integrated building energy simulation. *Appl. Energy* **2013**, *103*, 627–641. [CrossRef]

25. Koulamas, C.; Kalogeras, A.P.; Pacheco-Torres, R.; Casillas, J.; Ferrarini, L. Suitability analysis of modeling and assessment approaches in energy efficiency in buildings. *Energy Build.* **2018**, *158*, 1662–1682. [CrossRef]
26. Nguyen, A.-T.; Reiter, S.; Rigo, P. A review on simulation-based optimization methods applied to building performance analysis. *Appl. Energy* **2014**, *113*, 1043–1058. [CrossRef]
27. Aste, N.; Manfren, M.; Marenzi, G. Building automation and control systems and performance optimization: A framework for analysis. *Renew. Sustain. Energy Rev.* **2017**, *75*, 313–330. [CrossRef]
28. Østergård, T.; Jensen, R.L.; Maagaard, S.E. A comparison of six metamodeling techniques applied to building performance simulations. *Appl. Energy* **2018**, *211*, 89–103. [CrossRef]
29. ISO. *Energy Performance of Buildings—Assessment of Overall Energy Performance*; Technical Report No. ISO 16346:2013; ISO: Geneva, Switzerland, 2013.
30. ASHRAE. *Guideline 14-2014: Measurement of Energy, Demand, and Water Savings*; American Society of Heating, Refrigerating and Air-Conditioning Engineers: Atlanta, GA, USA, 2014.
31. Masuda, H.; Claridge, D.E. Statistical modeling of the building energy balance variable for screening of metered energy use in large commercial buildings. *Energy Build.* **2014**, *77*, 292–303. [CrossRef]
32. Paulus, M.T.; Claridge, D.E.; Culp, C. Algorithm for automating the selection of a temperature dependent change point model. *Energy Build.* **2015**, *87*, 95–104. [CrossRef]
33. Tronchin, L.; Manfren, M.; Tagliabue, L.C. Optimization of building energy performance by means of multi-scale analysis—Lessons learned from case studies. *Sustain. Cities Soc.* **2016**, *27*, 296–306. [CrossRef]
34. Jalori, S.; Agami Reddy, T.P. A unified inverse modeling framework for whole-building energy interval data: Daily and hourly baseline modeling and short-term load forecasting. *ASHRAE Trans.* **2015**, *121*, 156.
35. Jalori, S.; Agami Reddy, T.P. A new clustering method to identify outliers and diurnal schedules from building energy interval data. *ASHRAE Trans.* **2015**, *121*, 33.
36. Abdolhosseini Qomi, M.J.; Noshadravan, A.; Sobstyl, J.M.; Toole, J.; Ferreira, J.; Pellenq, R.J.-M.; Ulm, F.-J.; Gonzalez, M.C. Data analytics for simplifying thermal efficiency planning in cities. *J. R. Soc. Interface* **2016**, *13*, 20150971. [CrossRef]
37. Kohler, M.; Blond, N.; Clappier, A. A city scale degree-day method to assess building space heating energy demands in Strasbourg Eurometropolis (France). *Appl. Energy* **2016**, *184*, 40–54. [CrossRef]
38. Ciulla, G.; Lo Brano, V.; D'Amico, A. Modelling relationship among energy demand, climate and office building features: A cluster analysis at European level. *Appl. Energy* **2016**, *183*, 1021–1034. [CrossRef]
39. Ciulla, G.; D'Amico, A. Building energy performance forecasting: A multiple linear regression approach. *Appl. Energy* **2019**, *253*, 113500. [CrossRef]
40. Tagliabue, L.C.; Manfren, M.; De Angelis, E. Energy efficiency assessment based on realistic occupancy patterns obtained through stochastic simulation. In *Modelling Behaviour*; Springer: Cham, Switzerland, 2015; pp. 469–478.
41. Cecconi, F.R.; Manfren, M.; Tagliabue, L.C.; Ciribini, A.L.C.; De Angelis, E. Probabilistic behavioral modeling in building performance simulation: A Monte Carlo approach. *Energy Build.* **2017**, *148*, 128–141. [CrossRef]
42. Aste, N.; Leonforte, F.; Manfren, M.; Mazzon, M. Thermal inertia and energy efficiency—Parametric simulation assessment on a calibrated case study. *Appl. Energy* **2015**, *145*, 111–123. [CrossRef]
43. Li, Q.; Augenbroe, G.; Brown, J. Assessment of linear emulators in lightweight Bayesian calibration of dynamic building energy models for parameter estimation and performance prediction. *Energy Build.* **2016**, *124*, 194–202. [CrossRef]
44. Booth, A.; Choudhary, R.; Spiegelhalter, D. A hierarchical Bayesian framework for calibrating micro-level models with macro-level data. *J. Build. Perform. Simul.* **2013**, *6*, 293–318. [CrossRef]
45. Tronchin, L.; Manfren, M.; Nastasi, B. Energy analytics for supporting built environment decarbonisation. *Energy Procedia* **2019**, *157*, 1486–1493. [CrossRef]
46. Jentsch, M.F.; Bahaj, A.S.; James, P.A.B. Climate change future proofing of buildings—Generation and assessment of building simulation weather files. *Energy Build.* **2008**, *40*, 2148–2168. [CrossRef]
47. Jentsch, M.F.; James, P.A.B.; Bourikas, L.; Bahaj, A.S. Transforming existing weather data for worldwide locations to enable energy and building performance simulation under future climates. *Renew. Energy* **2013**, *55*, 514–524. [CrossRef]
48. Busato, F.; Lazzarin, R.M.; Noro, M. Energy and economic analysis of different heat pump systems for space heating. *Int. J. Low-Carbon Technol.* **2012**, *7*, 104–112. [CrossRef]

49. Busato, F.; Lazzarin, R.M.; Noro, M. Two years of recorded data for a multisource heat pump system: A performance analysis. *Appl. Therm. Eng.* **2013**, *57*, 39–47. [CrossRef]
50. Tronchin, L.; Fabbri, K. Analysis of buildings' energy consumption by means of exergy method. *Int. J. Exergy* **2008**, *5*, 605–625. [CrossRef]
51. Meggers, F.; Ritter, V.; Goffin, P.; Baetschmann, M.; Leibundgut, H. Low exergy building systems implementation. *Energy* **2012**, *41*, 48–55. [CrossRef]
52. PHPP. The Energy Balance and Passive House Planning Tool. Available online: https://passivehouse.com/04_phpp/04_phpp.htm (accessed on 9 January 2020).
53. Michalak, P. The development and validation of the linear time varying Simulink-based model for the dynamic simulation of the thermal performance of buildings. *Energy Build.* **2017**, *141*, 333–340. [CrossRef]
54. Michalak, P. A thermal network model for the dynamic simulation of the energy performance of buildings with the time varying ventilation flow. *Energy Build.* **2019**, *202*, 109337. [CrossRef]
55. Kristensen, M.H.; Hedegaard, R.E.; Petersen, S. Hierarchical calibration of archetypes for urban building energy modeling. *Energy Build.* **2018**, *175*, 219–234. [CrossRef]
56. Kristensen, M.H.; Choudhary, R.; Petersen, S. Bayesian calibration of building energy models: Comparison of predictive accuracy using metered utility data of different temporal resolution. *Energy Procedia* **2017**, *122*, 277–282. [CrossRef]
57. Tian, W.; de Wilde, P.; Li, Z.; Song, J.; Yin, B. Uncertainty and sensitivity analysis of energy assessment for office buildings based on Dempster-Shafer theory. *Energy Convers. Manag.* **2018**, *174*, 705–718. [CrossRef]
58. Antony, J. *Design of Experiments for Engineers and Scientists*; Elsevier Science: Amsterdam, The Netherlands, 2014.
59. Østergård, T.; Jensen, R.L.; Mikkelsen, F.S. The best way to perform building simulations? One-at-a-time optimization vs. Monte Carlo sampling. *Energy Build.* **2020**, *208*, 109628. [CrossRef]
60. EVO. *IPMVP New Construction Subcommittee. International Performance Measurement & Verification Protocol: Concepts and Option for Determining Energy Savings in New Construction*; Efficiency Valuation Organization (EVO): Washington, DC, USA, 2003; Volume III.
61. FEMP. *Federal Energy Management Program, M&V Guidelines: Measurement and Verification for Federal Energy Projects Version 3.0*; U.S. Department of Energy Federal Energy Management Program; FEMP: Washington, DC, USA, 2008.
62. Fabrizio, E.; Monetti, V. Methodologies and advancements in the calibration of building energy models. *Energies* **2015**, *8*, 2548. [CrossRef]

energies

Article

Enhancement of a Short-Term Forecasting Method Based on Clustering and kNN: Application to an Industrial Facility Powered by a Cogenerator

Giulio Vialetto * and Marco Noro

Department of Management and Engineering, University of Padova, 36100 Vicenza, Italy; marco.noro@unipd.it
* Correspondence: giulio@giuliovialetto.it

Received: 13 October 2019; Accepted: 18 November 2019; Published: 20 November 2019

Abstract: In recent years, collecting data is becoming easier and cheaper thanks to many improvements in information technology (IT). The connection of sensors to the internet is becoming cheaper and easier (for example, the internet of things, IOT), the cost of data storage and data processing is decreasing, meanwhile artificial intelligence and machine learning methods are under development and/or being introduced to create values using data. In this paper, a clustering approach for the short-term forecasting of energy demand in industrial facilities is presented. A model based on clustering and k-nearest neighbors (kNN) is proposed to analyze and forecast data, and the novelties on model parameters definition to improve its accuracy are presented. The model is then applied to an industrial facility (wood industry) with contemporaneous demand of electricity and heat. An analysis of the parameters and the results of the model is performed, showing a forecast of electricity demand with an error of 3%.

Keywords: data analytics; big data; forecasting; energy; polygeneration; clustering; kNN; pattern recognition

1. Introduction

Data management, machine learning, and artificial intelligence have been emerging themes in the energy sector during recent years, thanks to the increasing availability of data and the decreasing cost of sensors, storage, and data manipulation. Data analytics methods have already been used to analyze collected data to improve energy efficiency, for example in buildings [1,2], or combined with machine learning methods [3]. Different machine learning methods have been already defined [4], such as clustering, k-nearest neighbors (kNN), regression models, principal component analysis (PCA), artificial neural networks (ANNs), and support vector machines (SVMs). These methods are mainly used in the energy sector. In [5], ANNs are used to predict residential building energy consumption. In [6], SVMs and ANNs are applied to predict heat and cooling demand in the non-residential sector, whereas in [7] ANNs and clustering are used to predict photovoltaic power generation. PCA is considered to analyze and forecast photovoltaic data in [8] and [9], meanwhile, in [10] and [11], SVM is used. Data are also used to perform analytics on energy: In [12], open geospatial data are used to plan electrification, whereas in [13] social media data are proposed to better define energy-consuming activities. In another study, a methodology based on energy performance certification is defined to estimate building energy demand using machine learning (decision tree, SVM, random forest, and ANN) [14]. Ganhadeiro et al. evaluates the efficiency of the electric distribution companies using self-organizing maps [15]. Machine learning methods are implemented in different environments: MATLAB [16–18] and R [19,20] are the most famous ones for research. Recently, Fowdur et al. have provided an overview of the available platforms (mainly commercial, such as IBM solution, Hewlett-Packard Enterprise Big Data Platform, SAP HANA Platform, Microsoft Azure and Oracle

Big Data, but also open source software, such as H2O) on machine learning, and how it could be used for big data analytics [3]. Commercial environments have a more robust tool already developed and test with better documentation than open source software such as R. These environments would be preferred if the research aim is to develop commercial software. Open source software would be preferred for the aim of research, thanks to the possibility of full access to code and algorithms, in order to propose improvements and the free distribution of results.

In this paper, an enhancement of short-term forecasting based on clustering and kNN is proposed. In this context, "short-term" means few hours. When energy demand is sampled frequently (for example, every 15 min) and a dataset is available, data can be used to train a model to predict the energy request of the next few hours, with the scope of improving the operation strategy of the energy generation system and optimizing energy storage. Clustering is used to define the average curves, meanwhile kNN (the k-nearest neighbors algorithm) classifies each observation and forecasts the energy demand.

The clustering method has been already used to classify daily load curves [21,22] and to forecast energy demands [23–27]. Clustering and kNN are proposed in this study as forecasting methods and compared to other machine learning techniques, such as the previously cited ANN, SVM, or PCA. Such methods forecast data just by comparing the energy demand observed to the historical data. As a matter of fact, many industrial facilities collect energy demand data without considering other process variables that could be necessary to increase the accuracy of forecasts, for example weather conditions (air temperature, humidity, etc.). The complexity of the problem increases if the production process is a batch-type instead of continuous, as more variables are necessary (such as the properties of raw materials). As novelties compared to previous studies that proposed clustering and kNN for forecasting, an innovation on data normalization and an alternative criterion to define the most suitable number of clusters are suggested in this study in order to increase the accuracy of forecasts. Martinez deeply analyzed the use of clustering and kNN to forecast energy data using pattern similarity on historical data, silhouette criteria, and Dunn and Davies-Bouldin indices to define the optimum number of clusters (the clustering hyperparameter) [25]. Then, he improved forecasting accuracy with a weighted multivariate kNN algorithm in [27]. These papers present a similar algorithm to that proposed in the present article, even if improvements on the hyperparameter definitions are suggested here. The authors have already studied how to increase the efficiency by using innovative operation strategies and polygeneration systems, such as in [28–31]. Solid oxide fuel cells and heat pumps have been proposed to increase the efficiency on energy generation for the residential sector in different climates. Solid oxide fuel cells and electrolyzers have been suggested for energy generation in industrial facilities producing hydrogen [32]. In [33,34], some energy audits in industrial facilities have been performed to analyze the main inefficiencies, and to define which improvements are required. The main scope of this study is to define a method for improving the performances of short-term forecasting and, consequently, to use it in order to improve the operation strategy of the energy generation plant of an industrial facility. As a matter of fact, forecasting can help to optimize energy storage by giving suggestions on the energy demand of the next hours. The case study relates to a wood industry that requires low temperature heat to dry wood into steam-powered kilns by using a cogeneration plant. The industrial firm is organized with a batch process, and no data are available on the process (quantity of the wood into the kilns, properties and humidity of the raw wood, weather conditions, etc.). Energy demand data (both electricity and heat) are used to test the proposed improvements. The results show that the proposed methodology is able to predict the accuracy of the forecasting.

2. Method

In this section, the description of the method proposed in order to forecast the energy demand of an industrial facility is firstly reported. Successively, the method of the model training and the definition of its parameters are described.

2.1. Forecast Method Introduction

In this study, a forecast method based on clustering and kNN is proposed and applied to an industrial facility. Industry uses energy (thermal and electric) both for industrial processes and auxiliary purposes (lighting, compressed air, etc.). Generally speaking, energy uses related to the production processes are strictly connected to the variety and entity of the production output. If the production output remains constant in terms of the type and quantity of items, it is expected that the energy use does not vary significantly. Moreover, if the production output varies significatively (for example, because the industrial process is organized by batch), the complexity of the problem increases, and more variables are required. The aim of this study is to define a model based on a machine learning technique that allows the forecasting of energy demands for a short period (for example, the next hour) based on the demands observed and using a clustering approach without any other variables that could describe the process and/or the environmental conditions. In this study it is supposed that average profiles can be defined by using a dataset of at least one year of observation, in order to perform the forecast. Other machine learning methods, such as ANN or PCA are not proposed for this forecast problem due to the lack of variables which could describe the industrial process. Moreover, even if an ANN could be trained with only historical data to perform the forecast, the advantage of using clustering combined with kNN is the knowledge of the forecast process. If ANN would be used, a neural network is trained so a "grey box" model is defined, where the user knows the connection into the network, but it is unknown how the network works when varying the input variables. Instead, the methodology proposed here uses clustering to define similar patterns on historical data, where the user has a high control on the forecast process and may know each pattern proposed.

The first concept to introduce is the energy demand curve. It represents a temporal sequence of observations and forecasts of energy demand. Each curve can be split in two parts, namely, support and forecast. The former is the part of the data that will be provided to the model, constituted by the latest observations. The latter is the predicted data based on the support (Table 1). The length of the support (s) and of the forecast (f) is fixed by the user. In this model, it is proposed that $0 < f \leq s-2$. In the following discussion section, the performances of the model, varying f and s, for a real case study, will be described.

Table 1. Example of curves, definition of support and forecast (sample dataset).

				Support						Forecast	
$i = 1$	$i = 2$			...			$i = s = 8$	$j = 1$		...	$j = f = 4$
10	11	10	13	12	14	16	12	11	12	18	13

To perform the forecast, the model features a workflow (Figure 1) based on the following steps:

1. Model training: A dataset of observations is used to train the model. Observations define the average demand curves and train the classification model;
2. Classification: Observations are used to classify which is the most similar average curve;
3. Forecast: Average curve forecast is used to define forecast of the observations.

The model proposed is based on two machine learning methods, clustering and kNN. Clustering is a method used only in the training process to define the average curve, while kNN is used to classify the observations and to relate them with the average curves.

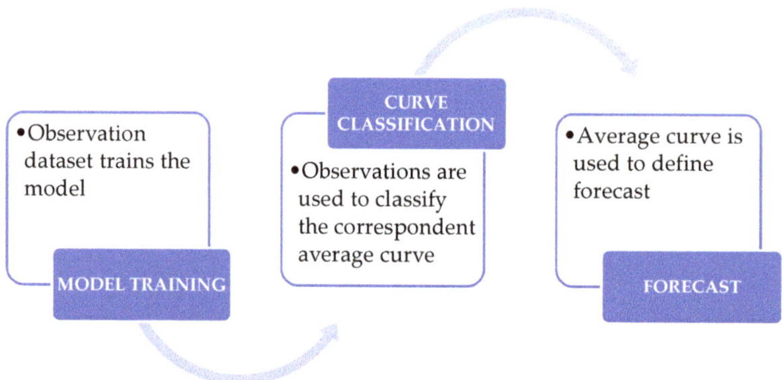

Figure 1. Workflow of the forecast method proposed.

2.2. Introduction to Clustering

Clustering is a data analytics method used to classify data and to perform data segmentation [4]. The samples are grouped into subsets or "clusters", where, in each cluster, objects are more likely related to one another than those assigned to different clusters. Clustering is strictly related to the concept of "degree of similarity" (or "degree of dissimilarity") between the objects of the same subset. A cluster method groups similar objects whereas similarity is defined, for example via a distance function.

K-means is a clustering method used when all the variables are quantitative, and the Euclidean distance between the objects is defined as a dissimilarity function, where the lower the distance, the greater the similarity ([4,35]). The Euclidean distance between each object x_a and x_b is measured by using the variable $i = 1...n$, which describes each object (Equation (1)):

$$d(x_a, x_b) = \sum_{i=1}^{n} \left(x_{a,i} - x_{b,i}\right)^2 \tag{1}$$

If a dataset with m objects is provided, K-means divides the dataset into N clusters, minimizing the Euclidean distance between each object of the cluster. The number of clusters, N, must be defined by the user as a hyperparameter. A hyperparameter is a value of a machine learning model that is defined before the training process. Silhouette [36], gap criterion [37] and other methods have been already developed and proposed to define the suitable number of clusters to divide a dataset. These methods try to define the minimum number of clusters to maximize the distance between the clusters themselves. For example, in [25], the performance of a forecasting method based on clustering and kNN with the silhouette, Dunn, and Davies-Bouldin methods, used to define the optimum number of cluster, is analyzed. In this paper, it the use of a criterion based on the clusters distance is not proposed, but instead to define the minimum number of clusters that minimizes the error of prediction under a threshold that is chosen by the user. In Section 2.7, discussing the hyperparameter definition, such a criterion will be described.

2.3. Introduction to kNN

kNN (k-Nearest Neighbors) is a machine learning method used mainly for classification and regression [4]. In the proposed forecast method, firstly, clustering training dataset is divided into N clusters, then an average curve for each cluster is defined. When a new observation occurs, it is necessary to classify which is its cluster. Here, kNN performs the classification task by analyzing how the k-neighbors nearest to the observation are classified, and the distances between them. In the model here proposed, kNN is used to define which is the cluster (and consequently, the average curve) defined with the training dataset closer to the new observation. kNN requires two hyperparameters, the number of neighbors (k), and the distance function. Section 2.7, discussing the hyperparameter definition, describes how they are defined.

2.4. Model Training

The main task to define the forecast model is the training process. The training process requires at least one year of observations. The observations are ordered and then used to defined curves with support and forecast. These curves define a dataset. The workflow of the training can be divided in the following steps (Figure 2):

1. Define dataset: Firstly, it is necessary to define and to normalize the dataset. Successively, it is randomly divided into three subgroups, namely, the validation, training, and test datasets. These subgroups represent 25%, 50% and 25% of the total observations, respectively. The validation dataset is used to define the hyperparameters of the model, whereas the training dataset is used to train both the cluster and kNN models. Finally, the test dataset is useful to verify the performance of the trained model.
2. Define hyperparameters: As previously mentioned, the proposed model defines both the cluster and kNN models. Both methods require the definition at least the distance function and the number of clusters (cluster model), or the number of observations for classification (kNN model). The Euclidean distance function is proposed for the cluster model, meanwhile, the number of clusters and number of observations for classification are defined using the validation dataset.
3. Train cluster model: When all the hyperparameters are set, the training dataset is used to train the cluster model and to define the average forecast curves.
4. Train kNN model: When both the cluster model and the consequently average forecast curves are defined, kNN is defined. kNN is used to forecast the observations.
5. Test model: The test dataset is used to test the trained model and to check its performance by using the mean absolute percentage error (MAPE) and root mean square error (RMSE) criteria.

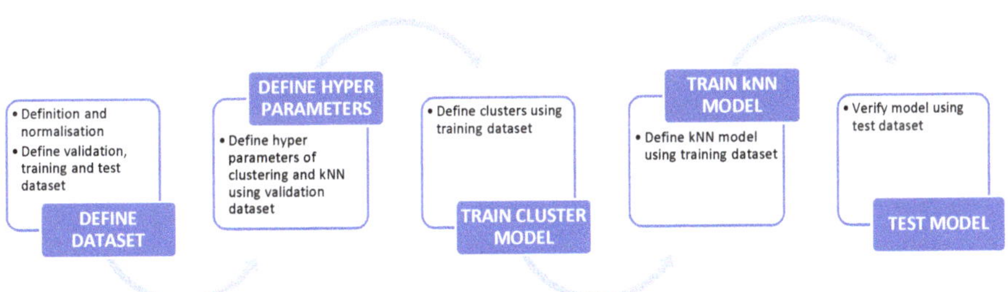

Figure 2. Workflow to train the model.

After the training process, the model can be used to forecast new observations.

2.5. Data Normalization

One of the first step of data analytics is data normalization. As datasets have different values and scale effect may occur, classification methods such as clustering will not work properly if data are not normalized. Usually, normalization is performed using standard score or minimum-maximum scaling [4,38]. The standard score normalizes the dataset (X) by using the average (μ) and the standard deviation (σ), as described in Equation (2):

$$\frac{X-\mu}{\sigma} \quad (2)$$

In this model, the authors propose to differently normalize dataset. As the goal of the model is to forecast energy demand curves, the idea is that different curves may have different scales but similar variation. The standard score would be normalized but the curves will still have a lower scale effect. Instead, in this study it is proposed to calculate the average of the observations for each curve, and then to calculate the variations between observations and average (Equation (3)):

$$n_{j,i} = \frac{o_{j,i}}{a_j} - 1 \tag{3}$$

where $o_{j,i}$ is the observation i of curves j, a_j is the average and $n_{j,i}$ the normalized observation. Figure 3 represents an example explaining the reason why this normalization is proposed. Curves 1 and 2 have different scales but similar variation. Firstly, the standard score is applied, then the average normalization follows. The average (avg) and standard deviation (std) for the standard score are calculated using all the support values. In the other case, the average of support of each curve is calculated and used for normalization. Forecast values are excluded because they only become known during the training process. As can be seen in Figure 3, curve 2 is 1.58 times larger than curve 1, and a noise is added. It is possible to appreciate that the proposed method (avg), that is based on the average of the curves, reduces the scale effect, but keeps the variation. As a matter of fact, the normalized curves 1 and 2 have similar values. Instead, the standard score method proposes normalized curves with different values because it normalizes not only the scale effect but the variation as well.

STANDARD SCORE	SUPPORT							FORECAST					
Curve 1	10.0000	10.1000	10.0000	10.0000	10.1000	10.0000	10.0000	10.0000	10.0000	10.0000	10.0000	avg	12.9369
Curve 2	15.8000	15.8000	15.9580	15.8000	15.8000	15.8000	15.9580	15.8000	15.8000	15.8000	15.8000	std	3.0187
norm, curve 1	-0.9729	-0.9398	-0.9729	-0.9729	-0.9398	-0.9729	-0.9729	-0.9729	-0.9729	-0.9729	-0.9729		
norm, curve 2	0.9485	0.9485	1.0008	0.9485	0.9485	0.9485	1.0008	0.9485	0.9485	0.9485	0.9485		

AVERAGE	SUPPORT							FORECAST					
Curve 1	10.0000	10.1000	10.0000	10.0000	10.1000	10.0000	10.0000	10.0000	10.0000	10.0000	10.0000	avg, curve 1	10.0286
Curve 2	15.8000	15.8000	15.9580	15.8000	15.8000	15.8000	15.9580	15.8000	15.8000	15.8000	15.8000	avg, curve 2	15.8451
norm, curve 1	-0.0028	0.0071	-0.0028	-0.0028	0.0071	-0.0028	-0.0028	-0.0028	-0.0028	-0.0028	-0.0028		
norm, curve 2	-0.0028	-0.0028	0.0071	-0.0028	-0.0028	-0.0028	0.0071	-0.0028	-0.0028	-0.0028	-0.0028		

Figure 3. Data normalization example.

2.6. Error Estimation

When a forecast method is proposed, it is necessary to estimate the error of the forecasting. As previously mentioned, error estimation is used also to define the hyperparameters. Here, MAPE- and RMSE-derived errors are suggested. *MAPE* is the acronym of mean absolute percentage error, and it is defined by Equation (4):

$$MAPE = \frac{1}{n}\sum_{i=1}^{n}(\frac{1}{l}\sum_{j=1}^{l}(\frac{p_{j,i}}{d_{j,i}} - 1)) \tag{4}$$

where n is the number of curves, l is the number of the forecasted values of each curve, $p_{j,i}$ is the model predicted value of the curve, and $d_{j,i}$ is the value observed. *RMSE* is the acronym of root mean square error. Here, it is proposed instead of mean square error (MSE) because it is possible to compare error using the same measurement unit of data. It is defined by Equation (5):

$$RMSE = \frac{1}{n}\sum_{i=1}^{n}\sqrt{\frac{1}{l}\sum_{j=1}^{l}\left(p_{j,i} - d_{j,i}\right)^2} \tag{5}$$

These errors are calculated on the entire forecast, meanwhile, the first forecasted value of each curve is the most important. *MAPE*1 and *RMSE*1 are calculated considering not all the forecasted values but only the first (l = 1).

2.7. Hyperparameters Definition

As previously mentioned, it is necessary to define the parameters for clustering and kNN. They are called hyperparameters. Clustering requires the "distance function" and the "number of clusters", while kNN requires the "number of the nearest neighbors" and the "distance function". Only the clustering distance function is defined a priori (Euclidean distance), whereas the other ones are defined using the validation dataset.

Firstly, the number of clusters is defined. As previously mentioned, different criteria have been already developed, and they usually try to minimize the number of clusters in order to maximize the distance between data. It is in the authors' opinion that a more suitable criterion for a forecasting method is to find the minimum number of clusters that minimize the forecasting error, for example under a threshold previously defined. The model proposed here clusters data to obtain average curves, and then it uses them to forecast the energy demand. It is proposed to vary the number of clusters (from 2 to N) and for each simulation to calculate *MAPE* between the data and average curves of the clusters. The parameter is the minimum n that has a *MAPE* lower than the average next three values:

$$\min(n)|\, MAPE(n) < \frac{MAPE(n+1) + MAPE(n+2) + MAPE(n+3)}{3} \tag{6}$$

Nevertheless, it is possible to define n as the minimum number of clusters associated with a *MAPE* lower than a defined threshold:

$$\min(n)|\, MAPE(n) < MAPE_{limit} \tag{7}$$

This method can be seen as an early stopping method, because the number of clusters increases by as much as the accuracy of the system is increased. Figures 4 and 5 report how this method is applied to a validation dataset of electricity and heat demand, respectively. Each curve has 8 observations as support, and 4 observations as forecast (data refers to the case study defined in Section 3.1). It is possible to appreciate that the curves have a *MAPE* decreasing rapidly between 2 and 10 clusters, whereas between 10 and 30 clusters they become more stable. With more than 30 clusters, the curves have very low gradient, and locally *MAPE* increases, even if the number of clusters increases. In this case, if the criterion described by Equation (6) is applied, then 10 clusters for heat and 13 for electricity are suggested.

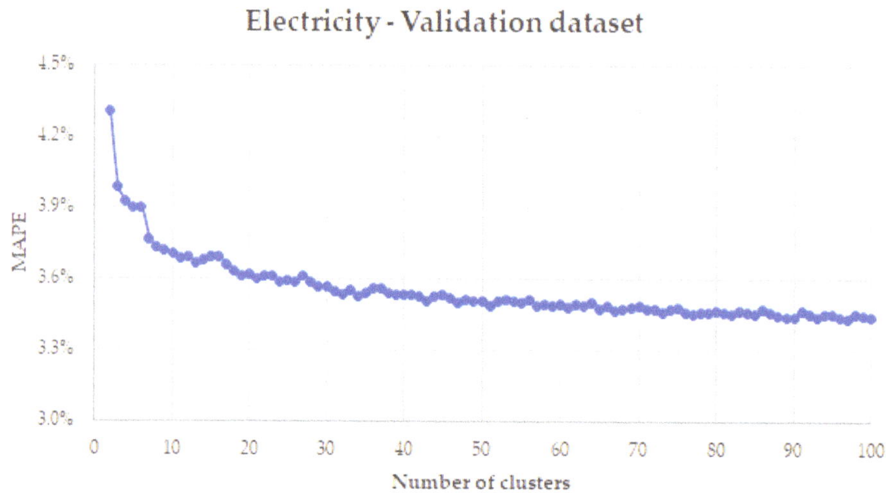

Figure 4. Electricity validation dataset mean absolute percentage error (*MAPE*) when varying the number of clusters from 2 to 100 for an 8-4 curve.

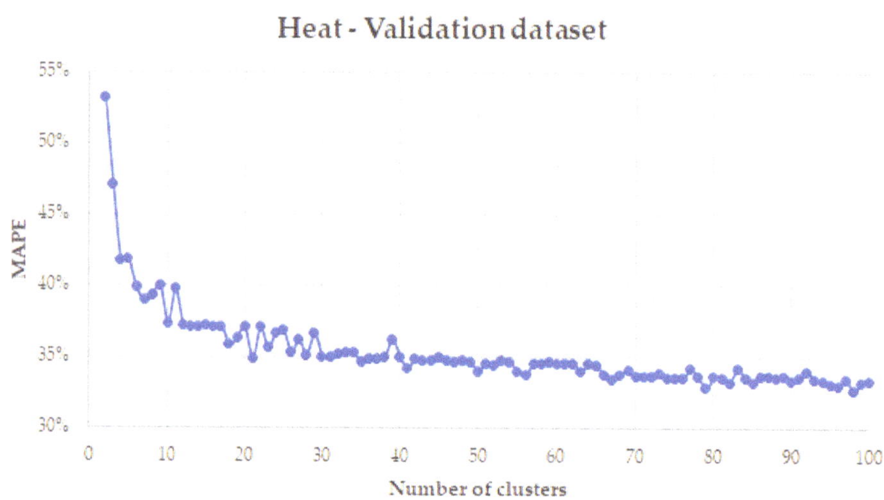

Figure 5. Heat validation dataset *MAPE* when varying the number of clusters from 2 to 100 for an 8-4 curve.

As previously mentioned, in other studies (such as [25]) where clustering and kNN are proposed for forecasting, the optimum number of clusters is defined by using a criterion such as silhouette or gap statistics. Here, the silhouette calculates the average distance between each member of a cluster from another cluster, and the minimum number of clusters that increases the distance is the optimum [36]. If the silhouette criterion was applied to the validation dataset (for both electricity and heat), the number of clusters suggested would be lower than the method proposed. In this regard, Figures 6 and 7 show that the number of clusters suggested is two in both cases. As a matter of fact, if this value was used, the *MAPE* would be the highest (Figures 4 and 5).

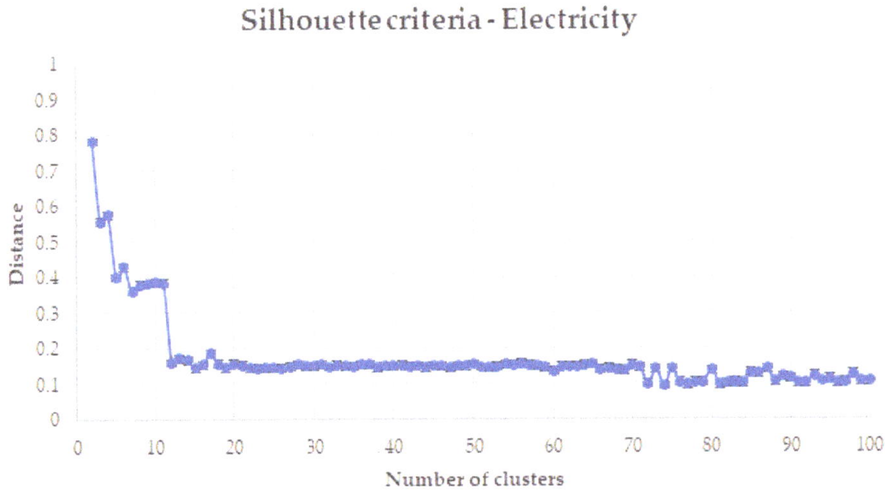

Figure 6. Silhouette applied to electricity validation dataset when varying the number of clusters from 2 to 100 for an 8-4 curve.

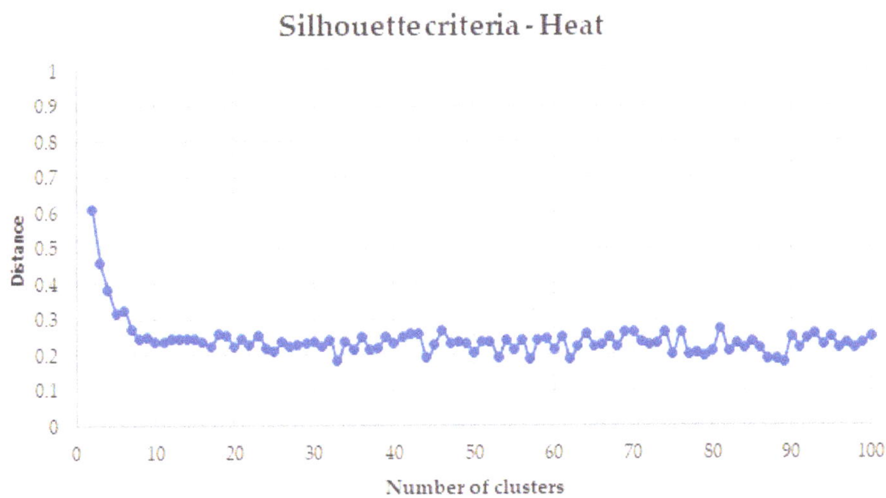

Figure 7. Silhouette applied to heat validation dataset when varying the number of clusters from 2 to 100 for an 8-4 curve.

The kNN hyperparameters are defined, instead, using MATLAB optimization with the 'Fitchknn' function. The latter optimizes the kNN model by choosing the distance function and the number of neighbors to decrease the classification error [39].

3. Results

The proposed method was applied to a case study based on an industrial facility characterized by a simultaneous demand of electricity and heat. The production process is organized by batch, and no data such as environment conditions, raw material properties, etc., were available. Data are used to predict the two types of energy separately by also using energy demand data, and the length of support and forecast was varied in order to verify the dependency of error. The aim was to verify the forecasting performances on energy demand (electricity and heat) of the proposed method. No improvements of the current energy generation system and/or industrial process are proposed.

3.1. Case Study Description

The energy consumption of an industrial facility selling wood (timber) laminated windows, plywood, engineered veneer, laminate, flooring, and white wood was analyzed. The industrial process requires heat to dry wood in kilns (working temperature of 70 °C), and to store it in warehouses. Electricity is used for the production equipment, offices, lighting in the warehouses, and to charge electric forklifts. Energy is generated by using two cogeneration systems (combined heat and power, CHP) based on internal combustion engines (ICE) to produce both electricity and heat. A natural gas fired boiler was present as an integration system for the kilns. Electricity is also exchanged with the grid when mismatching occurs between generation and demand. Figure 8 represents the energy fluxes and the interconnections between each component of the system.

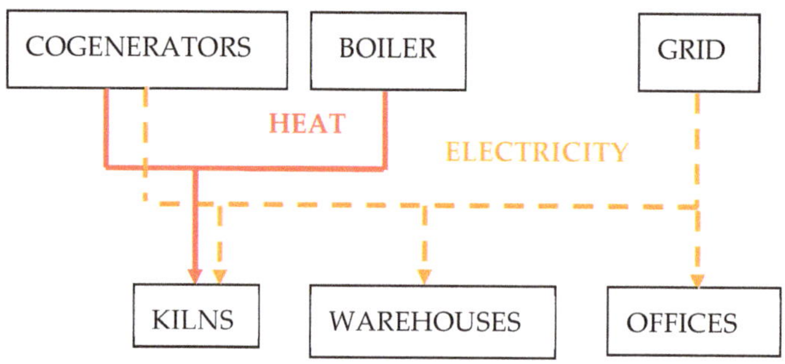

Figure 8. Electricity (yellow dot lines) and heat (red continuous lines) energy fluxes, connection between production (up boxes), and utilization (bottom boxes).

Energy use (both electricity and heat) was sampled every 15 min from 01/01/2015 to 25/09/2017. Electricity demand was available as mean power requested (kW). Heat demand, instead, was calculated by measuring the water flow rate (m^3/h) and inlet and outlet temperatures (°C) to heat the kilns. The data were stored in a structured SQL database. Here, we intended to use these data to define a curve with support and forecast, in order to train and to validate the forecast model. A dataset for heat demand, and another for electricity, has been defined.

As a matter of fact, these datasets can contain some sampling events with missing measurements or outliers. Missing measurements in a SQL database are managed with null values, so the events with at least one variable with a null value were not considered for the study, because the system was not able to sample the process, and the other variables could be affected by errors. Outliers could occur because the data were stored without any validation.

The data were plotted by a histogram (with a log scale on the x axis) and a probability plot of quartiles (QQ plot) to intercept outliers. The QQ plot was used to compare the dataset distribution with the normal distribution. The assumption here is that the data follow the latter, and if it does not, outliers are likely to be present. Figure 9 displays how the data were distributed. It is possible to appreciate that the outliers are present for both the electricity and heat demand. The electricity demand data were mainly between 100 and 1000 kW, while the maximum sampled value was higher than 106 kW. The same occurs for heat demand, where, in fact, the QQ plots show that the current dataset does not follow a standard distribution.

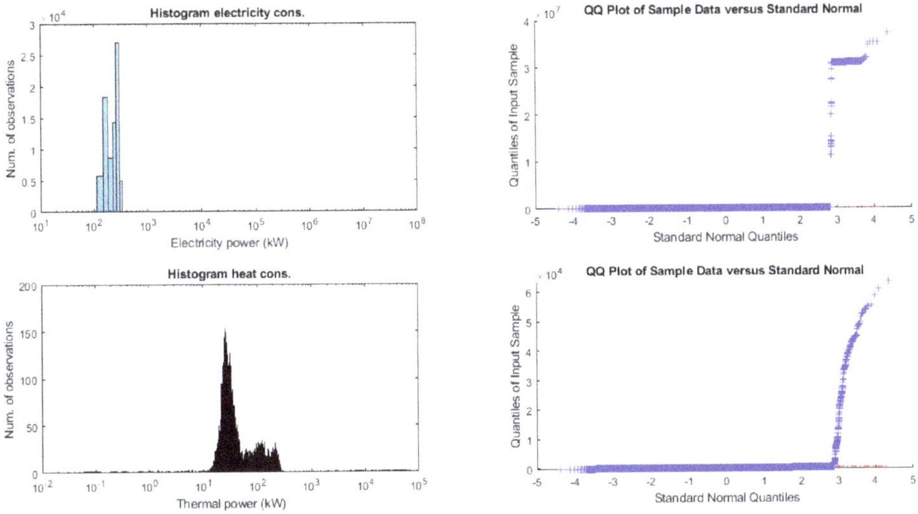

Figure 9. Representation of the dataset without filtering data, histogram, and probability plot of quartiles (QQ plot) of electricity (top) and thermal (bottom) power.

To filter the outliers, it was proposed to define an upper limit for each of the variables, both for electricity and heat. The limit was set considering the maximum demand of electricity and heat of the system. Figure 10 represents the filtered data, where the QQ plots show that the filtered dataset was closer to a normal distribution and that the range of the dataset decreased.

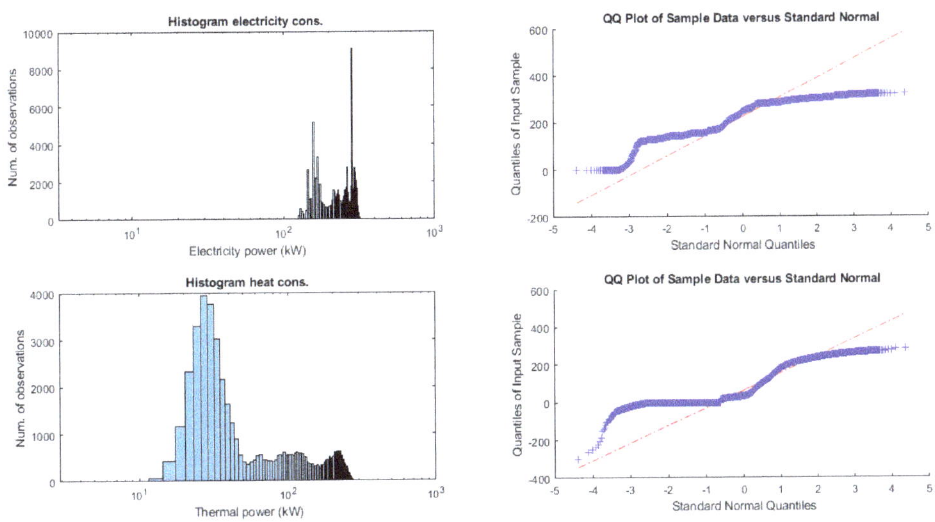

Figure 10. Representation of the dataset filtering data, histogram and QQ plot of electricity (top) and thermal (bottom) power.

3.2. Model Training and Test

Observations were used to define a dataset. The dataset was filtered by data related to null values or outliers. Here, it was randomly split into training, validation, and test datasets, representing 50%, 25%, and 25% of the entire dataset, respectively. The validation dataset was used to define the hyperparameters of the model, whereas the training dataset was used to train the model, and the test dataset was used to check the accuracy of the model. Accuracy was defined by calculating the *MAPE* and *RMSE* between the forecasted value of the model and the observed value of the dataset. Curves of different lengths for the support and forecast were defined in order to discuss the influence of definition on the hyperparameters, in particular, the number of clusters. Table 2 shows some simulations of the model, considering energy demand curves of different length (for example, an 8-4 curve represents a curve with 8 observations as a support and 4 observations as a forecast). *MAPE* was calculated using the test dataset (error between forecasted values and observed values), once for the first forecasted value (here, the test dataset is *MAPE* 1) and once for the entire forecast (here the test dataset is the *MAPE*). The *MAPE* value calculated with the validation dataset was also added in order to define the hyperparameter number of clusters (Section 2.5). It is possible to appreciate that the *MAPE* calculated with the validation dataset is a good predictor of the *MAPE* of the test dataset. For example, with an 8-4 curve with electricity, the *MAPE* calculated with the validation dataset was 3.60%, whereas the *MAPE* calculated with the test dataset was 3.58%. The results also show a difference between the electricity and heat datasets, where an 8-4 curve has a *MAPE* of 3.58% and 34.11%, respectively. The difference can be explained with a higher variation of heat values.

Table 2. Simulation of the model with different curves length.

Curve	Type of Energy	Validation Dataset	Test Dataset			
		Mean Absolute Percentage Error (*MAPE*)	*MAPE*1	*MAPE*	*RMSE*1	Root Mean Square Error (*RMSE*)
8-4	Electricity	3.60%	2.75%	3.58%	5.15 kW	3.82 kW
8-4	Heat power	35.41%	32.95%	34.11%	93.43 kW	55.43 kW
10-4	Electricity	3.71%	2.74%	3.57%	5.15 kW	3.82 kW
10-4	Heat power	35.23%	32.70%	34.95%	93.20 kW	54.82 kW
10-8	Electricity	4.79%	2.90%	4.47%	5.47 kW	3.53 kW
10-8	Heat power	36.66%	35.30%	34.12%	90.03 kW	41.99 kW
12-8	Electricity	4.69%	2.80%	4.47%	5.31 kW	3.53 kW
12-8	Heat power	39.00 %	32.10%	37.21%	95.14 kW	43.05 kW

4. Discussion

In this section, the influence of the curve size and the type of normalization are both analyzed.

4.1. Influence of the Curve Size

Observations were used to define the curves in order to train and test the forecast model. Support is the part of the curve that is used to classify observation, and, consequently, it defines the forecasted value (forecast part). The length of the supports (s) and forecasts (f) may vary the hyperparameter number of clusters and, consequently, the error on forecasting. By increasing the forecast length (equally with support length), the forecast error is expected to increase, because the model needs to predict more observations. It is unknown what the effect of increasing the support length (with the same forecast length) could be, that is, increasing or decreasing the accuracy of the classification of the curve. Figures 11 and 12 represent the value of the *MAPE* criteria for the validation dataset, varying the support and the forecast for electricity and heat, respectively.

	Electricity	FORECAST						
		2	4	6	8	10	12	14
SUPPORT	4	2.9%						
	6	2.9%	3.6%					
	8	3.1%	3.6%	4.1%				
	10	3.1%	3.7%	4.1%	4.8%			
	12	3.2%	3.8%	4.3%	4.7%	5.1%		
	14	3.3%	3.9%	4.4%	4.9%	5.3%	5.8%	
	16	3.5%	4.0%	4.5%	5.1%	5.3%	5.8%	6.3%

Figure 11. Heatmap of *MAPE* of electricity validation dataset with curves with different support and forecast length.

	Heat Power	FORECAST						
		2	4	6	8	10	12	14
SUPPORT	4	37.7%						
	6	36.7%	35.2%					
	8	33.8%	35.4%	39.2%				
	10	36.8%	35.2%	37.3%	36.7%			
	12	40.2%	37.8%	39.0%	39.0%	36.9%		
	14	35.0%	36.9%	33.1%	36.1%	39.2%	37.4%	
	16	35.4%	39.0%	37.8%	39.0%	39.3%	37.9%	38.8%

Figure 12. Heatmap of *MAPE* of heat power validation dataset with curves with different support and forecast length.

Firstly, it is possible to appreciate that the electricity validation dataset has a regular variation of *MAPE* in comparison to the heat validation dataset. When the electricity dataset is used, the *MAPE* increases when increasing support and/or forecast lengths. Here, it is supposed that the electricity demand varies differently from the heat demand. As expected, the electricity dataset shows that when increasing the forecast length of the curve the *MAPE* increases. Here, the *MAPE* increases from 3.5% for a 16-2 curve (4 support length, 2 forecast length) to 6.3% for a 16-4 curve. This shows that the error increases when the forecast period becomes longer. On the other hand, the increase of the support length is also related to the increase of *MAPE*, where it changes from 2.9% for a 4-2 curve to 3.5% for a 16-2 curve. Even if more observations are available to classify each curve, the error does not decrease.

4.2. *Influence of the Normalization*

As mentioned in Section 2.5, in this model, it is proposed to not use a normalization based on the standard score but instead on the percentage norm. Here, the aim is to reduce the scale effect of the curves but to maintain their variation. A representation of the *MAPE*, varying the number of clusters in the electricity validation dataset with a curve of 8 observations for support and 4 for forecast (Figure 13) and 10 for support and 4 for forecast (Figure 14) is reported. In both cases, it is possible to appreciate that the dataset normalized with the standard score has a higher *MAPE* with respect to the normalization with the proposed percentage norm.

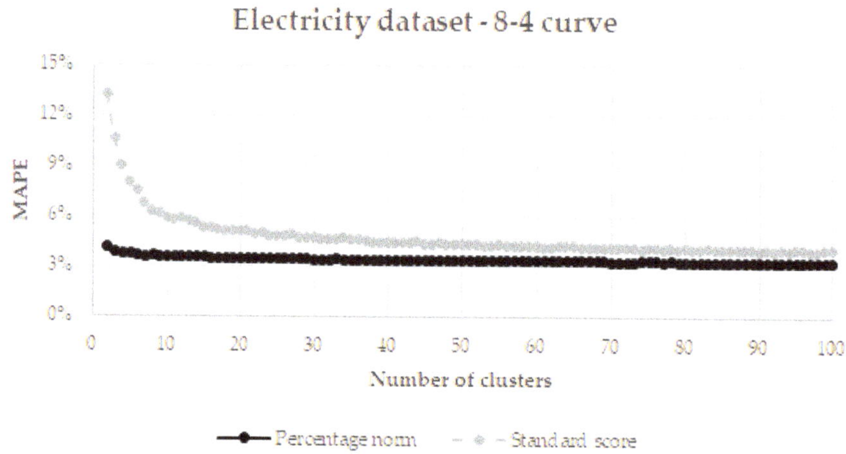

Figure 13. Comparison on *MAPE* with the electricity validation dataset, curve with 8 observation and 4 forecast values, normalization between percentage norm and standard score.

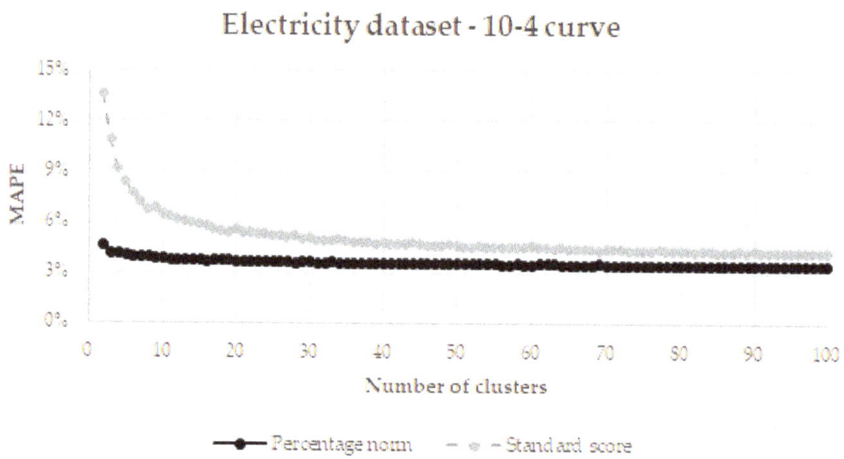

Figure 14. Comparison on *MAPE* with the electricity validation dataset, curve with 10 observation and 4 forecast values, normalization between percentage norm and standard score.

5. Conclusions

In this paper, the enhancements of a short-term forecasting method based on clustering and kNN machine learning techniques have been proposed and tested. A novel definition of hyperparameters (number of clusters) and data normalization compared to the state-of-art methods are presented here, in order to increase the accuracy on the forecast and to minimize errors. A dataset of observations is required to define the hyperparameters, in order to train the model and to test it. A case study based on an industrial facility with simultaneous electricity and heat demands was presented in order to apply the proposed energy forecast method. An analysis reported on how the length of the energy demand curves (numbers of observations and forecast) impacted the model performance. The industrial firm works with a batch process and only energy demand data were sampled and stored, as no other data on the process were available. The results show that the improvements suggested here, in terms of the definition of hyperparameters, decrease the error of forecasting compared to other criteria in the literature. An analysis of the effect of the length of the curves (both on support and forecast) on the error was performed as well. For the dataset used here, the longer the length (both on support and/or forecast), the higher the error. The validation dataset was not only used to define the hyperparameters, as it could be used to predict the error of the forecast as well. It is in the authors' opinion that further improvements on the methodology could be achieved by studying the most suitable distance function for the dataset and/or by weighting observations. Moreover, an investigation on how this forecast method could improve energy production and efficiency could be of interest, for example reducing the production of unnecessary heat and/or improving suitable operation strategy to decrease the cost of energy generation.

Author Contributions: Conceptualization, G.V.; Data curation, G.V.; Validation, M.N.; Writing—original draft, G.V. and M.N.

Funding: This research received no external funding.

Acknowledgments: Authors thank Corà Domenico and Figli S.p.A., who provided the dataset for the case study to test the proposed methodology.

Conflicts of Interest: The authors declare no conflict of interest.

References

1. Noussan, M.; Nastasi, B. Data Analysis of Heating Systems for Buildings—A Tool for Energy Planning, Policies and Systems Simulation. *Energies* **2018**, *11*, 233. [CrossRef]
2. Tronchin, L.; Manfren, M.; Nastasi, B. Energy analytics for supporting built environment decarbonisation. *Energy Procedia* **2019**, *157*, 1486–1493. [CrossRef]
3. Fowdur, T.P.; Beeharry, Y.; Hurbungs, V.; Bassoo, V.; Ramnarain-Seetohul, V. Big Data Analytics with Machine Learning Tools. In *Internet of Things and Big Data Analytics toward Next-Generation Intelligence*; Springer: Berlin, Germany, 2018; pp. 49–97.
4. Hastie, T.; Tibshirani, R.; Friedman, J. *The Elements of Statistical Learning*; Springer: New York, NY, USA, 2009.
5. Biswas, M.A.R.; Robinson, M.D.; Fumo, N. Prediction of residential building energy consumption: A neural network approach. *Energy* **2016**, *117*, 84–92. [CrossRef]
6. Koschwitz, D.; Frisch, J.; van Treeck, C. Data-driven heating and cooling load predictions for non-residential buildings based on support vector machine regression and NARX Recurrent Neural Network: A comparative study on district scale. *Energy* **2018**, *165*, 134–142. [CrossRef]
7. Cheng, K.; Guo, L.M.; Wang, Y.K.; Zafar, M.T. Application of clustering analysis in the prediction of photovoltaic power generation based on neural network. *IOP Conf. Ser. Earth Environ. Sci.* **2017**, *93*, 012024. [CrossRef]
8. Yang, D.; Dong, Z.; Lim, L.H.I.; Liu, L. Analyzing big time series data in solar engineering using features and PCA. *Sol. Energy* **2017**, *153*, 317–328. [CrossRef]
9. Malvoni, M.; De Giorgi, M.G.; Congedo, P.M. Photovoltaic forecast based on hybrid PCA–LSSVM using dimensionality reducted data. *Neurocomputing* **2016**, *211*, 72–83. [CrossRef]
10. Malvoni, M.; De Giorgi, M.G.; Congedo, P.M. Data on Support Vector Machines (SVM) model to forecast photovoltaic power. *Data Br.* **2019**, *9*, 13–16. [CrossRef]
11. Qijun, S.; Fen, L.; Jialin, Q.; Jinbin, Z.; Zhenghong, C. Photovoltaic power prediction based on principal component analysis and Support Vector Machine. In Proceedings of the 2016 IEEE Innovative Smart Grid Technologies-Asia (ISGT-Asia), Melbourne, VIC, Australia, 28 November–1 December 2016; pp. 815–820.
12. Korkovelos, A.; Khavari, B.; Sahlberg, A.; Howells, M.; Arderne, C. The Role of Open Access Data in Geospatial Electrification Planning and the Achievement of SDG7. An OnSSET-Based Case Study for Malawi. *Energies* **2019**, *12*, 1395. [CrossRef]
13. de Kok, R.; Mauri, A.; Bozzon, A. Automatic Processing of User-Generated Content for the Description of Energy-Consuming Activities at Individual and Group Level. *Energies* **2018**, *12*, 15. [CrossRef]
14. Attanasio, A.; Piscitelli, M.; Chiusano, S.; Capozzoli, A.; Cerquitelli, T. Towards an Automated, Fast and Interpretable Estimation Model of Heating Energy Demand: A Data-Driven Approach Exploiting Building Energy Certificates. *Energies* **2019**, *12*, 1273. [CrossRef]
15. Ganhadeiro, F.; Christo, E.; Meza, L.; Costa, K.; Souza, D. Evaluation of Energy Distribution Using Network Data Envelopment Analysis and Kohonen Self Organizing Maps. *Energies* **2018**, *11*, 2677. [CrossRef]

16. MATLAB. Deep Learning Toolbox. Available online: https://www.mathworks.com/products/deep-learning.html (accessed on 14 July 2019).
17. Paluszek, M.; Thomas, S. *MATLAB Machine Learning*; Apress: Berkeley, CA, USA, 2017.
18. Kim, P. *MATLAB Deep Learning*; Apress: Berkeley, CA, USA, 2017.
19. Ghatak, A. *Machine Learning with R*; Springer: Singapore, 2017.
20. Ramasubramanian, K.; Singh, A. *Machine Learning Using R*; Apress: Berkeley, CA, USA, 2019.
21. Amri, Y.; Fadhilah, A.L.; Setiani, N.; Rani, S. Analysis Clustering of Electricity Usage Profile Using K-Means Algorithm. *IOP Conf. Ser. Mater. Sci. Eng.* **2016**, *105*, 012020. [CrossRef]
22. Müller, H. Classification of daily load curves by cluster analysis. In Proceedings of the Eighth Power Systems Computation Conference, Helsinki, Finland, 19–24 August 1984; pp. 381–388.
23. Wahid, F.; Kim, D. A Prediction Approach for Demand Analysis of Energy Consumption Using K-Nearest Neighbor in Residential Buildings. *Int. J. Smart Home* **2016**, *10*, 97–108. [CrossRef]
24. Wu, J.; Kong, D.; Li, W. A Novel Hybrid Model Based on Extreme Learning Machine, k-Nearest Neighbor Regression and Wavelet Denoising Applied to Short-Term Electric Load Forecasting. *Energies* **2017**, *10*, 694.
25. Martinez Alvarez, F.; Troncoso, A.; Riquelme, J.C.; Aguilar Ruiz, J.S. Energy Time Series Forecasting Based on Pattern Sequence Similarity. *IEEE Trans. Knowl. Data Eng.* **2011**, *23*, 1230–1243. [CrossRef]
26. Talavera-Llames, R.; Pérez-Chacón, R.; Troncoso, A.; Martínez-Álvarez, F. Big data time series forecasting based on nearest neighbours distributed computing with Spark. *Knowl. Based Syst.* **2018**, *161*, 12–25. [CrossRef]
27. Talavera-Llames, R.; Pérez-Chacón, R.; Troncoso, A.; Martínez-Álvarez, F. MV-kWNN: A novel multivariate and multi-output weighted nearest neighbours algorithm for big data time series forecasting. *Neurocomputing* **2019**, *353*, 56–73. [CrossRef]
28. Vialetto, G.; Rokni, M. Innovative household systems based on solid oxide fuel cells for a northern European climate. *Renew. Energy* **2015**, *78*, 146–156. [CrossRef]
29. Vialetto, G.; Noro, M.; Rokni, M. Innovative household systems based on solid oxide fuel cells for the Mediterranean climate. *Int. J. Hydrog. Energy* **2015**, *40*, 14378–14391. [CrossRef]
30. Vialetto, G.; Noro, M.; Rokni, M. Thermodynamic investigation of a shared cogeneration system with electrical cars for northern Europe climate. *J. Sustain. Dev. Energy Water Environ. Syst.* **2017**, *5*, 590–607. [CrossRef]
31. Vialetto, G.; Noro, M.; Rokni, M. Combined micro-cogeneration and electric vehicle system for household application: An energy and economic analysis in a Northern European climate. *Int. J. Hydrogen Energy* **2017**, *42*, 10285–10297. [CrossRef]
32. Vialetto, G.; Noro, M.; Colbertaldo, P.; Rokni, M. Enhancement of energy generation efficiency in industrial facilities by SOFC—SOEC systems with additional hydrogen production. *Int. J. Hydrogen Energy* **2019**, *44*, 9608–9620. [CrossRef]
33. Lazzarin, R.M.; Noro, M. Energy efficiency opportunities in the production process of cast iron foundries: An experience in Italy. *Appl. Therm. Eng.* **2015**, *90*, 509–520. [CrossRef]
34. Noro, M.; Lazzarin, R.M. Energy audit experiences in foundries. *Int. J. Energy Environ. Eng.* **2016**, *7*, 409–423. [CrossRef]
35. MATLAB. K-Mean Function-MATLAB. Available online: https://it.mathworks.com/help/stats/kmeans.html (accessed on 19 January 2019).
36. Rousseeuw, P.J. Silhouettes: A graphical aid to the interpretation and validation of cluster analysis. *J. Comput. Appl. Math.* **1987**, *20*, 53–65. [CrossRef]
37. Tibshirani, R.; Walther, G.; Hastie, T. Estimating the number of clusters in a data set via the gap statistic. *J. R. Stat. Soc.* **2011**, *63*, 411–423. [CrossRef]
38. Lantz, B. *Machine Learning with R*; Packt Publishing: Birmingham, UK, 2013.
39. MATLAB. FitchkNN Function-MATLAB. Available online: https://it.mathworks.com/help/stats/fitcknn.html (accessed on 23 June 2019).

MDPI
St. Alban-Anlage 66
4052 Basel
Switzerland
Tel. +41 61 683 77 34
Fax +41 61 302 89 18
www.mdpi.com

Energies Editorial Office
E-mail: energies@mdpi.com
www.mdpi.com/journal/energies

www.ingramcontent.com/pod-product-compliance
Lightning Source LLC
Chambersburg PA
CBHW061446210926
51850CB00008B/1625
9783039362189